P9-CJU-597

Like, Comment, Subscribe

Like, Comment, Subscribe

Inside YouTube's Chaotic Rise to World Domination

Mark Bergen

VIKING

VIKING
An imprint of Penguin Random House LLC
penguinrandomhouse.com

Copyright © 2022 by Mark Bergen
Penguin Random House supports copyright. Copyright fuels creativity, encourages diverse voices, promotes free speech, and creates a vibrant culture. Thank you for buying an authorized edition of this book and for complying with copyright laws by not reproducing, scanning, or distributing any part of it in any form without permission. You are supporting writers and allowing Penguin Random House to continue to publish books for every reader.

Library of Congress Cataloging-in-Publication Data

Names: Bergen, Mark (Business journalist), author.
Title: Like, comment, subscribe : inside YouTube's chaotic rise to world domination / Mark Bergen.
Description: New York : Viking, [2022] | Includes bibliographical references and index.
Identifiers: LCCN 2022002106 (print) | LCCN 2022002107 (ebook) | ISBN 9780593296349 (hardcover) | ISBN 9780593296356 (ebook) | ISBN 9780593653098 (international edition)
Subjects: LCSH: YouTube (Firm) | Google (Firm) | Internet videos—Social aspects. | Internet industry. | Internet entertainment industry.
Classification: LCC HD9696.8.U64 Y6834 2022 (print) | LCC HD9696.8.U64 (ebook) | DDC 338.7/6102504—dc23/eng/20220706
LC record available at https://lccn.loc.gov/2022002106
LC ebook record available at https://lccn.loc.gov/2022002107

Printed in the United States of America
1st Printing

Set in Warnock Pro
Designed by Cassandra Garruzzo Mueller

For Annie, my love

So much has been done, exclaimed the soul of Frankenstein—more, far more, will I achieve; treading in the steps already marked, I will pioneer a new way, explore unknown powers, and unfold to the world the deepest mysteries of creation.

MARY SHELLEY, *FRANKENSTEIN*, 1818

It was going to be a joke. This was all going to be a joke. Why did it become so real?

LOGAN PAUL, "WE FOUND A DEAD BODY
IN THE JAPANESE SUICIDE FOREST,"
YOUTUBE, 2017

Contents

Part III

Part IV

Like,
Comment,
Subscribe

PROLOGUE

March 15, 2019

Haji-Daoud Nabi, a grandfather with a thick white beard and a bright smile, met the man who would end his life on a sunny Friday afternoon in Christchurch, New Zealand. Nabi stood at the entrance to his mosque. As the younger man approached, Nabi assumed this visitor was coming to join in worship and greeted him warmly, "Hello, brother."

Before his arrival the younger man had sent an email with the subject line "On the attack in New Zealand today." The email began with a confession—"I was the partisan that committed the assault"—and included a long manifesto. It landed in in-boxes of the nation's newspaper editors and television producers, professionals who, not too long ago, determined what was broadcast to the world. The email was dismissed as spam or the ramblings of a crank.

Then the calls flooded in. Gunshots heard everywhere around Christchurch's Hagley Park. Lifeless, bloodied bodies lay across two mosques, sacred sites on either side of the park's rich green. At least fifty dead, including a three-year-old child and Nabi. A reporter who landed on the scene, Kirsty Johnston, saw only carnage and wounded survivors frantically waving down taxis to the hospital. She had grown up in this placid island nation, where police usually didn't carry guns, where violence and vitriol happened overseas, on the news. Not here.

But the nation had changed, and so had the news. Everyone soon learned that the terrorist behind the killings, a twenty-eight-year-old white man, had strapped on a body camera and broadcast seventeen minutes of his rampage live on the internet. And so those same newspaper editors and television producers combed through his footage and manifesto for any clues behind the country's gravest mass shooting ever. They waded through arcane references to Serbian politics, sixteenth-century warfare, and internet subcultures. Little made sense. But one bit stood out as recognizable: the name of a YouTube star. "Remember, lads," the terrorist said on-screen, moments before opening fire. "Subscribe to PewDiePie."

• • •

In the days before, halfway around the globe, YouTube employees soaked in a gigantic, heated resort pool. They had arrived in shuttle buses, as always. The buses had come north, through San Francisco and Berkeley, through tony suburbs and towns, through forested state parks dotted with soaring redwoods, and into the heart of California wine country to Indian Springs, to a quaint hotel situated atop a natural hot spring in Calistoga. In 1859, that city was declared settled by California's first millionaire, a man made rich by promoting the gold rush to anyone who would listen.

Claire Stapleton, a YouTube manager, unpacked her bags in a hotel cottage. She had done these corporate retreats many, many times before. This particular one was only for members of YouTube's marketing division, those responsible for the upkeep of its public image, its brand. It would also be her last; she didn't know it for certain then, but she had some suspicions.

Stapleton was pale with dark brown hair, nearly black. Normally carefree in demeanor, she could pull off stern, as she had in *The New York Times* four months earlier, where she was photographed in a solemn black turtleneck as the face of a rebellion inside Silicon Valley. At the resort, Stapleton walked out of her cottage and past an outdoor spill fountain, a lovely trellis garden, and a meditation circle to the small conference rooms named River and Reflection.

Overnight stays there for business groups cost around $350 a room. No issue for YouTube, which earned more than $11 billion in sales the prior year. Though these guests signed in under the name of another corporation: Google, YouTube's owner and parent since 2006. Google posted sales of more than $136 billion in 2018. Yet the technology Goliath was trying to be more cautious with its wealth. A financial chief had come from Wall Street and pinched pennies at a company famed for spending freely. Also, two years into Trump's America, Google and its Silicon Valley peers, used to being celebrated as innovators and underdogs, suddenly found themselves reviled as greedy, irresponsible, and too powerful. As *the establishment*. Even some of Google's own employees had started seeing the company this way.

To limit such unwanted attention, Google now held fewer retreats at lavish, exclusive spots. Indian Springs was perfectly understated. From the outside the two-story, Spanish-mission-style property looked like a simple 1950s motel, but there were subtle touches of luxury inside: organic shampoos, "Be Well" filtered-water stations, and faux fireplaces. The resort had managed, by some miracle, to pipe hot springs water into a gentle, calming Olympic-sized pool.

Stapleton and her colleagues were encouraged to enjoy themselves. It had been a stressful year or two. Google had, everyone knew, some of the happiest employees on earth. Yet its regular survey of staff satisfaction, called Googlegeist, had recently returned troubling results: more employees were reporting shrinking faith in the company's leadership and priorities, and nearly half felt their pay was not "fair and equitable." That fall, Stapleton had led thousands of employees in a protest against Google's handling of sexual harassment charges. Google had a long-established alarm system for its vast computer network. "Code Yellow" meant software programmers should work overtime to fix a flaw or bug. "Code Orange" was close to an emergency. "Code Red" rang when Google's search page or its email service stopped working. *Fix it now.* The company had extended this alarm system to nontechnical affairs, such as the satisfaction levels of staff.

This YouTube retreat at Indian Springs was unofficially designated a "Well-Being Code Red."

Stapleton's marketing team went to wine-tasting sessions in town and took pizza-making classes. They strolled through the resort's manicured agave garden and its Buddha Pond. They roasted marshmallows outside at a fire pit. They practiced self-care. They drank. YouTube staff expensed treats at the Chaise Lounge, an old-fashioned soda shop, and reclined under the shade of retro blue-and-white-striped awnings. A big American flag waved at the pool's entrance above an old-fashioned clock inscribed with "Pepsi-Cola." Indian Springs described this design as "Old Hollywood," a poetic irony for its guests from YouTube, a business that, more than any other, took the "old" of Hollywood—the studios, the agents, the groomed stars, the entertainment people paid for—and blew it up.

Once Stapleton arrived in the River and Reflection rooms, she sat for a mandatory screening of a YouTube video they had all seen before. Someone pulled it up online and pressed the iconic triangular Play button.

YouTube: "Our Brand Mission." June 22, 2017. 1:48.

The video begins with a small boy in his bedroom, too cute with his too-big electric guitar. It cuts to another child, somewhere in Asia, herding sheep; then a woman crying; a man landing a skate trick. "Look at these moments," a female voice-over begins. "All of these stories. Secrets. Revelations. From every corner of the world." It is a montage of inspiring YouTube footage. Babies, athletes, random acts of kindness, a woman in a hijab, group dancers, group huggers, more criers. "This is the rarest, purest, most unfiltered portrait of who we are as a people," the narrator says. "This is what happens when you give everyone a voice, a chance to be heard, a stage to be seen."

The video closed with familiar text, YouTube's brand mission: "Give everyone a voice and show them the world." *So strange,* Stapleton thought. So much of the world had changed since her team first made this motivational montage back in 2017. So much of *YouTube* had changed. She had been in countless conversations about reworking its brand mission since then, and they were still playing the same old video as motivation. She kept these thoughts to herself.

Fourteen years after its founding, YouTube.com was still a marvel of the modern world. In under two decades, on-demand lightning-fast internet television had gone from an impossibility to a simple fact of life. YouTube had become *the* place for free moving pictures online. "The video scaffolding of the internet," one employee called it. More than 2 billion people visited every month, making YouTube the second most frequented website on earth (behind Google). It was also the world's second most popular search engine (behind Google). By mid-2019, some 1.7 billion people—more than a third of the world's internet population—visited YouTube every day. They used YouTube for entertainment, instruction, and comfort. Polls showed that a quarter of Americans relied on YouTube for news, and more people visited the site regularly than Facebook, Instagram, or any other social media. An entire generation of children had ditched television for YouTube. In many countries, YouTube *was* television. YouTube reinvented supermarket tabloids and instruction manuals. In parts of Silicon Valley, futurists imagined YouTube replacing college professors and doctors.

And unlike virtually any other website open to the masses, YouTube paid its contributors. That novelty had birthed an entire creative industry, a stable of performers, personalities, artists, influencers, instructors, and franchises, a new media as revolutionary as radio and TV, all in just a few short years. YouTube made everyone broadcasters. YouTube has given us "Gangnam Style," Charlie biting a finger, "Baby Shark," *It's Friday, Friday, gotta get down on Friday.* Yoga with Adriene, lunch doodles with Mo, professional *Minecraft* players. A diverse sea of enormous talent old media had ignored or overlooked. Thousands of microcelebrities you might not know, but millions of young fans certainly do, watching with a fervor they never show for movie or TV stars.

YouTube was formed in the same stretch as a crop of flashy consumer internet upstarts, and has outlasted nearly all of them, except Facebook. But while Facebook has since struggled to remain relevant to young people, YouTube never has—it draws in younger and younger audiences, year after year. No company has done more to create the online attention economy we're all living in today. YouTube started paying people to make videos when Facebook was still a site for dorm-room flirting, when Twitter was a techie fad, and a decade before TikTok existed. All those companies ripped from Google their philosophy—more information online is better—and their business model: get a free service to as many people as possible and mine their clicks, habits, and data to sell ads. Influencers, preteen millionaires, fake news, internet addictions, con artists—Google and YouTube enabled all these messy wonders and ills of social media, usually first. "Google made the wheel," one veteran of Google and Facebook said. "Facebook and every other internet company copied it."

YouTube was as reliable as running water and as voluminous as an ocean. Before the Indian Springs retreat, the company publicized this unfathomable stat: 450 hours of video were uploaded every minute to its site. Imagine the longest movie you've ever seen. A *Lord of the Rings*, maybe. Now imagine watching it one hundred times in a row, and you still have not sat through the footage added to YouTube every sixty seconds. Since 2016 people have watched more than a billion hours of video there every single day. The mind numbs. Want a video on any topic imaginable? It's probably there. *Type, type, click. Like, comment, subscribe.* Billions did that every day, with little knowledge of how the video that popped up before them wound up there.

Everyone knew YouTube. But few knew how it works—who runs it, what decisions they make, and why those decisions matter. This book was written to remedy that. It is the story of a business that transformed from a money pit into a raging commercial success, a pillar of the internet that made Google one of the world's most profitable and powerful companies. It's the story of a new mass media programmed not by editors, artists, or educators but by algorithms. And it's the story of some very consequential, and very strange, recent events that most people who visit YouTube.com know nothing about.

For most people used YouTube as a utility, a well of information, or a harmless pastime.

But YouTube contained *far more* than that. YouTube's marketing team was all too aware of a different, more disturbing side. And as they sat in the resort, in River and Reflection, watching their own marketing footage, touching and warm, they couldn't help but notice how distant it felt from the Nightmare Fuel.

Nightmare Fuel was a grim moniker some on the team used to refer to a daily email that monitored press coverage and online chatter about YouTube, normal fare. But the emails also exposed the site's underbelly, its bizarre corners and horrors. The term "Nightmare Fuel" was born after a YouTube star broadcast a dead body hanging in a Japanese forest. Because the marketing team handled all of YouTube's official online accounts, they needed to be made aware of the site's constantly shifting controversies lest they accidentally wade into one. Like when teens started posting videos documenting their consumption of laundry detergent pods, marketing had a new edict—*don't post anything about laundry.* When a thirteen-year-old girl generated press for making ASMR videos on YouTube with troubling sexual connotations—*best to avoid ASMR, and maybe teenage girls.* Or when a news story noted the preponderance of horse bestiality on YouTube—*nothing horse related.* This deluge of videos and critical coverage piled into their in-boxes every morning. Stapleton worried that Nightmare Fuel made her team think YouTube reflected the worst of humanity.

Most of the retreat offered a nice escape from the Nightmare Fuel. Except one agenda item: the team was told they were to end a long moratorium and begin using YouTube's corporate accounts to promote PewDiePie. This order, marketing staff there understood, had come from on high, "from Susan." This was Susan Wojcicki (pronounced wo-JIT-ski), Google's first marketer and YouTube's chief executive since 2014.

Each person at the retreat knew the tale of PewDiePie.

Real name: Felix Kjellberg (pronounced SHEL-burg), a Swede, not yet thirty, fond of shrieking his online persona as a customary greeting in his videos. *PEW dee PIIIIIIIIE!* He was YouTube's biggest star by a metric the company invented: "subscribers." Viewers clicked on a tiny red button to subscribe to video makers (as they would for a magazine or cable package, only this was free). When YouTube introduced this feature, its inventors never imagined that any given broadcaster would get more than a few million subscribers, tops. By March 2019, PewDiePie had nearly 100 million, on par with the Instagram followings of celebrities like Miley Cyrus and Katy Perry. Kjellberg had debuted as PewDiePie nine years earlier, eons in YouTube years, streaming himself playing video games and publishing his reels, like everyone else, at his own discretion, straight into the great big video tapestry. Over time, he had amassed a ferociously loyal fan base.

Kjellberg and Google made considerable money together. Google knew this. Google counted everything; it named itself after a number (googol, an unbelievably large sum) and existed on the 0s and 1s of computer code. Google counted every minute of footage consumed on YouTube. For a seven-year stretch starting in 2012, humanity consumed 130,322,387,624 minutes of PewDiePie videos, according to internal company documents. During this period Kjellberg earned $38,814,561.79. A lion's share (95 percent) came from commercials YouTube inserted before and during his videos. Every time an advertiser spent a dollar to run a commercial, YouTube handed fifty-five cents to the video broadcaster and kept the remaining forty-five—its fee for hosting footage and providing the enormous machinery that made its site work. So, from PewDiePie's videos in that stretch, YouTube made, roughly, $32 million. Given the size of his audience and the salaries silver screen stars commanded, one could argue Kjellberg was underpaid.

And yet the headaches. All the marketing staff knew them. How, in 2017, *The Wall Street Journal* printed a story with an image from a YouTube video of PewDiePie next to a sign that read DEATH TO ALL JEWS. How, in the ensuing chaos about bad jokes, neo-Nazis, and things taken out of context, Google ended a broader commercial deal with Kjellberg. How, seven months later, he had dropped the n-word casually in a video game stream (and apologized) but then went on to direct his viewers to a YouTuber who spouted anti-Semitic bile. How critics called PewDiePie "dangerous" and "flirting with the alt-right." How one headline about him read, WHEN DID FASCISM BECOME SO COOL? How people compared him to Donald Trump.

But the marketing team *also* knew that all this only strengthened the resolve of PewDiePie's fans—his "Bro Army," he called them. That past autumn, when a YouTube channel featuring Bollywood songs looked set to surpass PewDiePie's subscriber total, his army had formed a rallying cry to defend their king against critics and the threat to his crown. *Subscribe to PewDiePie!*

That cry rang out from every corner of the internet. It appeared on signs at the Super Bowl in Atlanta and during a basketball game in Lithuania. A British political party tweeted it. A colorful cast of characters spread the cry: HackerGiraffe, Goose Wayne Batman, Elon Musk. It was an antiestablishment mantra, a screw-you to corporate internet overlords, a cultural phenomenon. A meme. And the cry, like YouTube.com, grew beyond anything YouTube, the company, ever imagined.

Since the DEATH TO ALL JEWS incident, YouTube had kept its biggest star at arm's length, making no public displays of support or promotions as it did for other broadcasters. But the company had decided to change its stance. The marketing team discussed this decision at the retreat. Claire Stapleton's boss, Marion Dickson, wrote an email that Thursday, March 14, to her staff at the retreat on the edict to "re-engage" with PewDiePie. "I want to make sure we put some proper guidelines and principles in place on how we do this so we can manage any potential backlash," Dickson wrote, noting the importance of being "clear on how this aligns with brand values and messaging."

Staff soon boarded their buses in the retreat parking lot, drove past the palm trees and redwoods, out of wine country, and back home. They pondered these questions of proper guidelines, principles, and brand values. By that evening it was Friday in New Zealand. Stapleton and her colleagues watched news alerts light up their phones and emails pour in. A terrorist had broadcast mass murder online, posting footage of his rampage that looked like a first-person-shooter video game. With children as shooting targets. It appeared first on Facebook Live, a YouTube competitor, then soon on YouTube, where accounts kept reposting copies of the video even as the company scrambled to take it offline. One of the only clues of the murderer's intent was the PewDiePie rallying cry.

Not long ago virtually no one took YouTube that seriously. No one cared about its rallying cries. In the intervening years the company had pursued a massive business opportunity so swiftly, driven so much by a blind faith in its technology, that YouTube had unwittingly built a machine that exposed the vilest parts of human nature. Some outside the company thought its business flourished because of that vileness. By 2019 the world was beginning to grapple with social media's real impact—how a few computer science companies in California suddenly controlled most pathways of information and speech. YouTube, when it could, liked to fly below the radar of these debates. But in so many ways, YouTube had set the stage for modern social media, making decisions throughout its history that shaped how attention, money, ideology, and everything else worked online.

By that Thursday in March, when Stapleton and her colleagues left the retreat, their company had been through two hellish years of firestorms with unruly stars, wackos, conspiracies, child abusers, and an existential business crisis. They were eager to put that behind them. Mending public ties with PewDiePie felt like a first step. Then the Christchurch shooting happened, a tragedy that unspooled on the company's website and, it seemed, involved the influence of its biggest broadcaster. A nauseous feeling Stapleton had about her company and its role in society got even worse. As she processed the horror, she thought of the marketing video they had just rewatched hours before. Was *this* gruesome act actually, as the video said, the "most unfiltered portrait of who we are as a people"?

No one at the company wanted it to be. But this would hardly be the first time YouTube's creation careened forward in a direction the company behind it couldn't control.

Part I

CHAPTER 1

Everyday People

Chad Hurley wanted to create *something*; he just wasn't sure what.

It was early 2005, and Hurley spent most of his time hunched over a computer screen in Northern California. Hurley didn't look like the brainy Poindexters around Silicon Valley. He had broad shoulders and a high forehead, a high school jock's physique, and dirty-blond locks that he swept back surfer-dude style. He liked beer and the Philadelphia Eagles and considered himself an artist of sorts. With a friend, a kindred spirit, he had recently started a menswear line making laptop bags after finding most on the market ugly and dull.

But Hurley, a web graphic designer, knew the real money was in computers, not bags, so that's where he and his two programmer pals, Jawed Karim and Steve Chen, hoped to strike gold. At twenty-eight, Hurley was the eldest, by a year, and de facto leader. He had a toddler son and had married into Silicon Valley royalty—his father-in-law was Jim Clark, a famed internet entrepreneur. Hurley began dreaming of his own company at the dawn of Web 2.0—websites filled with the work of regular folk, not professionals. Web surfers rushed to post online diaries, photo albums, poems, recipes, screeds, whatever they liked. "Everyday people," Hurley would call them. For months, Hurley and his pals had batted around proposals for a new internet business, meeting at his house in Menlo Park or a café nearby, where they discussed popular Web 2.0 fixtures to emulate, like Friendster, a social network, and the blogging websites growing like weeds. More often they talked about Hot or Not, a skeletal site that let people upload photographs of a face and vote on its attractiveness. Crude, but so popular. The trio knew one of Hot or Not's creators from a coffee shop they frequented at their old jobs, and they knew he was making decent money. That was cool.

The three finally settled on an idea for a website to let people share and watch video. On Valentine's Day they had stayed up way too late, working in Hurley's garage with his dog, and settled on a name for their idea. Hurley tried words that evoked personal television, riffing on old slang for the medium, "the boob tube." A tube for you. They typed it into Google. No results. That evening, they bought the web domain YouTube.com, a first step on solid ground.

Eight days later Hurley opened an email from Karim with the subject line "Strategy: please comment."

> **The site should look good, but not too professional. It should look like it was thrown together by a couple of guys. Note that hotornot and friendster, while easy to use, don't look professional, and yet they've had enormous success. We don't want to look too professional because it scares people off . . .**
> **The most important aspect of the design is ease of use. Our moms should be able to use this site easily.**
>
> **Timing/Competition:**
>
> ---
>
> **I think our timing is perfect. Digital video recording just became commonplace last year since this is now supported by most digital cameras.**
> **There is one site I'm aware of: stupidvideos.com, that also hosts videos and allows viewers to rate them. Luckily the site hasn't caught on very much. We should discuss why this is the case, and why we expect our site to gain more traction.**

Hurley read on.

Site Focus

Our focus should implicitly be dating, just like hotornot. Note that hotornot is a dating site without seeming too much like one. This puts people at ease. I believe that a dating-focused video site will draw much more attention than stupidvideos. Why? Because dating and finding girls is what most people who are not married are primarily occupied with. There are only so many stupid videos you can watch.

Hurley was married but agreed that dating could motivate people to make and watch videos. "People want to see and be seen," he would write weeks later. Karim's email ended with a target date for YouTube's launch less than three months away: May 15, 2005.

They set to work. Hurley tinkered with the way YouTube.com would look to a visitor. Chen and Karim engineered the code to bring the site to life. Then, on March 20, Yahoo announced it was buying Flickr. Yahoo was a web titan, a valuable "portal" for online activity that raked in billions a year. Flickr, a sleek Web 2.0 service, let people upload digital photos. The press reported that Yahoo had paid as much as $25 million for the acquisition. The Flickr deal lit a fire. Karim sent another email, with the subject line "new direction":

Chad and I were discussing today that the focus of the site should be more like flickr. Basically a repository for all kinds of personal videos on the internet.

In the coming weeks Hurley, Chen, and Karim doubled down and debated anew how their site should work. *Like the dating site or the photo site?* Hot or Not appealed to "hip college kids with raging hormones," Chen wrote in one email, while Flickr reached "designers, artists, and creative folks." *Who would use YouTube? Should they make two sites?* Hurley wavered on the Flickr model, concerned that video was more difficult to upload and edit online, but he also didn't want to pigeonhole the site. Late on Sunday, April 3, Hurley emailed the others that they should just get their site out into the world; they could "figure out where we are going down the road."

Ten days later an obstacle appeared on the road. Google put out a call online for people to submit amateur videos, which the company might then post for the world to see. Hurley would later recall his reaction: "Ah, fuck." Google was scarier than Yahoo. When Google started, it was one of many web search engines, but it quickly pulverized all its competitors. Now Google was starting to reveal its true ambition. On April Fools' Day 2004, it launched an email service, Gmail, with so much unheard-of free data storage that people assumed it was a joke. Then it announced a massive, free digital map of the planet. And now Google, spigot of cash, hoarder of brilliant programmers, was coming for YouTube.

When Hurley and his pals next met, a new agenda item appeared: *Should we give up?*

• • •

Chad Hurley grew up in Reading, Pennsylvania, and had landed in California the customary way: on a living room floor. After college, a small school in Pennsylvania where he took odd jobs cobbling together websites, Hurley had moved back in with his parents. He was aimless and a little bored. One day he was leafing through an issue of *Wired* magazine and came upon an article on Confinity, a California company trying to move currency through early handheld computers called PalmPilots. Confinity needed a designer. On a whim Hurley sent a résumé. He heard back the next day. *Could he come for an interview tomorrow?*

This was 1999, when Silicon Valley was flush with cash and desperate for warm bodies. Confinity asked Hurley to design a logo for their new payments service, PayPal, and hired him quickly. He flew to Northern California, the new epicenter of innovation and commercial success, and slept on a mattress pad on the floor. His host, Erik Klein, a programmer from Illinois, had also started in California this way. Practically all the rookie hires at Confinity initially crashed on pads or couches near the office; with little job and life experience, the twentysomethings waited for a broker they knew who rented apartments without checking references.

At Confinity, Hurley soon met another new hire, a young coder named Steve Chen who had round cheeks, spiky black hair, and an easy laugh. Chen had arrived on a one-way ticket from Chicago, deserting college one semester before graduation, a shock to his parents. Born in Taipei, Chen moved to the United States with his family at the age of eight. On the flight over he didn't know how to ask the stewardess for water in English. Much of his youth in suburban Chicago was devoted to learning the language. Then, at fifteen, he went off to boarding school, the Illinois Mathematics and Science Academy, where he found his real fluent tongue: computerspeak. Chen got his own hulking desktop machine, and with no parents around he stayed up late, downing coffee, toying with code that animated screens. He went on to study computer science at the University of Illinois but frequently skipped classes. Assignments asked for outcomes from code—a program or an algorithm optimized to perform a task. *If* this, *then* that. Chen could figure that out with just a book and a keyboard. And he already had a contact in the industry. One of Confinity's founders, Max Levchin, a University of Illinois alum, liked to recruit from Chen's high school; Levchin once told a reporter that the Illinois academy churned out "hard-core smart, hardworking, non-spoiled" coders, perfect for a start-up. Chen started at Confinity on a Sunday, walking into its office to find four other coders playing video games. *Heaven.*

Chen liked to work late, fueled on cappuccinos and cigarettes, sometimes stumbling into the office past noon. Colleagues teased him for being a mischievous joker; he took frequent smoke breaks and frequent programming "shortcuts," inelegant technical work-arounds others avoided. Chen loved writing code in Python, an obscure computing language nobody else used. He liked it because it was open source, built and maintained by contributors around the world with an unrestrained spirit he saw in himself.

Occasionally, he worked alongside Jawed Karim, another gifted, mischievous immigrant from the University of Illinois. Karim also loved the internet for its rule-bending openness. In college Karim had invented MP3 Voyeur, a file-sharing service tech-savvy college students used to rip music online, months before another named Napster debuted.

Chen, Karim, and Hurley lived through turbulent years at Confinity. Start-ups back then were sharks circling for blood, willing to turn on a dime at the smell of money. Confinity ditched its first idea, security software, for another—mobile payments. It managed to survive the dot-com crash, when bursting markets demolished young internet businesses, and changed its name to PayPal. Emerging from the rubble, PayPal went public and, in 2002, sold itself to the auction site eBay.

PayPal's early staff were a tight-knit group of type A overachievers, and after the eBay purchase several of them went on to join blue-chip investment firms and start marquee companies: Yelp, LinkedIn, SpaceX. The press would christen this cadre (of mostly men) "the PayPal Mafia." At PayPal, the YouTube founders had a reputation as more of the B team. Hurley had left soon after the eBay acquisition, chafing at its stuffy corporate culture. Chen worked on PayPal's expansion into China but grew to hate the culture too, believing that it valued financial gain over programming gusto.

When the trio started talking up their new idea for a video website in early 2005, few took them that seriously. In April, Chen sent over their test website to an old colleague.

"Nice," the PayPal alum wrote back, "it works really well. how you gonna keep out the parn?" (He mistyped "porn.") Chen reassured him that they would, then asked, "Want to put a video in??????"

The internet was not yet an enormous public stage, not yet the automatic place for people to share and overshare. Posting unpolished personal stuff felt weird. Chen's ex-colleague wrote back, "I'm not sure I have any."

• • •

Ultimately, a lukewarm reception didn't deter the YouTube dudes. Neither did Google's entrance into amateur web video. Google wasn't the only one after all. Microsoft had a video site and so did a litany of start-ups, like Revver and Metacafe, and crass shock portals, like Big Boys and eBaum's World. Each featured footage on their own websites or applications, but they lacked a way to let videos play everywhere else on the web. YouTube had found a way.

Jawed Karim showed it off to a PayPal coder he knew, Yu Pan, at a house party. "It's Flash," Karim explained. With Flash, a software system for rendering text, audio, and video graphics, YouTube could embed its video player box on other websites. That would ultimately be the trio's most brilliant move, an innovation that enabled YouTube to leapfrog all its competitors. At the party Karim pulled up a test video for Pan, who had toyed with Flash at PayPal and grasped its technical potential. On-screen Hurley had sketched a simple rectangle and tiny triangular Play button in pixels, a miniature TV that could plonk down anywhere online.

Getting videos to play in Flash was easy. Getting them to sync with sound was not. Chen made countless four-second films of himself just talking, going back each time into the code to make sure his lips and words moved together. Once the dudes were confident it worked, Karim published their site's first proper video, an eighteen-second clip with a crude wink.

jawed: "Me at the zoo." April 23, 2005. 0:18.

Karim is in a black ski jacket at the San Diego Zoo. Kids chattering in the background nearly drown out his voice, but his mouth moves in sync. "All right, so here we are in front of the elephants," he says, looking straight at the camera. "The cool thing about these guys is that they have really, really, really"—pause—"long trunks. And that's pretty much all there is to say."

To populate the site, Karim uploaded clips of 747s taking off and landing. Chen posted clips of his cat, PJ, with his handle, tunafat.

But they needed more material. They were still toying with the dating idea, and for that to work, they had to have plenty of videos of women. "YouTube is in need of creative content!" Chen wrote in a Craigslist post. "If you are a female or an extremely creative male between the ages of 18 to 45 and if you have a digital camera that can create short video clips, please follow these steps to earn $20." Instructions followed: visit YouTube.com, start an account, upload three videos of yourself. Visitors could pick from a drop-down menu: "I'm a FEMALE seeking MALES between 18 to 45."

They posted these listings around Las Vegas and Los Angeles. No one responded.

Back at the drawing board, Hurley decided these calls for creativity were too daunting, arguing that they should encourage "real personal clips that are taken by everyday people." For him the site's handicap was that it lacked a clear purpose. Was it a place to flaunt your opinions or your sex appeal? "I keep getting mixed signals from both of you," he wrote in one email, exasperated. "Are we moving towards blogging or dating?" Karim wrote back, "Screw blogging. we should just be a site where you can post videos of yourself. Broadcast yourself. That's it." An early slogan the three had used for their site—"Tune In, Hook Up"—was a dud. Karim now proposed that phrase—"broadcast yourself"—as the site's motto, and it stuck.

Karim's bluntness would soon start to grate on his co-founders. But his decisiveness and motto certainly helped their site once it went live to the public that May. Still, YouTube's real fuel came from some prescient tweaks the trio made a month later. They added features for people to leave comments and a small button to easily email friends links to clips. And when someone clicked on a video, a row of related ones appeared on the right of the page, prompting them to watch more.

• • •

In the years before YouTube was formed, America's broadcast networks, the lords of TV—NBC, ABC, CBS, and Fox—had just emerged alive, bruised and bent *but alive,* from a bloody bout with their first nemesis: cable. Unleashed by regulators in the 1990s, fleets of cable stations took over the airwaves, seizing viewers and advertisers from networks that had dominated TV since its inception. Networks fought back. They started competitive 24/7 stations (MSNBC, Fox News), and they consolidated: Viacom, owner of CBS, bought TNN, BET, and MTV within a few short years. But their winning weapon was reality TV. Shows with amateur casts and scant scripts (or none at all) were cheap to make, and audiences ate them up.

By 2005, though, that novelty had worn thin. *The Real World* was in its sixteenth season, *Survivor* in its tenth. Reality TV's artifice had become all too obvious; audiences now knew that its characters and dramas were staged and manicured and that stardom there led to only fifteen minutes of fame, if that. Networks adapted. They turned to C-list celebrities (ABC had *Dancing with the Stars*) or promised real, lasting fame: Fox's new show *American Idol* topped TV ratings, drawing twenty-six million viewers an episode. A year earlier NBC brass, fretting about the end of its megahit *Friends,* lucked out with the surprise success of *The Apprentice,* a reality contest starring a washed-up real estate heir, Donald Trump. TV moguls had seen what the internet did to music. Punk pirates at a site called

Napster gave it away for free, obliterating the industry. Thankfully, boomers still soaked up reality TV.

But younger audiences, that slippery cultural bellwether, were already drifting away from broadcast again. During George W. Bush's presidency, the kids got their news from Jon Stewart on Comedy Central. They spent their time on MySpace. By the summer of 2005, MySpace was the hottest thing online; it reeled in sixteen million visitors a month, making it the fifth most popular web destination. News Corporation, Fox's parent company, paid $580 million in July to buy MySpace and its young audience.

By then Steve Chen had found a shortcut on MySpace to juice YouTube's growth.

MySpace had a messy, abundant menagerie of blogs, music, and classifieds. But not video. Chen thought MySpace devotees were an ideal target for YouTube. They already shared photos. *Why not videos?* Thanks to Flash, YouTube was able to run its video files directly on MySpace pages, enticing visitors over to its own website. Newcomers arrived, posting footage from family vacations, cat clips, and oddities not found on TV. Traffic from MySpace, coupled with the new features for comments and related videos, brought a steady flow to YouTube that would never relent.

Chen may have had help from another social network. After leaving PayPal, he had taken a job with Facebook, then a new start-up from Harvard students. One early Facebook staffer recalled Chen showing off YouTube.com inside Facebook's offices. After Chen left for YouTube, he held on to his Facebook corporate computer, and some at Facebook suspected that he authored code for YouTube on that device, which might entitle Facebook to the intellectual property. (Chen denied doing this and said he simply didn't take the time to return the device to Facebook's offices. Facebook never raised the issue publicly.)

Chen was busy that year. With the rush of viewers and uploaders from MySpace that summer, he quickly realized YouTube needed help keeping the site afloat. So he turned to his PayPal friends. Chen lured over Yu Pan, a proficient coder colleagues described as a "mad scientist." Erik Klein left PayPal on a Thursday, talked to Chen Monday morning, and had a new laptop and job by the afternoon. Several other PayPal alums soon joined.

Each day the team had a simple edict: don't let YouTube.com crash. Sometimes the website would keel over from bugs or the sheer weight of video uploads. Coders took the train to a small office YouTube had rented and whipped out bulky laptops and wireless modems to work on their commutes. Chen, a night owl, read through user email complaints past midnight and left a fresh pile of fixes for his team each morning. For each video to play online, YouTube needed tons of computing bandwidth and gear. Chen bought 42U racks—hulking boxes, larger than fridges, made to hold servers—by the truckload. But that solution was short-lived. By September YouTube's site was playing more than 100,000 video views a day. Eventually, Chen found a company in Texas that rented out server space, putting all the bills on his credit card, which he frequently maxed out.

At PayPal, programmers had used cash instead of their own service in their daily lives—a half joke in response to the fact that their system, exploding in usage, was so fragile that one transaction could push it over the edge. They continued the tradition at YouTube. The men who built the machinery behind the site rarely watched any of its videos.

But the founders in charge tried to watch. Very quickly the quirky home videos had begun to mix in with footage that looked an awful lot like TV. In July, Hurley found a few videos titled "budlight commercials," pirated rip-offs of TV spots. He voted to pull them. The beer maker owned the copyright on its commercials, and if those videos appeared without its consent, Hurley knew YouTube could be in legal trouble. Karim disagreed, restoring twenty-eight removed videos. Those clips could spread widely, exposing more people to the site, he argued in an email. *Worth the risk.*

Hurley replied, ominously, "Ok man, save your meal money for some lawsuits! ;)"

Hurley was less casual the following month when he found YouTube clips of a NASA shuttle landing taken directly from CNN. "If the boys from Turner"—CNN's owner—"would come to the site, they might be pissed?" he emailed. "These guys are the ones that will buy us for big money, so lets [*sic*] make them happy."

While Hurley fretted over the boys from Turner and Chen tried keeping the site's lights on, Karim was on his way out. Like Chen, Karim had dropped out of college to join an unknown internet company. He completed his degree online, but his parents were both scientists and, even after PayPal's success, urged him to go to graduate school. He left YouTube that fall. Chen felt abandoned, down one expert programmer when their site needed that expertise the most. Later Karim's role would come under dispute. He told people that he had birthed the initial idea for YouTube after getting frustrated with missing major live TV touchstones, like the 2004 Indonesian tsunami or Janet Jackson's Super Bowl nipple flash. *Once they aired, why couldn't you watch them again?* Chen would say the idea had come to him from a conversation during a dinner party at his apartment.

• • •

After Karim left, YouTube began to truly take off, but not from anything his co-founders did. Instead, a group of young, creative oddballs had started to use the site religiously to build cultural touchstones all on their own.

Brooke Brodack was ten when she unwrapped her first video camera on Christmas. Or maybe eleven—she doesn't remember. She does remember the grandiose video tours she shot in her backyard and the holidays when she filmed her family sitting around eating turkey, corralling them together to play back her footage. At thirteen she began writing skits, learning to plan and edit shots. This hobby stuck through college, where she studied broadcasting and worked as a hostess at the "Lucky 99," the local name for a surf-and-turf eatery in Worcester, Massachusetts.

In the autumn of 2005, in between hostess shifts, she found a new film canvas.

Brookers: "CRAZED NUMA FAN !!!!" 4:03.

> A warning appears in all caps: "This video contains a very catchy song and dance. Some viewers may be susceptible to influence and attempt to copy this for fame." A young woman appears, lopsided ponytails, big gap between her front teeth. She wears a paper with #1 NUMA FAN, an internet in-joke, stapled to her shirt.

Once she joined YouTube, Brodack, then nineteen, posted a flood of madcap home videos under the handle Brookers. In one she lip-synched to a song from the band Chicago while wielding a katana sword. "Crazed Numa Fan!" cast her as an obsessive devotee of another bedroom webcam performer: Gary Brolsma, who posted his footage in 2004, pre-YouTube, to a site called Newgrounds.com. In that video Brolsma's headphones are wrapped over his round face, which stared off camera. Electronic music thumped, a Euro-pop song with a chorus, in Romanian, that sounded like "numa numa." Brolsma mouthed along, growing more animated as the song unfurled. Viewing it was like intruding on someone's private moment of joy. Brodack's version, a huge YouTube hit, was longer and wilder, her limbs flapping around like Gumby.

Most people stumbling upon YouTube in 2006, especially young people, knew Brolsma's clip. They had gathered around computer screens to enjoy his lip-sync routine. Brodack found its enthusiasm infectious, like seeing an introvert crawl from his shell. Her tribute became emblematic of YouTube's emerging aesthetic: re-creation, conversation, inanity. It was a copy of a copy, each as famous as the one before.

CHAPTER 2

Raw and Random

Rats. The place was teeming with rats. In nooks between the ventilators and the ceiling, beneath the floorboards. God, more every day.

By early 2006, YouTube had moved into a bigger office to accommodate its rapid, surprising growth. It was kind of a mess. The start-up occupied the second floor above Amici's pizza in San Mateo, a satellite city near San Francisco and train lines, and the twentysomething software engineers rarely cleaned up after their meals, resulting in an endless rodent problem. Someone placed two plush toy rats on YouTube's reception desk, company mascots. The office was horseshoe shaped, with a stairwell in the middle and rows of makeshift desks, tacky fluorescent lights, and gray carpets. Hurley and Chen shared a corner table near the few windows. Hurley had commissioned an artist to paint spiraling red and gray stripes on the walls to represent bandwidth and maybe spruce up the place. They hung cheap white sheets from the ceiling as dividers. They bought bulk Costco snacks for employees, as Google did, some of which rotted in the back of the fridge. Each new hire had to assemble their own IKEA desk and chair, a ritual signaling the start-up's scrappiness.

Hurley and Chen had formed a board and recruited business operators to join their coders, but the company kept its raffish quality, even as it drew more attention. MC Hammer, a bygone-era star, visited in February to check out the trendy new site, and Kevin Donahue, one of the recent hires, gave him an office tour. Staff recorded it and posted the resulting clip on YouTube (title: "Hammer Time!"). A *Forbes* reporter came by to watch a popular clip of a stepfather playing a cruel joke: while his kid was deep into a video game, he flashed a sinister face on-screen, prompting terrified sobs. "That's just wrong," said Chris Maxcy, another new YouTube business hire. Others laughed. Donahue described their site to *Forbes* as "raw and random."

Just a few months earlier YouTube's operational costs were still loaded on Chen's credit card, and the company was barely hanging on. Mercifully, YouTube found a savior. Chen and Hurley, who were still tracking new accounts on their site, spotted his name that summer: Roelof Botha. Roelof had money. At PayPal he was *the* money guy—he served as chief financial officer—and had since moved to Sequoia Capital, a famed venture capital firm that once backed Google. A tall, pragmatic South African with a business degree and a geek's fetish for tech novelties, Botha had recently taken a new digital camera on his honeymoon in the Italian countryside. He posted some footage to the site he had heard about through the PayPal network. Buzz outside the network was also building. In August, Slashdot, an influential tech news site Google's founders read, gave a shout-out to YouTube, bringing in waves of fresh traffic. Botha reconnected with his former PayPal colleagues and drafted a memo that month persuading other partners at Sequoia to invest.

By late August, YouTube had eight thousand visitors, who generated more than fifteen thousand videos. Botha crunched its numbers: YouTube was paying around $4,000 a month to serve videos, with each play costing a tiny fraction of a cent in computing oomph. YouTube could conceivably charge people for features, like special video effects, or it could make money with ads, as Google did. In his memo Botha noted the recent success of "user-generated content" Web 2.0 companies like Flickr and Tripadvisor, a crowdsourced travel site that had sold for more than $100 million. YouTube could net at least that much.

The memo worked. In November, Sequoia announced an investment of $3.5 million, praising the marvel of YouTube holding eight terabytes of footage, "the equivalent of moving one Blockbuster store a day over the Internet"—an astounding stat then, when Blockbuster still mattered. One Sequoia partner, Michael Moritz, would later call YouTube "the fourth horseman of the internet" alongside Amazon, Microsoft, and Google. Sequoia placed Botha on YouTube's board and took 30 percent ownership.

The investment gave the startup enough money to rent its office in San Mateo above the pizza shop, where the rodents roamed freely. Once, Chris Maxcy, the fresh business hire, followed an awful smell to find a dead rat decaying in a trap near the office ceiling. Maxcy, who preferred slacks and button-downs, was the only one in the office to tuck in his shirt—Hurley and Chen, when they had to dress up, preferred untucked—but he was still playful. He held the dead rat at arm's length in a trash bag. Micah Schaffer, a twenty-five-year-old newcomer with unruly hair, hipster jeans, and flannel, started recording a YouTube video.

"Let's show this to Heather," he quipped.

Heather was Heather Gillette, one of YouTube's earliest hires and one of the few women in the company. She grew up nearby in Palo Alto, where her parents always rented their home, and as an adult she dreamed of buying her own. She wanted space for her dogs, cats, and two horses, and a flock of chickens she was thinking of buying. That summer she had found a perfect spot—a vacant lot beneath green rolling hills and a forest of redwoods. Elated, she pulled out a small handheld video recorder to share the perfection with her family. When she scoured the web for a way to send the footage, she found only a seedy-looking site, MPEG Nation, demanding $25. She had not heard of YouTube until she visited the home of Kathy, her childhood friend. Kathy's husband, Chad Hurley, sat silently hunched over his computer in the dining room. Kathy explained Chad's latest venture, a free video-sharing site, and Gillette walked over to see what he was up to. She had worked in customer service roles but was now between jobs and, thinking of the sprawling lot for sale, wanted a new one. "We're not going to need customer service," Hurley told her. "It's free."

But weeks later, on Gillette's next visit to her friend's home, Hurley had changed his mind. He needed an office manager.

And as office manager, Gillette had to deal with the rodents. She couldn't stomach killing them. The animals she owned were like her "children." Men on staff teased her for her affection.

• • •

But Gillette had more complicated problems than rats. Hurley had also asked her to handle screening.

To avoid having YouTube become a crass shock site, its founders had outlawed uploads of porn and extreme violence. Still, those came. Initially staff took turns on moderation duty, scouring for offenders during the workday; Chen scanned during his late-night, coffee-fueled binges. A software system they built let viewers flag troubling or rule-breaking videos. But the daily barrage of footage demanded a better solution.

Gillette hired ten moderators for a team called SQUAD (Safety, Quality, and User Advocacy)—some of the internet's first frontline workers. They sat at computers that displayed a steady, unending queue of videos viewers had flagged. Four buttons appeared on their screens' top right corner. "Approve"—keep the video up. "Racy"—mark the video as suitable for eighteen and older. "Reject"—remove the video. "Strike"—remove the video and add a penalty to the account. Too many penalties, and the account went down. Gillette recruited more reviewers for night and weekend shifts and purchased industrial-sized protectors to shield screens from other onlookers. At first, these reviewers sat right by the office entryway, but soon YouTube decided it might not be ideal for visitors to first encounter a row of people scanning the internet's underbelly. So they were moved.

Micah Schaffer was tasked with writing screening guidelines. He printed out one "Reject" rule and taped it above his desk: "Just to clarify, if the only reason that genitals are not visible is because they are inside someone else's genitals—that's not racy, that's a strike." Porn caused problems: when YouTube courted Disney for a partnership, Disney suits complained about easily spotting still frames of adult videos on YouTube. A reporter once called on a Friday morning asking about the abundance of salacious material. Gillette stood up during YouTube's lunch hour and told everyone in the office they needed to spend their weekend scrubbing the site of genitalia. It worked: the reporter never published a story.

But moderation was rarely as simple as merely spotting private parts. Schaffer and Jennifer Carrico, an attorney volunteering at YouTube, sat together in a room one week to work on the guidelines. After watching some of the borderline, surreal footage pouring in, the pair settled on a tactic: they pointed at each body part and wondered aloud where it could go. *Can someone put a thumb in there? What do we do with videos showing that?* At one point Carrico wondered, "What kind of Pandora's box have we opened?"

Julie Mora-Blanco, a college classmate of Schaffer's, joined the SQUAD in the summer of 2006. YouTube paid her $45,000 a year with health benefits and equity, which she found incredible. Colleagues warned her about all the evils she might encounter, or so she thought. One morning, early in her tenure, she saw a video that would haunt her for more than a decade to come. "Oh, God," she cried once it began. Later she would only describe it as involving a toddler and a dimly lit hotel room. A co-worker talked her through next steps: hit "Strike," nix the account, and feed it to a nonprofit that monitored child exploitation and alerted federal authorities.

But those haunting videos, along with explicit porn, were really the SQUAD's only clear-cut zones. When moderators had questions, footage often went to Gillette. She could watch most evils with a level head and found adult porn amusing. Anything bleak involving animals—fetish clips of women in heels stepping on creatures, cats being boiled alive—Gillette had to hand off. She also had nightmares for years about the few she did see.

Mora-Blanco and her colleagues used dark jokes as a coping mechanism for this traumatic work—a running one involved an octopus and sexual consent—but they also felt deeply proud of their jobs keeping YouTube's nascent web community safe and sanitized.

Soon Gillette's job extended to keeping YouTube in good legal standing. Hurley approached her one day with a small slip from a company that housed YouTube's computer servers; it said that YouTube had broken the law. "Do you think you could handle copyright?" Hurley asked.

From the early garage days, Hurley knew the legal risks of running pirated footage without approval from copyright owners. As YouTube grew, so did takedown requests from old media. (In instant message chats, later surfaced in a lawsuit, staff complained about these requesters variously as "copyright bastards" and "fucking assholes.") But Hurley also knew that *if* the copyright owner *did* approve, and YouTube simply wasn't aware, it would be stupid to take videos down. In October 2005 a user named joeB uploaded a three-minute, mesmerizing clip of Ronaldinho, the soccer superstar who had signed with Nike. Was it pirated? The video went gangbusters, and YouTube kept it up. They soon learned that joeB belonged to Nike's marketing department. And that drove home a key lesson: YouTube could be a threat to copyright owners but it could also be a very valuable tool for businesses looking for audiences.

At the start of 2006, YouTube had another potential joeB on its hands: "Lazy Sunday."

Saturday Night Live, NBC's legendary show, was entering its fourth decade and growing stale. In a revival attempt it began airing "digital shorts" from new cast members like Andy Samberg, a floppy-haired comic with a Disney-prince jawline. His troupe's skit "Lazy Sunday"—two white guys rapping about cupcakes and *The Chronicles of Narnia*—appeared on YouTube in December and went viral. Hurley sent an email to NBC: *If you didn't put it up, we'd happily take it down. Just let us know.* For weeks no one answered, and "Lazy Sunday" continued to rack up views. Then, on February 3, an NBC lawyer finally replied in a stern letter demanding YouTube remove the sketch *and* all videos tagged "Saturday Night Live" or "SNL." Kevin Donahue, YouTube's new vice president, tried to convince NBC of the promotional value of keeping such viral content up. The value to YouTube was clear: that month most web searches bringing people to its site were the two words "lazy Sunday." Eventually, YouTube did pull the clip, but most visitors the skit drew in stuck around. For other copyright claims Gillette tried to comply swiftly, which irritated the engineers, who worried that too many video removals would turn people off from uploading.

By 2006 programmers had most of YouTube's frequent outages under control, though there were occasional hiccups. An annoyed viewer once called into YouTube's office line and left a voice mail. "I need to goddamn masturbate, and I can't do that when you don't have all those videos up," he shouted into the phone. "Get your shit together, you goddamn whores." Staff laughed nervously. Unnerving messages like that, the NBC drama, the sprawling, unfettered moderation mess—it all made one thing terribly clear: YouTube needed a full-time lawyer.

• • •

Zahavah Levine loved music more than anything. When she was nine, she took the subway to the Philadelphia Spectrum to see Kiss, screaming along, "Shout it out loud!" She had had an enormous blues record collection until some asshole stole it during law school. Levine had wavy brown hair and a fighting spirit; at school she protested South African apartheid and the contras in Nicaragua. After Berkeley Law she was swept into the dot-com explosion happening across the bay, an industry plowing into digital frontiers and desperate for lawyers.

This was the roaring 1990s. Once the province of colleges and computer nerds, the internet had come to the masses. In 1995 sixteen million Americans went online from their homes; that number had jumped nearly tenfold by 1998. Surfing, shopping, banking, screwing—everything was moving online. Congress and Bill Clinton's White House faced immense pressure from all directions to regulate the web. Religious right scolds wanted to wipe out sex and other ills, free-market champions wanted fewer impediments to commerce, and media lobbyists wanted protection from intellectual property theft. The resulting flurry of muddled laws included two whoppers that defined the modern internet. In 1996 the Communications Decency Act went after "obscene and indecent" materials online and included a short provision, Section 230, that gave websites permission to remove smut and shielded them from liability for posts users wrote. In 1998 the Digital Millennium Copyright Act (DMCA) provided ways for owners of intellectual property, like songs and movies, to claim rights online. In theory, the laws protected websites from lawsuits and copyright trouble.

In practice, the laws were far from clear. During their passage Levine first worked for a law firm that drew up "hyperlink contracts," deals between companies that linked to one another's websites. Such things seemed necessary until everyone got more comfortable online, and then they weren't. In 2001, Levine found a job involving her true love: music. Listen .com ran the web music service Rhapsody, and one of Levine's first tasks there was to explain how copyright law had knocked out Napster.

Napster was a fallen internet star. Music fans loved its free file-sharing system; the music industry hated it. Eighteen record labels sued Napster for copyright theft. Napster argued that it worked like a VHS player: people didn't hold the VHS manufacturer responsible for videotapes they played. Courts did not agree. In 2001, Napster lost a California ruling, effectively sweeping it into the dustbin. Levine's new company cut licensing deals with record labels to offer aboveboard on-demand streaming. She became an expert in the byzantine laws governing digital music. Within a few years, though, her tiny company risked obsolescence. Microsoft introduced a streaming service, and Apple released iTunes (ninety-nine cents a song).

Chris Maxcy had been a Rhapsody dealmaker before he left for YouTube, an upstart Levine had never heard of, and he started to send her messages: "We really need you. It's doing well, you have to believe me." Levine checked out YouTube and saw similar terrain to Rhapsody: scores of videos using popular songs as soundtracks, or just playing entire tracks with static images. But YouTube's library contained far, far more than music. She interviewed with Hurley and Chen and faced a tougher grilling from Botha, the investor. YouTube extended an offer, but Levine was ambivalent, and much of her uncertainty stemmed from the DMCA. She called a friend and fellow lawyer, Fred von Lohmann, and asked him to meet after work at her usual watering hole, the Rite Spot Cafe, a dive bar in San Francisco's Mission District. Levine arrived with pages printed out from the law, slapping them down on a table in front of her friend. Under poor bar light, she read directly from Section 512.

A website with copyright-infringing material—say, a clip from *SNL*—would *not be held liable* if one of three things happened: (1) the website lacked "actual knowledge" that the material was infringing; (2) the website didn't get any "financial benefit directly" from the infringing material; or (3) the website, once it was notified of the material, took it down "expeditiously."

"What does that mean?" Levine asked. *What did "actual knowledge" mean for YouTube?* The company didn't even know who uploaded what, let alone whether copyrighted material was posted with authorization. YouTube had started running ads on pages where videos played, but they weren't yet targeted based on the content. *Was that ad money "directly" attributable to "infringing material"?* So much of the DMCA was ambiguous. *Newsweek* had recently run an article calling YouTube the "video Napster." *Was that true?*

Finally, Levine looked up at von Lohmann and asked, "Should I take the job?"

"Hell, yes," he replied.

"But are they going to get sued out of existence?"

"Who cares?"

Few knew the DMCA like Fred von Lohmann. He had worked on one of its first prominent cases, defending Yahoo in a lawsuit over bootleg video game sales on its site. (Yahoo won.) Now he was a copyright gun for hire for the Electronic Frontier Foundation, or EFF, a prominent Silicon Valley civil liberties group. One of its founders, John Perry Barlow, an eccentric former lyricist for the Grateful Dead, railed against government and corporate attempts to rein in the web. In a 1994 essay Barlow predicted the internet's coming ubiquity and laid out the philosophy that would define Silicon Valley:

> Once that has happened, all the goods of the Information Age—all of the expressions once contained in books or film strips or newsletters—will exist either as pure thought or something very much like thought: voltage conditions darting around the Net at the speed of light, in conditions that one might behold in effect, as glowing pixels or transmitted sounds, but never touch or claim to "own" in the old sense of the word.

The future was on YouTube's side. Von Lohmann also knew that the site had hired at least one comrade: Micah Schaffer, the young staffer with the frank genitalia guidelines. Before YouTube, Schaffer had hung around with EFF staff and Cult of the Dead Cow, a hacktivist coding collective of rabble-rousers that supported dissidents. When Kevin Mitnick, a famed jailed hacker, was released from federal prison, Schaffer was there with friends filming a documentary. He had worked for Rotten.com, a repository of morbid and gross-out images created, in part, as an up-yours to the Communications Decency Act. (Many images came from medical texts.) "We love you, Micah," von Lohmann would tell him. "But there are some things that we can't unsee." If YouTube was Napstered, von Lohmann then told Levine, she would get an enjoyable front seat to one of the most important legal cases of the decade.

Levine took the offer. She arrived at YouTube's dinky San Mateo office feeling, at thirty-seven, like a grandmother. A torrent of legal issues immediately rained down. One record label honcho, friendly before, now screamed at her that YouTube owed the label "hundreds of millions of dollars!" Even peaceful talks had bumps. Schaffer had set a cheeky placard above his desk after YouTube received multiple requests from German officials. (Germany had strict laws against displaying Nazi imagery, but YouTube, which had no office there, didn't have to comply.) The placard read, DO NOT APPEASE THE GERMANS. It came down after Levine hosted a group of German record executives who did not get the joke.

But one of Levine's strangest cases came after less than two weeks on the job. PETA, the animal rights group, suddenly demanded YouTube remove a video of a truck running over a fish. "Oh my God," Levine said to a friend. "Is this cruelty? Where should we draw the line?"

• • •

freddiew: "Aces." February 22, 2006. 1:22.

The title appears, followed by a twang and a whistle, an ode to spaghetti westerns. Two guys sit at a card table next to cheap dorm-room furniture. Camera zooms on the faces, cuts to the poker chips. Faces, cards, chips. Then the gimmick: there are eight aces on the table. The smaller guy flips the table, pulls out a gun, and leaps through the air, firing, pure Quentin Tarantino.

Freddie Wong, the smaller guy, had filmed this footage at the break room in New North, a freshman dorm at the University of Southern California. Everyone on his floor was a film student or wanted to be one. They scripted and shot their own videos on cheap Flip cameras, plugged them into laptops with a FireWire, spent hours editing, then showed them off. The school gave them only fifty megabytes of free computing storage, so Wong, prowling for more space to store his creations, stumbled upon YouTube. There he found like-minded young strivers: nigahiga, a Hawaiian high schooler who uploaded spirited lip-synching clips; Little Loca, the "Mexican-American homegirl," an alter ego of a twenty-two-year-old from rural California. They were producers, directors, stars. They competed in an informal brinkmanship to see how quirky, irreverent, and attention grabbing their videos could get. Unlike *American Idol,* YouTube had no judges, only audience. It felt as if everyone watching YouTube were also creating YouTube videos.

In San Mateo, YouTube staff, enamored with this explosion, tried to keep pace. Once, after noticing how often visitors conversed with one another through video, Chen ordered up a new feature on a Friday. Over the weekend coders created a simple button to add a video response posted beneath another. Uploaders then flooded popular clips with replies to get attention.

YouTube's systems started to reward persistence. Mark Day, a Glasgow transplant in San Francisco, tried his stand-up act before countless middling, half-attentive audiences at BrainWash, an open-mic city café that doubled as a Laundromat. On YouTube he stood before a bright yellow wall in his house, and the audience came to him. He posted a gush of clips in the emerging video blogging style—talk directly to viewers and talk fast, editing out any gaps between words. It was an instant dopamine rush hit when his first clip crossed fifteen thousand plays.

DeStorm Power, a physical trainer and musician in Brooklyn, had tried MySpace and obscure music sites to get noticed. A training client asked him if he could run a workout online, so Power posted grainy footage of his push-ups and leg routines on YouTube and saw that people were viewing them. Akilah Hughes, a college student in Kentucky who dreamed of becoming the next Oprah, watched "Lazy Sunday" go viral; here was a way to create a portfolio if you lacked the right connections. *I might as well get an account,* she told herself. *I'm on this website every day.* Young video makers of color, like Power and Hughes, were some of the earliest YouTube adopters, in part because they knew how much old media was stacked against them.

These pioneers felt an easy camaraderie on the site. "It was a cool club for the kids that really weren't that cool," recalled Justine Ezarik, a graphic designer from Pittsburgh who started posting as iJustine in 2006. After gaining some traction on the site, she moved to Los Angeles to room with Brooke Brodack, the gap-toothed teen who uploaded as Brookers. The club kids got a name: YouTubers. By spring of that year Brookers had become the first certifiable YouTuber star. More than a million watched her "Numa Numa" lip sync. YouTube had created its subscription feature the prior October, and by that summer Brodack had more subscribers than anyone else. She moved west after the NBC late-night host Carson Daly found her clips and reached out, offering a job with his show. "I just love that no middleman is involved," Daly cooed about YouTube. "There's no agent, nothing."

In his dorm room, Freddie Wong obsessively watched these stars rise, tracking the elements and formulas that made their footage go viral and spread as easily as breathing. On his own account he ran tests, posting clip after clip. His jackpot: *Guitar Hero.* At a friend's apartment, Wong filmed himself for five minutes playing the popular video game as a shameless blowhard. Wong had boxcar glasses, a mop of black hair, and instant nerd charisma on-screen. And he performed the game with phenomenal mastery, snapping its toy instrument in half at the end like Jimi Hendrix. The video broke through and put freddiew on YouTube's map.

Years before Instagram influencers and TikTok stars, these young creatives invented an entirely new model of fame, luring in audiences not yet trained to spend hours of their days absently flipping through the internet.

Back then, no YouTuber went as viral as Bree.

lonelygirl15: "My Parents Suck . . ." July 4, 2006. 01:01.

Bree, a teenager, sits close to the camera in a maroon shirt, clutching a stuffed animal in her lap. Curtains of brown hair frame a small heart-shaped face with a delicate mouth and magnetic, swooping eyebrows. "I'm really upset right now," she laments.

Mesh Flinders knew Bree would be a hit the first moment he saw her. Flinders, a struggling screenwriter, had pitched scripts all over Hollywood featuring a character he invented: a nerdy homeschooled girl who obsessed over theoretical physics and boy problems. But he had no takers. He had nearly given up on his entertainment dreams when he found YouTube, a website full of nerdy kids escaping, posting hours of strange, confessional, experimental videos. At a karaoke bar he met a collaborator, Miles Beckett, a plastic surgeon in training who wanted to make movies instead. Flinders pitched his idea. "I want to make a show about a girl that disappears," he told his new partner.

They opened a YouTube account, lonelygirl15, and held a casting call, where they found Jessica Rose, a nineteen-year-old straight out of acting school, to perform as Bree, vowing to pay her once the experiment succeeded. (Rose, when she first heard the pitch, understandably thought it was porn.) Web cameras back then used a slightly distorted lens, a fish eye. When Rose leaned in, her features were magnified, "a face made for a browser screen," *Wired* would write. She appeared as Bree in her bedroom, doing what other YouTubers did: lip-synching, making carefree chatter. She talked about her friend Daniel. She replied to comments. In her July 4 video she complained about her parents prohibiting a trip with Daniel, dropping hints that her family was mixed up in a cult. Behind her was a bed with soft peach pillows and a nightstand covered in pink faux fur.

The video was shot in Flinders's apartment off Pico Boulevard. He had spent a few hundred bucks at Target to outfit the room as a teenage girl might. While Beckett filmed, Flinders punched out scripts for their next YouTube video in the corner. The pair had by now teamed with an entertainment lawyer, whose wife played Bree offscreen, writing to fans and keeping up the ruse. Flinders had scripted an elaborate story line involving the occult, which would unfold bit by bit like an epistolary novel. But he soon realized that letting fans direct his plot worked way better. One suggested a romantic spark between Daniel and Bree, so he wrote it that way. It felt as if they were inventing the future of entertainment.

Within two days the July 4 video surpassed half a million views, an audience on par with a cable TV hit. Flinders called his partner. "Holy shit," the screenwriter said. "This is working even faster and better than we thought."

CHAPTER 3

Two Kings

Robin Williams—*the* Robin Williams—strutted out onstage, and Google employees in the front rows erupted in glee.

The audience behind them at the Las Vegas Hilton auditorium was filled with tech fanatics (mostly men) wearing lanyards around their necks. Google had the marquee slot at the Consumer Electronics Show, or CES, a gadget industry bonanza, in the first week of 2006. Larry Page, one of Google's co-founders and its chief visionary, stood onstage to showcase his company's inventions. Midway through his presentation Williams joined him, firing off a fusillade of off-color jokes about the adult film convention in town and Asian companies at the trade show.

George Strompolos, a fresh Googler—what company staff called themselves—sat in the crowd waiting for Google Video to come up. He knew this presentation had been cobbled together at the last minute. A year earlier Google had begun its first experiment with digital video. At the time Google was riding high: it had just listed on the Nasdaq as a public company worth north of $23 billion, cementing the search engine as one of few survivors of the dot-com bust. Google had a voracious appetite for innovation and, like others in Silicon Valley, sensed that TV, film, and home videos were moving online, a trend that the company had to seize to remain relevant in decades to come.

Only, it wasn't sure how. Google Video had premiered as an internet service to convert captions from TV shows into a searchable database, as Google.com had done with web pages. Google toyed with user-generated videos, but Google brass had decided the top priority would be getting professional media online. Once Robin Williams left the Vegas stage, Page unveiled that project: a new Google Video store, the internet's answer to the musty cable box. Page promised that people could visit Google's store to watch everything from NBA games to *Rocky and Bullwinkle*. When CBS's chief, Les Moonves, appeared onstage to announce that some shows would be coming from his network, Strompolos applauded in genuine surprise. But, still, the greenhorn Googler had a nagging feeling. He admired the renegade, amateur media bubbling up online. He saw "Lazy Sunday" on YouTube and knew young people were flocking there in droves; they weren't coming to Google Video and wouldn't start coming there for prime-time CBS. What he and nearly every other Googler didn't know was that Larry Page had been thinking the same thing.

Back in November a deputy had sent Page an email exchange from Google Video staff about the upstart YouTube. Page scanned it and hit reply four minutes later. He noted the Sequoia investment and wrote, "I think we should look into acquiring them."

Then, two weeks after the Las Vegas show, another email was sent, this one to ideas@google.com, an address to which any Googler could pitch. Dan O'Connell, a sales associate, sent the email late on a Sunday. A guitar player and snowboarder, O'Connell had discovered he could watch clips of both hobbies on this new website, even upload his own easily. "We should look to either strike a significant partnership with YouTube or simply acquire them before someone else (namely Yahoo!) does," O'Connell wrote.

The next morning Page forwarded the email to his top lawyer, adding one line: "Where are we at on this?"

• • •

Larry Page invented Google with Sergey Brin, and everyone called them Larry and Sergey—just first names at Google, another way the technologists were upending tired hierarchies of corporate America. By 2006 the story of "the Google boys," as the press still dubbed them, was enshrined in Silicon Valley lore. At their Stanford PhD orientation Page, a recalcitrant brainiac from Michigan raised by two computer scientists, met Brin, a rambunctious math savant whose family had fled Soviet Russia. They hit it off. "You can't understand Google," Marissa Mayer, an early deputy, once told a reporter, "unless you know that both Larry and Sergey were Montessori kids." Both attended schools rooted in the pedagogy of letting children chase their own interests and buck authority. They brought this sensibility to their company. "Why aren't there *toys* at work?" Mayer said. "Why aren't snacks free? Why? Why? Why?" The boys had taken an academic idea for web search and formed a business. They rented space in a two-thousand-square-foot house of a mutual friend, Susan Wojcicki, YouTube's future chief, who watched them cram her garage full of computer servers, gear, and more brainiacs.

There were many other search engines at the time, but these listed web results mechanically, relying on just the text used within a page. Google .com ranked websites based on the number of links they received from others, using the web's recursive logic, the internet itself, to improve its results. Every Google search made Google work better. Page and Brin, not yet thirty, pinned a ludicrously broad mission statement to their founding documents: Google would "organize the world's information and make it universally accessible and useful." By 2006, Google was applying this mission to anything the internet touched. Email. Imagery. Maps. Streets. Legally treacherous fields. A lawyer who joined in 2004 was asked, with a straight face, "Do you want to work on the project where we're scanning all the books in the world, or the project where we're recording all the TV in the world?"

The lawyer froze. "What?"

The Google boys created or championed these schemes to will the future into being, but they rarely did the grunt work. They were visionaries, not managers. Investors persuaded them to hire Eric Schmidt, a veteran software engineer and executive, as CEO—someone who "made the trains run on time and broke the ties" between the founders, explained an early investor. Schmidt and the founders relied on a tight inner circle of deputies. Most grunt work fell on two women, Marissa Mayer and Susan Wojcicki, who, after lending her garage, soon joined as employee number sixteen. Mayer, another Stanford graduate, led Google Books; Wojcicki looked over Google Video. Employees nicknamed them "mini-founders"—smart, ambitious operators who could channel the wishes of Page and Brin. As Google grew, the women were among the few who could reliably get the Google founders' attention.

While Google Books had a straightforward strategy—make a digital replica of every printed page—Google Video struggled to get its footing. After TV captions, the unit had added amateur video, putting out the call for submissions in April 2005. Wojcicki gathered around with a group of colleagues to watch the first entrants. Furry purple puppets came to life onscreen, singing in a language Wojcicki didn't understand. Thirty-six, petite, and practical, with a no-fuss brunette bob, Wojcicki wore few visible signs of being a newly minted millionaire (which she was, thanks to Google's public listing). She had little experience in media production, and these purple befuddling puppets confirmed her initial suspicions of amateur video. *What type of person would want strangers to watch their videos online? Who would watch strangers?*

Years later Wojcicki would say her opinion changed after she played the footage for a test audience: her two young children at home. They erupted in giggles and asked to see it again.

But back in the spring of 2006, Wojcicki moved gingerly with amateur video and had a very clear problem: YouTube was trouncing Google. The site from the dinky San Mateo crew was raking in forty million video plays a day and growing at a tremendous rate. On *Slate*, the tech writer Paul Boutin praised YouTube and MySpace as the "next-generation Internet where people contribute as easily as they consume." After Google's January show in Vegas, Boutin had predicted that Google's video store would soon replace cable. "I was wrong," he wrote four months later, "and I think Google has failed to take off for the simple reason that it's more annoying to use than YouTube."

Schmidt, Google's CEO, was not pleased. He emailed the article to Wojcicki, adding, "Perhaps this is why YouTube and MySpace are cleaning our clocks."

Wojcicki typed out a response quickly on her BlackBerry. Most of those annoying bits were going away, she assured her boss. Google Video had initially required people to install a separate application. No more. In two weeks they would have a web system faster than YouTube with YouTube-style features like "tagging" and video sharing. The main hiccup, according to Wojcicki, was Google's wager that people would pay to download marquee movies and shows in huge numbers. They were ditching that strategy. "I think we are doing the right things now to win but we fell behind," she wrote. She followed up when she got to her desk and reassured Schmidt of an upcoming deal with Viacom giving Google shows like *SpongeBob SquarePants* and *Punk'd*.

And she had another card to play: "leverage Google.com." That is, put a link to its video store on Google's search home page.

For some of her staff, though, this didn't feel like winning. Shiva Rajaraman, a young Texas native, had taken a job with Google Video after reinventing himself. Internet companies offered two tracks up the corporate ladder: *engineering*, coding that made technology work; and *product*, design and strategy that let people use that technology. Product managers were mini-CEOs. After working for middling software firms, Rajaraman went to business school to emerge as a product leader and landed at Google, the major leagues. *Now this?* Google Video did not inspire. A hip-hop devotee, he noticed artists were rising not on MTV, the old kingmaker, but on YouTube, the new cultural barometer. He pulled up Google Video and saw Charlie Rose.

Google had another apparent handicap: the company was reluctant to publish videos without screening them first. (The Authors Guild had sued Google over its book-scanning project in 2005, which left Google lawyers nervous.) That screening process worked well during the week, when Google Video teams were clocked in, but not on weekends, when they weren't. If someone posted a video Friday night, why would they want to wait two days to see it? Especially if they could see it immediately on YouTube. As a plucky startup, YouTube could take all sorts of risks that Google, a public company with a big target on its back, could not.

Rajaraman also found some of his colleagues equally uninspired. Google Video wasn't a profit center for the company, so it felt like an easy place to goof off, as one co-worker did. Google paid a bonus for referring young interns, so this co-worker built a simple tool to scrape websites listing collegiate résumés. Every free hour he would peel emails off that list and plug them into the referral system in the blind hope that one grad would land a role. It wasn't the mark of people working on "the next great thing," Rajaraman thought. He began to wonder if he'd made a mistake joining Google Video.

• • •

Things for Chad Hurley, however, looked very different. Even when he thought he'd blown it, he struck gold.

It happened in Idaho, in July 2006, when he stood before the glowering stares of entertainment's most powerful moguls in Sun Valley, site of the exclusive deal-making, glad-handing summit that the investment bank Allen & Company put on every year. All the balding TV and movie executives were there, in matching blue vests the bank handed out. Each summer the bank picked a few lucky upstarts to attend and present. In 2006 it chose YouTube.

Onstage Hurley cycled through the presentation slides his marketing and publicity manager, Julie Supan, had assembled. Months before, YouTube had referred to itself in a letter to NBC as a simple "online forum." Hurley, not a natural gloater, now tried upping the ante. "YouTube is a consumer media company," he began. "The YouTube brand is synonymous with online video, and we are focused on building a sustainable, profitable business." (That month YouTube would actually bleed over $1 million, but he didn't share that.) Hurley flipped through flattering stats: eighty million videos watched a day; more than two million videos uploaded; 60 percent of all online video in the United States piped through YouTube. He showed Google Video's number: 17 percent. YouTube was the "birthplace of a new clip culture," a place for moguls to market their movies and shows. *A partner.* He mentioned the Brookers deal with Carson Daly, then tried going out on a high note: "People want to be seen, and YouTube is a stage for everyone to participate." *Polite applause.*

He walked off and immediately apologized for a lackluster performance to a banker standing nearby. The banker would later remember Hurley on the verge of tears. *God, no,* the banker thought. *It was frickin' genius!* The moguls were bracing for a slick punk kid, a Napster type, to lecture them on how ill-prepared they were for the future. Hurley, with his nonchalant, surfer-dude vibes, looked unthreatening. A wolf in sheep's clothing. A few bigwigs—most notably Mark Cuban, an internet billionaire and movie studio owner—chastised YouTube as a bunch of pirates. But most in the industry showed benign interest, even sympathy. A Time Warner executive visiting YouTube that summer noticed how its sixty-person operation shared only ten phone lines, telling a reporter he "almost felt bad for them."

After Hurley's Sun Valley appearance, the press christened him "Silicon Valley's golden boy." He chatted with Bill Gates at one event and showed Martha Stewart how to set up a YouTube account. He met George Lucas, the *Star Wars* creator, who was asked to join YouTube's board. (Lucas declined.) In August, Hurley's sales team cut a huge deal for advertising on the site that enabled YouTube to turn its first profit. (A short-lived one: it went back into the red the following month.) They began talks with record labels, TV networks, and mobile companies. Hurley and Chen donned black suit jackets (over jeans, shirts still untucked) in front of cheap red IKEA curtains at the office for a glossy magazine photo shoot.

Even scandals boosted YouTube. That fall a hawk-eyed blogger discovered the truth about lonelygirl15 and its fake starlet, Bree. *The New York Times* outed the aspiring filmmakers as the account's puppet masters. Yet even with the artifice exposed, viewers didn't flee. They still wanted to know what would happen next to Bree.

Behind the curtains at YouTube, though, Hurley was close to losing his title.

As the business grew, Botha, YouTube's driving investor from Sequoia, felt it was time to bring in another CEO. In his investment memo a year earlier, Botha had written that YouTube needed to "quickly hire" one. Sequoia had done this with Google, where it brought in Schmidt alongside the founders. Botha hunted for a seasoned manager to come in above Hurley and Chen and nearly settled on one: Mike Volpi, an executive from the networking gear provider Cisco, a company where another YouTube board member worked. Hurley and Chen met Volpi at an Italian restaurant near their offices that fall. Volpi, like others in Silicon Valley, believed YouTube could make money from subscriptions, charging for streaming licensed shows and movies. He thought the meeting went well: the YouTube dudes were easygoing but passionate about what set their site apart, the ease with which people could upload videos. But afterward YouTube fell silent.

Chess pieces were moving even faster in Mountain View, Google's home.

• • •

None of the Google Video features Susan Wojcicki had laid out for her CEO in the spring were working. Then, in August, Google played its ace: it added a link to Google Video on Google.com, its most valuable real estate, the place a gazillion people visited every day. Traffic barely budged. One Google Video employee recalled being stunned: "One of our secret levers, the spigot, seriously did nothing."

After that Shiva Rajaraman received a tap on his shoulder at his desk from a slight, soft-spoken man everyone at Google knew: Salar Kamangar, a walking company legend. He had started at Google as a volunteer, pledging to work for free for his Stanford classmates Page and Brin. Kamangar, a programmer born in Iran, went on to create AdWords, the auction system for selling search ads that became Google's eternal oil well. Staff had labeled him their third influential "mini-founder" and "the secret president of Google," and he had quietly been tracking YouTube's rise.

"I need to put a few slides together," Kamangar told Rajaraman. "Maybe you can help with that." They were documents to justify buying YouTube.

For everyone involved, the coming weeks were a blur. Hurley courted acquisition offers right and left, and YouTube's mounting copyright headaches and the soaring costs of hosting web video bandwidth made these offers impossible to ignore. Fox's News Corp, which had paid half a billion for MySpace, inquired. R. H. Donnelley, a telephone directory publisher, even made a bid. But only two offers really mattered: Yahoo and Google.

Google recruited one of its go-to banks, Credit Suisse, to assess the deal. Bankers gathered in a Google war room to watch YouTube footage, amused and a little bewildered as to why the silly clip site was worth so much. Earlier that year Hurley and Chen had talked to Google about an acquisition but dismissed Google's offer of $50 million as too low. Google was now proposing $615 million. One banker pondered the thankless pitch Google would have to make to its shareholders about YouTube: "The good news: it has no revenue. The bad news: you've never heard of it."

Hurley and Chen met Yahoo brass at a Denny's near their office, a discreet spot chosen to avoid recognition. Google brass had scheduled time the following day. *Wouldn't it be funny*, the YouTube dudes joked, *if we met at the exact same Denny's?* They did.

Back at their office Hurley and Chen held a boisterous staff meeting on Friday, October 6. It would be the last above the pizza shop. Now nearing seventy people, YouTube had scouted a larger facility north in San Bruno, a sleepy San Francisco suburb. The group cracked beers and made a champagne tower to celebrate the move. A few flipped over empty watercoolers for a makeshift drum circle. Schaffer shot another video. "Over here we have engineers and designers crammed in like cattle," he narrated, panning across the office. "The content SQUAD diligently reviewing pornography." Trash cans overflowed with pizza boxes and Chinese take-out bins. Bathrooms lacked paper towels. One person who visited the office after YouTube left recalled seeing piles of cease-and-desist letters stacked dormant on a fax machine.

Nerves were running high, too. Yahoo, despite its courtship, had recently filed a lawsuit against YouTube for poaching a group of its sales staff, who now had to unceremoniously pack up to leave YouTube. Many employees had heard acquisition rumors and wondered about their futures. Each Friday, YouTube had a running gag: Hurley would climb on a chair to start the staff meeting with the same line: "Everyone, I have some exciting news: we're getting bought by eBay." (The founders' antipathy toward eBay, PayPal's acquirer, was well known.) On that Friday, Hurley climbed on a chair and announced, "We have exciting news." A pause. *Was this the actual acquisition?* "We're moving to San Bruno." Laughter and groans.

Over the weekend a few employees were let in on the real plans. A small group of early engineers, who were given special voting shares, had to sign off on the deal ASAP. Some first thought the news was a prank.

Not everyone was told. Heather Gillette, YouTube's circus ringleader, arrived unawares that Monday morning at the new San Bruno offices kicking mud off her shoes. She had found a bank willing to offer funds for her dream lot—one of those shady "no document" loans from the housing bubble—but could only afford to stay in an RV there and hadn't paved over the dirt. Worried about falling behind on her mortgage, Gillette had started talking to a neighbor in case she had to sell. At around noon she was still settling into the office, far bigger than expected, when she saw a camera crew rush backward through the doors. Hurley followed them in. Then Eric Schmidt, the Google CEO. *Aha.*

The men huddled in a conference room, soon joined by Chen, Botha, and a gaggle of Google lawyers. A few YouTube employees were out scarfing sandwiches nearby and had to be corralled back inside for the official news: they were now part of Google.

Sergey Brin, Google's co-founder, showed up in a tight black shirt and sleek matching pants, billionaire chic. Hurley and Chen were in their formal wear: loose black jackets tossed over jeans and untucked shirts. Chen wore a choker necklace. Schmidt, in a pink tie and tailored suit, joked about his own appearance: "I dressed up because I was going to be on YouTube." Brin, who with Page once handpicked every Google new hire, greeted the YouTube founders and tried small talk. "So, you guys just moved here recently?"

Chen smiled and said, "About four hours ago."

"Oh," Brin replied, surprised.

Few there knew that the acquisition almost didn't happen. Google had a secret way to peek on the performance data of websites that used its web advertising system, and during the talks one Google manager peeped at YouTube's numbers, irking Hurley so much that he threatened to walk. Schmidt managed to calm him and close the deal.

At YouTube's office Schmidt stood to address his new employees. In eighteen months, YouTube's staff had carried their site from a wobbly forum hanging on Chen's credit card to one of the most high-priced, talked-about, unreal acquisitions in business. "First, you guys hit the ball out of the ballpark," Schmidt began. "You did it quickly, and you did it well. I don't exactly understand why you did it so well and so quickly, but I want you to continue because I know winning when I see it." He assured the staff that they would all keep their jobs, and Google would keep operating YouTube as a "separate facility, separate brand, separate site." Brin and the YouTube dudes spoke briefly, then opened the floor to questions.

"So," one person asked, "what are you going to do with Google Video?"

• • •

That Monday, Schmidt walked into the Google Video offices in Mountain View for a much less ceremonious delivery of the same news. Those working on indexing television captions for search, one part of the unit, were staying put, he said. Everyone else was told to pack their desks for San Bruno to join YouTube.

George Strompolos, the young operative smitten with amateur video, was in shock. It felt as if Google were throwing in the towel. But, on the other hand, he suddenly belonged to the newest tastemaker in pop culture and tech. He looked around the room. Everyone was still trying to wrap their heads around the sum Google paid for YouTube, for by now they had all read the release: *$1.65 billion?!* Schmidt was later commended for the foresight of being willing to add a cool billion to the offer to box out other suitors. Page and Brin, Google's founders, had largely been enamored with YouTube as a place people used to search.

Back in San Bruno, once Google's executives and cameras cleared out, everyone went to a TGI Fridays nearby to drink and register their fortune. Gillette cried; she could keep her home. Legal filings would later show that Hurley and Chen earned more than $300 million apiece from the deal; Jawed Karim, the third co-founder who was no longer there, made $66 million.

Before Hurley and Chen could join the festivities, a staffer reminded them that Schmidt, now their boss, had asked the newest millionaires to make a video. Schaffer shot it outside the restaurant.

YouTube: "A Message From Chad and Steve." October 9, 2006. 1:36.

> Hurley begins. "Today we have some exciting news for you. We've"—pause, a step back from the camera—"been acquired by Google." Chen jumps in to thank the audience for contributing, reassuring them this will help address technical issues. It is clearly unscripted. Hurley vamps, "This is great. Two kings have gotten together." At that, Chen giggles uncontrollably, turning from the camera. "I don't know. Just keep going!" Hurley continues. "The king of search, the king of video have gotten together. We're going to have it our way." They both laugh. Hurley looks back at the camera. "We can't do that. Cut."

Schaffer quickly edited the footage and then uploaded it onto YouTube's official account.

CHAPTER 4

Stormtroopers

Everyone had to see this. Sadia Harper called her YouTube colleagues to her desk early in 2007 to watch. On her screen a tiny preteen with a crew cut and an oversized dress shirt was belting a song by the R&B singer Alicia Keys. "This kid is amazing," Harper said. The singer's mother had been badgering her with emails to feature her son, Justin Bieber, on YouTube's home page.

Harper was one of YouTube's "coolhunters"—a team tasked with curating videos that appeared on YouTube.com and keeping their fingers on the site's pulse. Right before joining Google, YouTube had cut a deal with Verizon Wireless to put its video player on select mobile phones; app stores weren't yet a thing. The carrier wanted a handpicked selection, not the site's normal free-for-all. Apple wanted the same thing for its new device, the iPhone; during a meeting, Steve Jobs had scolded the YouTube crew, "Your videos are shit." So YouTube hired another alum from the music service Rhapsody, Mia Quagliarello, as an editorial director. She recruited Joseph Smith, or Big Joe, as everyone called him, a graveyard-shift video screener who posted his own YouTube clips and was remarkably adept at spotting budding viral hits before they exploded in popularity. Officially, the members of this team were "community managers," but a colleague had dubbed them the coolhunters, a more resonant title.

By then YouTube was already a grand, expanding estate—full of aspiring comics, filmmakers, musicians, performers, hobbyists, and enthusiasts in every niche and of every age imaginable. (Briefly in 2006, one of the most popular YouTubers was Peter Oakley, a well-dressed British retiree who went by geriatric1927.) Many people landed on YouTube from a link sent by a friend or a web search. Increasingly they watched clips from the "related videos" panel sidebar next to the main footage. But a fair number still came in through its front door, YouTube.com. The coolhunters tended the front, selecting videos for everyone to see on YouTube.com. Sadia Harper, a high school friend of Chen's, joined a few months after the Google acquisition. Each morning, she scoured a list she assembled of blogs on music, entertainment, tech, and architecture, searching for fresh clips to place on the home page. Small frames of each one, called the thumbnail, appeared on the site stacked in a row of ten beneath a "Featured Videos" banner. Her team swapped these slots every four hours, giving YouTubers behind the videos they selected a guaranteed cascade of views. Some leaped immediately into wider pop culture. Harper featured a music video with a catchy, whistling hook from a band called Peter Bjorn and John. A week later, Drew Barrymore wore the band's T-shirt on *SNL*.

Quagliarello, the coolhunter boss, encouraged her team to make videos introducing themselves. Harper shot hers in her bedroom and added clips of do-it-yourself crafting, another emerging YouTube subculture. She solicited videos to be sent to her email address, which is where the little Canadian singer's mother kept emailing. Harper had to politely tell Bieber's mom that they preferred to feature original songs, not covers. Still, even when the coolhunters passed on a technicality, YouTube minted stars—a year later, a record executive would find Bieber's videos on YouTube and make him a pop sensation.

More editors joined, each handling a select vertical on the site. A radio programmer came to curate music. Mark Day, the Scottish stand-up who found an audience on YouTube, was hired to handle comedy. Steve Grove, a young, earnest journalist from Minnesota, joined for "news and politics."

Blogging had resurrected noble ideals of citizen journalism—everyday people who could use the internet to document their communities, check facts, hold power to account. Grove started a channel, CitizenTube, spotlighting the genre's practitioners on YouTube. "What do you think about Iraq or Social Security or abortion or health care?" Grove, donning a white YouTube T-shirt, asked viewers. "What do you think about the pothole outside your front door?" (Hurley, who rarely micromanaged, did make it clear he didn't love political videos on the home page.)

YouTube's editors were proud of discovering the undiscovered, those who were willing to post experiments and post often. They took risks themselves. In 2007 one video exploded on the site featuring a chintzy keyboard and a baby-faced singer named Tay Zonday, with a stunning baritone and bizarre, poetic lyrics. Dozens of YouTubers started taping their own versions of his song "Chocolate Rain." As a lark the editorial team planned their first "takeover": they filled the entire home page with tributes to the song. An engineer rushed over, panicked, assuming YouTube had been hacked. But the gimmick worked, and they orchestrated other takeovers periodically. If the editors felt uninspired, one colleague joked, they couldn't go wrong with cats. Steve Chen loved cats, and so did the internet.

With Google's cash YouTube began 2007 on a hiring spree, bringing in dozens of newcomers. Jasson Schrock, a web designer from upstate New York, arrived for his interview in a suit, unaware it was YouTube's office "pajama day." Once hired, he strained to work out the messy startup code behind YouTube's video player, which had been strung together "like spaghetti."

Then there were the Googlers. Just weeks earlier, some had been trashing YouTube as a band of pirates. Before the acquisition, as Google considered taking YouTube's no-screening approach, a Google manager worried in an email that the company was inviting a "giant index of pseudo porn, lady punches and copyrighted material." The first meeting in YouTube's offices between its staff and the newcomers from Google Video was an awkward standoff. Steve Chen wasn't sure if he should shake hands or throw a punch. It felt as if busy parents had dropped their kids off and then just left. *You kids are living together now. Make it work.* The enormity of the task they faced quickly dissolved any tension. Chen and Hurley went into endless calls with team after team at Google, working to integrate various back ends and business plans together. Googlers delved into YouTube's code base and numbers. YouTube's team filled out piles of Google's new employee paperwork. It reminded Erik Klein, the YouTube engineer, of *Brazil,* the great cinematic parody of bureaucracy. (On acquisition day, Klein had told Brin that most YouTube engineers had failed their Google interviews or hadn't bothered applying.) At times YouTube staff felt they had the wrong pedigree. Google recruiters routinely looked at SAT scores and Ivy League stamps, and the company bragged about an acceptance rate lower than Harvard's. "It was this weird sea of bland, same-degree people," recalled Julie Mora-Blanco, the YouTube moderator. "Like, 'Oh, you got your MBA at Stanford, too?'" The YouTube crew, full of state school alums and dropouts, joked that they were townies on a college campus.

Some of the crew had more awkward encounters. On multiple occasions, Google's CEO, Schmidt, who was married but openly dated other women, brought Kate Bohner, a former TV anchor he was dating, into YouTube's offices and asked staff to give her advice on growing her YouTube channel.

Many Googlers arrived begrudgingly at the new video unit: YouTube didn't yet have catered meals and the other perks Google lavished on its staff. One Google transplant, Ricardo Reyes, felt as if newcomers were the Empire's stormtroopers marching into YouTube's Rebel fortress. Reyes, a former Bush White House operator, didn't have much of a choice about joining YouTube. He was a fixer for Google, the go-to publicity handler for crises. On a Friday that February, he had taken some Google staff out for a break, an afternoon showing of the new *Spider-Man* movie. When the movie ended, he saw his phone light up.

"Where are you?" a colleague on the other line demanded.

"Uh," Reyes copped, "I'm at *Spider-Man*."

"Get back to the offices. We've just been sued for a billion dollars."

• • •

Viacom didn't see YouTube coming.

Sumner Redstone, Viacom's chairman and legendary mogul, had turned his father's drive-in movie chain, formed in 1952, into a dynastic, backslapping, backstabbing media conglomerate. His holdings included assets as disparate as *South Park*, *Survivor*, SpongeBob, and Al Gore. Redstone had once dismissed the internet as "a road to fantasyland," but by 2006 Viacom needed that fantasyland badly. Its primary business, pay TV, had peaked in 2000 at 83 percent of American households and then began an alarming decline. Viacom tried to buy MySpace but lost to its archrival, News Corp, the owner of Fox ("a humiliating experience," Redstone admitted). Viacom offered $1.6 billion to Facebook, but the social network rebuffed it, another humiliation. Some at MTV, a Viacom station, had been keeping tabs on YouTube, and the suits knew clips from its shows were appearing on the video site without their permission. But Viacom was mostly preoccupied with another upstart: Grokster, a file-sharing website and subject of industry scorn. (Before buying YouTube, a Google executive derisively called YouTube in an email a "video Grokster.")

Also, YouTube's business didn't make any sense to Viacom. The entertainment, the expensive productions, the artistry—they gave that away for free! And then placed ads beside it. "It's like handing people keys to cars on the car lot, and selling hot dogs," recalled a Viacom executive, who then thought of YouTube as "a couple kids in the basement engaging in piracy."

Then Google paid $1.65 billion. Suddenly the kids belonged to an adult company. What followed is disputed. Shortly after Google's announcement, Eric Schmidt sat down with Viacom executives and proposed a deal to guarantee up to $500 million in ad sales to indemnify YouTube against copyright claims. Viacom had proposed a figure closer to $1 billion, and the talks stalled over this gap and "other technical questions," according to *The King of Content,* Keach Hagey's book on Redstone. Another person involved from Viacom recalls a handshake agreement closer to $800 million. Then, according to this person, the Viacom team flew out to Google's offices for a follow-up meeting around the holidays, at which point Google proceeded to undo everything. "It's a funny company," the Viacom executive observed. "When the CEO agrees on something, everyone else sees it as a suggestion."

For its part, YouTube had seen Viacom coming with Zoey Tur.

Tur was well known as the most accomplished aerial photojournalist in Los Angeles. Before the Rodney King verdict in 1992, Tur had scoped out South Central, chatted with the neighbors and local Crips gang members, sorting out exactly where to park her helicopter as riots and police brutality unfurled. She did the same two years later, floating above O. J. Simpson's white Bronco at just the right moment. The chopper cost $2 million, but payments from news channels for air footage like that more than covered costs. A decade passed, and Tur happened upon a video-sharing website. A few clicks, and there were her proprietary shots of the L.A. riots and O.J.'s famed Bronco playing on YouTube. Some clips even showed ads beside them. YouTube managed to keep nudity off its website. Why couldn't they do the same for her footage? *They never spent a dime or put their lives at risk to take these videos,* she thought. Outraged, Tur sued YouTube in the summer of 2006 for violating DMCA copyright laws. YouTube removed the videos Tur flagged but argued that it lacked the tech and manpower to find every new upload. The suit lingered, and in October, Tur sat in traffic listening to radio news when she heard about the $1.65 billion acquisition. "Wow," she blurted out. "Crime pays."

Most of YouTube's rank-and-file staff knew nothing about Tur's lawsuit or the backroom meetings with Viacom. Micah Schaffer, the former hacktivist, prepared to head out for a weekend in February—a drinking weekend, because the Super Bowl was that Sunday—when Zahavah Levine stopped him. Viacom had just sent over 100,000 links to videos it claimed were uploaded without its consent. "Can you handle that?" she asked. Schaffer and colleagues had to delete them in batches to avoid overloading YouTube's computers. This was only a prelude. Viacom had interpreted Google's negotiation response as a sign that the search company was prepared for a legal fight. And Sumner Redstone, a lawyer by training, loved legal fights. He had installed another lawyer, Philippe Dauman, as Viacom's CEO. Years earlier the two men had orchestrated a takeover of Viacom by threatening to sue its board as a bartering tactic, maneuvering, an admirer swooned, "like Leonard Bernstein and Stephen Sondheim writing *West Side Story.*" *When you're a Jet, you're a Jet all the way.*

Viacom claimed $1 billion in damages from Google on March 13, 2007. The opening lines of its lawsuit read like an indictment of entertainment's greatest criminal mastermind:

> YouTube has harnessed technology to willfully infringe copyrights on a huge scale, depriving writers, composers and performers of the rewards they are owed for effort and innovation, reducing the incentives of America's creative industries, and profiting from the illegal conduct of others as well.

Viacom's lawsuit claimed it had found, even after the Super Bowl weekend notices, some 150,000 clips on YouTube of copyrighted material, which had been viewed "an astounding 1.5 billion times." Viacom wasn't the only company suing YouTube. A year later, another lawsuit would come from a motley crew that included a French tennis team, several record labels, and Zoey Tur, who had dropped her suit to join larger forces.

The lawsuit sent immediate shock waves within YouTube.

David King, another refugee from Rhapsody, was just settling into his first assignment at YouTube when the stakes changed. Levine had struck a landmark deal with some record labels: they agreed to upload their songs into a database, which YouTube would use to find identical matches on its site using a "fingerprinting" technology. Labels could then ask YouTube to remove those matches, *or* they could reap money from ads on those videos (at least the portion that didn't go to YouTube). King's task was to manage a similar system for everything on YouTube, not just music. Interesting work, sure, though not something that brought him into contact with corporate honchos. But after the Viacom lawsuit his project to appease copyright holders became a matter of life or death. Suddenly King was invited into secretive meetings with Google's general counsel and CEO, eager to hear his plans.

Steve Grove, the politics editor, had posted several CitizenTube clips a week when he first started his job. He imagined turning his account into a weekly show summarizing current events and political chatter on YouTube, à la *Meet the Press*. After the lawsuit Hurley advised Grove to tone it down; YouTube didn't want to look too much like a TV network.

Viacom's lawsuit hung like a sword of Damocles over YouTube, and Levine and other lawyers began devoting considerable amounts of time to the company's defense. Every decision about every video now carried more weight. The gravity extended beyond just the issue of copyright. YouTube soon encountered more and more unexpected complications in an online world it had rushed to embrace. Schrock, the new designer, had started working on research studies, where YouTube invited regular people in for surveys. Once, Schrock monitored a session when a colleagued asked respondents what videos they liked to watch. "I like cage fighting," a man replied quickly. "Watching animals fight." Schrock stood slack-jawed. No one expected that. The man left and, like all research subjects, was asked to sign a nondisclosure agreement.

Even darker moments occurred, including a tragedy that echoed the horrors to come in Christchurch years later. In the fall of 2007 an eighteen-year-old walked into a high school in Finland with a semiautomatic pistol and murdered eight people before shooting himself. He was a YouTuber. He uploaded under the handle Sturmgeist89, posting about metal music and the Columbine school shooters. In videos he held a gun and wore a black shirt that read "Humanity Is Overrated." He described on camera how his shooting would proceed, but his footage wasn't flagged by viewers or YouTube's machines, so moderators never saw it. Soon after the tragedy an email arrived in Schaffer's in-box. The shooter's father wanted to see videos of his son, an attempt to piece together the horror that had unfolded. YouTube had pulled them off-line and sent them to authorities. The father kept emailing YouTube his request.

Schaffer felt torn between helping the grieving father and keeping YouTube's privacy standards that prohibited sending deleted videos. Unsure what to do, he never responded. Eventually, the emails stopped.

CHAPTER 5

Clown Co.

The first time Evan Weiss saw his new star he got a splitting headache. Weiss, a journeyman Hollywood agent, had been producing TV shows for household names—Tyra Banks, Pamela Anderson—but had grown tired of the network grind and began scoping out smaller screens. One day a friend showed him a rising star on YouTube.

Fred: "Fred on Halloween." October 30, 2006. 4:32.

> A young kid's face, far too close to the camera. A cheap green wig, witch's hat, braces covering his teeth. His voice many registers too high—and sped up, fast and frantic, like the editing. "My mom took me to our school counselor a few times. She said I have, I have temper problems." Pause. *Scream.* "I don't have fuc—" *Chaos.* The reel spins around violently for a few seconds, then back to the kid. "I think I broke my mom's camera," he squeaks.

"It's in your face," Weiss concluded. "Over the top." But his friend insisted. He worked in licensing and had just taken this kid to a mall in the suburbs to sign T-shirts outside a Hot Topic, and he had to shut it down when hordes of young fans mobbed them. Weiss watched some more.

Lucas Cruikshank loved to goof around with his cousins and a camera in his small Nebraska town. It beat middle school, where he never fit in. For inspiration he watched *Mad TV* and Brookers, the madcap YouTube comic, then invented his own outlandish characters and practiced stretching his elastic face. Fred Figglehorn, his YouTube persona, was born after Cruikshank discovered how to speed up his voice with editing software. In mere months Fred had surpassed lonelygirl15 and every other account on YouTube in viewership. Older YouTubers (barely out of their teens) were pissed about the usurper. People within the company were flummoxed, ascribing his popularity to a youth zeitgeist they couldn't comprehend.

Weiss rethought his first instinct. He considered the four markers of a good comic he had learned over the years: (1) The comic must have an original voice; Fred had that. (2) The comic must be popular. Check. (3) The comic must be subversive. (4) The comic must have a point of view. Chipmunk squeal aside, Fred's shtick was surprisingly dark and heavy. His skits touched on bullying, gender, mental health, abusive parents, the complex inner life of an adolescent. (Cruikshank, who was twelve, had lied about his age to break YouTube's rules prohibiting those under thirteen from opening accounts.) "This is *it*," Weiss concluded. He had to sign Fred.

• • •

Google, for its part, was not at all certain what to do with YouTube's constellations of broadcasters or how to make money from them. Some clearly had commercial potential, netting audiences on par with those for prime-time TV. But commercialism was anathema to many other YouTubers. At the CES conference in January 2007, YouTube's business executive, Kevin Donahue, remarked briefly that the site was "noodling" with turning its popular clips into a TV channel. This did not play well. boh3m3, a popular YouTuber with a goatee and a sordid opinion on everything, posted a video titled "Dear Kevin Donahue," tearing into the executive. Fans echoed the sentiment in reply videos, with one inviting boh3m3 to check out some YouTube clips of cats on acid. These sorts of videos fueled concerns for some at Google that wide swaths of their new video service would not play well with advertisers.

A few YouTubers tested commercial waters on their own. DVRs were all the rage at the time, letting people skip TV commercials on recorded broadcasts. The ad industry, panicking, had invented product placement, which put marketing messages directly inside shows. lonelygirl15 ran a product plug for Neutrogena. In late 2006, Brendan Gahan, a twentysomething working at an agency in San Francisco, overheard a partner pass on an offer to run a TV commercial because the production budget was too low. Gahan proposed instead approaching some teenagers blowing up on YouTube: Smosh, an account from Ian Hecox and Anthony Padilla, two recent high school grads in Sacramento with swooping emo hair, who adored "Lazy Sunday" and emulated its style in lo-fi sketches. Smoothie King once gave them $500 to make a video, a princely sum. (Hecox still dressed up as Chuck E. Cheese part time for minimum wage.) Gahan, the adman, invited the Smosh teens to their first-ever office meeting and made his pitch: produce a video that mentions the Zvue, a squat portable iPod rival, and a check is yours.

Smosh uploaded "Feet for Hands," a three-minute absurdist sketch with a premise that's exactly what its title suggests. The Zvue appeared on-screen for roughly eighteen seconds. The YouTubers received $15,000. A cottage industry of influencer marketing turned in its womb.

Chad Hurley initially opposed paying YouTubers. "We didn't want to build a system that was motivated by monetary reward," he explained at a conference. YouTube's mandate from Google didn't involve monetary rewards, either. "You guys get to completely run the ship," Eric Schmidt told Steve Chen during acquisition talks. "As long as we agree on this simple checkbox": *grow users, videos, and views.* Another executive recalled Schmidt saying, "Just grow this thing. Don't worry about how much money you spend."

But growth required keeping people motivated to upload. Revver, a rival amateur video site, paid uploaders, and popular YouTubers sometimes touted this fact in videos. Hurley finally came around, devising a project called Apple Pie to develop ways to fund YouTubers—as in, paying people to make video was as American as apple pie.

And yet Hurley continued to resist Google's bailiwick: ads. He particularly disliked "pre-rolls," commercials that appeared before a video, which he found disruptive to the viewer experience. The few times colleagues ever saw Hurley grow animated involved interruptions to people's experience of his site. Hurley once shot and uploaded a clip during a meeting and counted the second-hand ticks as they passed, demanding, "Why hasn't this posted yet?"

YouTube ran tiny billboard ads on its home page, but, unlike TV, it didn't have an audience accustomed to regular commercial breaks. Sales staff tried hawking "brand pages," custom accounts for companies, like YouTube.com/Coke, that played videos automatically when someone landed on the site, but there were few takers. YouTube tried award shows for funniest videos, sponsored by Sierra Mist. And it tried clever pop-up ads—Homer Simpson chasing a pink doughnut across the screen to promote a movie premiere. But those were time intensive and didn't "scale," transferring easily and widely—two big strikes as far as Google was concerned, because it worshipped speed and scale. The hodgepodge approach didn't take. For its first few quarters at Google, YouTube's advertising sales team failed to meet goals.

To better approach the problem, YouTube's business unit divvied up its massive corpus as if it were a beast. There was the Head, consisting of top-shelf, quality stuff—footage from TV networks, studios, and musicians signed to labels. Then the Torso, amateur YouTubers like Fred and Smosh with prospects of going pro or at least having commercial appeal. The Long Tail was the bottomless heaps of clips where Google, for now, saw little economic value. (Google ranked these, as it did with everything. Videos categorized as nine and ten went to the Head, six through eight the Torso, the rest to the Long Tail.)

George Strompolos, the Google Video transplant, sat at YouTube's Torso unit. As a lanky teenager in Denver, he would set a camera on the pavement with his friends, hit record, and attempt skateboarding feats, gathering around TVs to watch the results. At Google staff were encouraged to set aside a day a week to work on harebrained ideas, a "20 percent project," and Strompolos used his harebrained days to sketch out commercial concepts for video producers. At Google Video he had persuaded a sponsor to back a pair of broadcasters who wore lab coats and exploded Diet Coke and Mentos. Strompolos had a perma-scruff and an easy, approachable manner, particularly for a Googler. After the acquisition he sent out introductory emails to handfuls of YouTube's most popular accounts. Half didn't reply. They had never heard from anyone at YouTube before, so they mistook it for spam. But Strompolos managed to wrangle a few to participate in YouTube's first grand economic experiment. He assembled thirty popular accounts willing to host ads next to their videos or as tiny pop-up banners within them, in exchange for a share of the take. The group included screwball comics (Smosh and Ask a Ninja, a single-shtick troupe), indefatigable video bloggers (sxephil and What the Buck?), and, of course, lonelygirl15.

The men behind lonelygirl15 asked Strompolos for an advance to fund their show, a normal arrangement in Hollywood. Strompolos took the request to his boss, who turned it down. This would be a simple ad split, no advances. More views, more money. For every dollar of advertising that ran with the videos, Google would take forty-five cents and give the balance to the YouTuber. YouTube's "partner program" launched in May. Those paying close attention saw it coming: back in January, Hurley, hungover, on a stage next to Bill Gates at Davos, had blurted out plans to share revenue with producers.

• • •

YouTube's ability to mint money also suffered from a technical hiccup unrelated to its broadcasters. The company was using somebody else's machines.

In San Bruno, Shiva Rajaraman, the Google product manager, had joined YouTube's team determining where and how ads appeared on the site. Early on Susan Wojcicki called him into a meeting. In addition to Google Video, Wojcicki oversaw much of Google's advertising technology, and she had a blunt question, "Why aren't you using our stuff?" Rajaraman had to explain that YouTube's software pipes were already tied intricately to the plumbing of another company, DoubleClick.

DoubleClick Incorporated was the Mad Men of the internet age. Created in 1995, the Manhattan-based company devised a clever name—a nod to the computer mouse tap-tap—and a shrewd business model. The web was then searching for ways to stand on its legs. DoubleClick invented software that could place digital billboards online and pay website owners for the space. The company also used cookies, invisible bits of code that clung to someone's browser whenever they visited a web page, tracking them nearly everywhere they surfed. Browse a few sites on home decor, then click on a news article and boom: waves of "banner" ads for furniture. Thank cookies for that. DoubleClick set up an exchange for buying and selling these "behavioral" ads, a Nasdaq for web marketing. In 2006, DoubleClick raked in an estimated $300 million. Its true grit was its Madison Avenue wheeling and dealing, a sales force of old-media types who convinced marketing chiefs and ad agencies their commercials belonged online. One banker who worked with the firm called them "bald dudes banging on phones."

Before Google bought YouTube, someone at DoubleClick had phoned the hot video site and arranged to place banner ads there. YouTube was DoubleClick's single largest account when YouTube joined Google. But DoubleClick had gone through the wringer—probes from regulators over its data-collection habits, a messy sale to private equity—and was now on the auction block again. And Google was looking for a way to get more of a foothold in another sector to complement search ads. Its executives disagreed with Hurley and thought YouTube videos were a natural place to serve banner ads. So just months after spending an eye-popping sum on YouTube, Google prepared to spend more on DoubleClick.

Wojcicki got involved, and a banker who worked on the sale process recalled an odd meeting that offered a glimpse into her mind. Everyone present walked through DoubleClick's financial figures before Wojcicki spoke up. "The real question is," she said, "would I buy this for free?" *Blank stares.* Price tags floated in the press were about $2 billion, and everyone knew Microsoft was also gunning for the company. *Was she really proposing . . . paying nothing?* Wojcicki continued to think out loud. Buying DoubleClick would bring instant revenue, sure, but it also meant years of operational integration—sales teams to mix, software to meld, human resource headaches, and so on. If someone just handed her the business, would it be worthwhile? Finally she concluded, "Yeah, I would take it for free." In April, Google announced plans to purchase DoubleClick for $3.1 billion.

Over a decade later, many U.S. lawmakers would question why Google was permitted to buy DoubleClick, a mammoth property that, paired with YouTube, helped Google dominate online advertising. But, at the time, YouTube's business outlook looked so paltry that few cared.

• • •

While Google closed its DoubleClick purchase, two Googlers walked into a brand-new building on Avenue of the Stars in Los Angeles, with dark slabs of glass and a big rectangular hole in its center that made it look like a giant movie screen. The building housed the headquarters of Creative Artists Agency (CAA), the talent firm, and other makers and breakers of dreams. Later, in Hollywood circles, the office complex would be known as "the Death Star."

Kevin Morris, a Hollywood lawyer who represented edgy TV fare like *South Park*, had convened a gathering at his office there—part mixer, part creative brainstorm, but mostly a détente between old media and new. A year after Google's YouTube acquisition, Hollywood's power brokers, aware of how digital music had destroyed the recording industry, had no interest in suffering the same fate. From one camp Morris had invited familiar faces—TV network execs, a screenwriter, and a sprinkling of movie stars. The other side included a Sequoia investor; Marc Andreessen, the software wunderkind behind the Netscape browser; and two guys from Google.

Jordan Hoffner, a YouTube director with a horseshoe hairline and fashionable glasses, sat on tech's side of the soiree. He knew the other side well; he had spent twelve years at NBC before joining YouTube's Head unit. But since entering the tech world, Hoffner had seen a yawning language gap between his new industry and his old one. At the mixer he tried a routine to explain YouTube that many of his old-media peers had already heard. Hoffner went to the whiteboard. "Let's all agree," he began, "a YouTube video that gets a million views, that's a hit?" Nods all around. ("Charlie Bit My Finger," a charming home video a British father uploaded in May, would hit a million views in a matter of months.) "Let's assume," Hoffner continued, "that a commercial on that can sell at $20 CPMs." (TV sold ads on rates called CPM, cost-per-thousand viewers.) Everyone agreed twenty bucks was an acceptable rate for TV. YouTube had begun showing more banner ads on videos, but a big chunk of the site's audience came from overseas, where YouTube didn't run ads yet, Hoffner explained. For simplicity's sake he put U.S. viewership on that million-view hit at 50 percent. He did the math on the board. At $20 rates, with only half the audience seeing ads, this YouTube hit would generate $10,000.

Dead silence.

Mad Men, which debuted that year, drew in about a million viewers per episode. It cost around $2.5 million *just to make* each one. Of course, popular TV shows could recoup those costs with cable subscribers and commercial breaks. YouTube was free, and its brass weren't sure how many commercial breaks viewers would tolerate, if any. When YouTube first let pop-up ads run on Smosh, the account's teenage creators disabled most of them, worried that viewers would be turned off. People still called them "sellouts" in comments. Hoffner wanted media types to adjust their business forecasts to the world of YouTube, where talent like Smosh owned their intellectual property and a new commercial model needed to be built. His economics lesson raised unsettling questions for those in the room: *How long would TV's old-fashioned model hold? With TV viewership slipping away to the web, what the hell happened next?* The next year Jeff Zucker, NBC's CEO, would urge his industry to hastily find a workable business for the internet lest they "end up trading analog dollars for digital pennies."

Google believed the internet was coming for TV no matter what. But company brass were less confident in the potential of Smosh or Charlie's finger biting to make YouTube's business work. They wanted more esteemed media, what they called "premium content." Members of YouTube's Head team flew to New York, Los Angeles, Tokyo, everywhere, pitching networks and studios on placing their materials on the site. Tim Armstrong, Google's jocular sales chief, lobbied heads of sports leagues, TV's big moneymakers, on making the jump. Almost immediately after buying DoubleClick, Google hunted down another ad company, Donovan Data Systems, which sold electronic invoices to TV networks. It was, essentially, the DoubleClick of TV, and Google arranged to buy it and plug it into YouTube. The companies batted around a price as high as $2 billion but couldn't agree on terms, and talks fizzled.

Still, by late 2007, Google unleashed what it thought would be a huge selling point: the "fingerprinting" service, which YouTube's lawyers had created with record labels, was ready for prime time. Google called it Content ID, and it did precisely that: it identified copyrighted content on YouTube, letting rights owners automatically delete it or (Google hoped) leave it up and reap money from any ads that ran on the footage. YouTube had devised tools to recognize reuploads of audio *and* visual files and, eventually, snippets of material that suited the music industry's commercial complexity, forming the most robust media rights systems ever. Of all YouTube's inventions, Content ID eventually contributed the most to making its business survive and thrive.

And yet, at first, the Head had few takers. Streaming, on-demand video was not yet a thing. Netflix still shipped DVDs in the mail. Media companies that didn't regard YouTube with outright malice (Viacom) were, at best, deeply uneasy. "None of them wanted to be first out of the foxhole," recalled one Google executive. Mostly the unease came down to money. For decades TV networks had been collecting licensing commissions for their shows, called carriage fees. A cable provider, like Comcast, paid fees to ESPN, packaged ESPN in a cable bundle, and charged consumers for ESPN on their cable box. Naturally, TV networks expected this fee from YouTube. Some at Google argued they should pay it, but the c-suite disagreed. Paying violated a central tenet of Google; it did not pay fees to the millions of websites its search engine indexed to show on Google.com or Google News. If not there, why pay for content on YouTube? If it *did pay* for shows on YouTube, what was to stop all the websites, newspapers, and blogs from demanding the same? So, no carriage fees.

This stance wasn't always clear to Hollywood. After Jordan Hoffner finished his presentation inside CAA, an A-list actor in attendance approached him. Something about YouTube's magic stuck, if not its economics. The A-lister had an idea for a series of sprightly shorts, à la Charlie Chaplin, in which he would star on YouTube. "Great!" Hoffner enthused. "Who's going to fund it?"

The movie star looked at the man from Google, a company worth more than $100 billion. "You."

"No," Hoffner replied, taken aback. "We just run the ads."

• • •

The Chaplin homage never happened. And when YouTube did manage to get A-listers on board, the outcome wasn't always pretty. Damon Wayans, an edgy comic star, created a YouTube account, WayoutTV, tailored to young male audiences. In one clip, "Abortion Man," a young man, informed of his girlfriend's pregnancy, rushes to the window to yell, "Help!" The titular superhero swoops down in a cape to punch the girlfriend in her womb until a prop bloody fetus pops out. "Are you ready for the most disgusting, insensitive, unfunny, and offensive video you've ever seen?" the website Jezebel asked. YouTube had recruited Toyota to sponsor WayoutTV videos, and staff had to orchestrate an apology and salvage its multimillion-dollar deal.

Google's business teams adjusted on the fly to this "unfettered upload world" of YouTube, recalled Patrick Keane, a former sales director. "How quickly can you take stuff down? How quickly can you defend stuff? Is ad targeting really going to work?" No one knew. Meanwhile, Google was working to settle in both YouTube and DoubleClick, its $4 billion prizes. "You have these real pillars-of-the-internet-type acquisitions trying to be integrated," said Keane, "at a time when the chaos was still very real." The competition was real, too. Media conglomerates that weren't suing YouTube were plotting to take it on. For months executives at NBC, Fox, and

Sony Pictures planned a web video service to host their shows, movies, and advertisers. Word of the project made it to Google, where it was met mostly with scoffs. The music industry had tried this Traveling Wilburys approach to the web and totally stepped in it. Besides, old media lacked technical chops to pull off streaming. Someone at Google coined a pet name for the media endeavor: "Clown Co."

The cocky Googlers underestimated the competition. When "Clown Co." finally launched, as Hulu, in March 2008, it sent shock waves of paranoia across Google. Hulu's press release mentioned the word "premium" five times. Inside YouTube, discussions began over whether it should launch its own "premium" service and where, exactly, that premium stuff would come from. The Head redoubled its efforts. When an executive based in London, Patrick Walker, heard about Hulu's planned launch in the U.K., he called contacts at TV networks and persuaded them to stay off Hulu, kneecapping the rival's launch. Twice, YouTube neared significant American deals. Officials had a series of negotiations with CBS to bring shows like *The Amazing Race* online, flying to meet the network chief, Les Moonves, in Manhattan and hammering out proposal after proposal. Once, after a meeting with CBS in Las Vegas, a YouTube exec told a deputy, "The deal's done." It wasn't. Talks fell apart after the two sides couldn't agree on terms for splitting ad sales. It did not help that CBS, while a stand-alone company, was still controlled by Viacom's Redstone. As YouTube courted CBS, Viacom higher-ups made their displeasure known, according to a former network executive.

Still, YouTube didn't have Viacom mucking up a deal with the remaining Hulu holdout, ABC. Discussions advanced with plans to give the network guaranteed promotions on YouTube, and some YouTube staff were pumped that their favorite show, *Lost,* might join their site. At the last minute, ABC signed with Hulu instead. (Google's "Clown Co." jab at old media, per one person involved, soured negotiations.)

Not every media entity sat out. WWE, the faux-pro-wrestling channel, embraced YouTube. At first, fans uploaded shaky footage of WWE's pricey pay-per-view shows to YouTube, taking elaborate steps to get around copyright rules. Viewers also loved a digital series from "The Miz," the old *Real World* star turned wrestler. WWE leaned into this fandom, posting promotions on YouTube for its shows, including 2007's "Battle of the Billionaires," which starred a fading reality star, Donald Trump. WWE also understood a programming philosophy that would later dominate YouTube, summed up in the title of a WWE host's biography: *Controversy Creates Cash.*

Even Oprah Winfrey, the queen of TV, joined in. She invited Hurley and Chen on her show in November 2007. YouTube publicists had to drag them to Chicago for the taping; they shied away from publicity and didn't feel that Winfrey's audience matched theirs. During the show Winfrey stood next to the YouTube dudes and recorded them with her pink handheld camera, noting YouTube's 200 million regular visitors and its astounding tally of videos. "Are you able to watch them yourselves?" the queen asked.

"That's why we don't have time for these interviews; it's because we're watching too many videos," Hurley joked. Then he added, seriously, "It's hard to keep up."

But the more memorable guest on *Oprah* that day was Tyson, a hit YouTube skateboarding bulldog. That cute, lowbrow act—dogs on skateboards!—defined YouTube in mainstream culture, and the company would spend years trying to prove that it offered much more than that reputation.

• • •

While YouTube struggled to bring in ads and premium shows, commerce, like life, found a way. Fred, the squeaky-voiced YouTube sensation, finally met Evan Weiss, the agent hunting him, in a lobby of the Las Vegas Mandalay Bay. Weiss arranged chairs for Cruikshank, Fred's creator, and his parents at a Starbucks, pitching the Nebraska family on turning their son into a media franchise starting with a Christmas record, like the Chipmunks. Cruikshank wanted to make a movie. "Definitely," Weiss said. "I'll guarantee it."

Other YouTubers searched elsewhere for money or exposure. Some tried blip.tv, a rival service that paid uploaders. iJustine, an early YouTube regular, began to "lifecast," streaming her entire daily existence with a laptop and a portable camera strapped to her baseball cap. (No bathroom trips or nudity. Friends teased her as being the "PG Princess.") She broadcast this on a new site called Justin.tv, which would try out various business models before renaming itself Twitch.

Still, the biggest commercial forces emerged right on YouTube. Michelle Phan, an art student in Florida, opened an account in 2007 and shot videos narrating and staring steadily into the camera as she applied makeup, pausing to show particular products and techniques. She transformed herself into a Barbie doll, an anime character, a "seductive vampire." Viewers could not turn away. Phan helped birth a media phenomenon that would upend fashion. Others followed suit in different industries.

kravvykrav: "iPod nano Video (3g): Complete Hands-On Review and Unboxing." September 8, 2007. 6:18.

> In grainy footage inside a home office, Noah Kravitz, shaved head and tinted glasses, introduces himself. He hoists the tiny device. "New Apple product hits the streets. People going nuts."

Noah Kravitz wrote reviews for PhoneDog.com, a gadget website, when personal gizmos were all the rage. He began posting his reviews as videos, and his exhaustive commentaries on new devices were a hit. YouTube sent him branded tube socks as a prize for his fandom. Viewers were particularly enamored when Kravitz slowly unpackaged items on film. Once he exchanged messages with a fan to ask why. "You have the phones. We don't," the fan replied. "We can't afford to go buy all of them. It's the closest thing we can get to that feeling of bringing the phone home and unboxing it on our own."

Unboxing would soon take another commercial turn that neither Kravitz nor anyone at Google expected.

CHAPTER 6

The Bard of Google

"If it wasn't implicit, my life has completely changed!"

Claire Stapleton typed this email to her friend Chloé on a Saturday in late July 2007, soon after arriving at Google. This was over a decade before her "well-being" stay at the Indian Springs resort and light-years away from the crushing cynicism of that trip. Those weeks before sending her email passed like a whirlwind—graduation from college, a cross-country trip, starting a proper adult job. Stapleton grew up in Oakland, where she ran track and acted in theater, and had a casual, sometimes dorky air that put people around her at ease. She arrived at Google orientation as a twenty-one-year-old on a picture-perfect Northern California day. She donned the Noogler hat—a goofy cap, fitted with a blue propeller, handed out to all new employees—and thumbed her corporate-issued BlackBerry. ("Who'd have thunk," she wrote her friend.) There was a blitzkrieg of protocols and principles to absorb, a flurry of new people to meet and analyze.

"Google! it's pretty incredible," she wrote. "A strange utopia, really, everyone is either an ivy leaguer who accrued the world's most ridiculous resumé after graduating, or an ivy leaguer who got plucked for potential or whatever. everyone's very smart, very privileged in a certain way, but also cool . . . the place just zings with dynamism, and at times the ambitious nature of it is infectious."

During her freshman year at the University of Pennsylvania, the college gained access to a new website begun by Harvard students, thefacebook.com. Stapleton posted photographs there with her friends sipping from red Solo cups at parties, splaying in parks, posing for goofy, candid photos. Mostly good shots, sure, but never too much polish—no one thought people outside college campuses would ever use the social network. She thrived in her classes, majoring in English and tearing through works by writers like Thomas Pynchon. She studied with Kenneth Goldsmith, a poet who fashioned himself a "text-based sculptor" and a "radical optimist" about the web. A T-shirt he wore on campus read, "If it doesn't exist on the Internet, it doesn't exist." Senior year came, and Stapleton considered applying for Teach for America, a popular program for new grads, until she spotted a Google table at a campus job fair. *Why not?* she thought.

In her application essay, she cited Goldsmith's T-shirt and questioned the contempt Pynchon and other intellectuals showed for the "Age of Computers." They were wrong, she argued, because Google kept "channels of interconnected intellectual thought, well, flowing." She was hired.

It was a watershed year for Google's college recruitment. By late 2006 the company was adding new components, like YouTube, at a faster pace than its lean public relations team could handle. Leaders there decided to fill the gap with fresh graduates, training them to field press queries and thread corporate spin. They picked thirty recruits, mostly from Ivy League schools, with Stapleton in the first batch. Her initial assignment felt like a disappointment. She was placed on "internal communications," a unit assigned to handle messaging within the company, not the more thrilling task of clashing with reporters and TV anchors. But she ended up with a fascinating role: writing for Google's sacred ritual, TGIF.

Each Friday around 4:30 p.m., Google's entire staff convened for a meeting in Charlie's Café, an office eatery that was filled with beer, soda, and snacks. Google's co-founders, Page and Brin, held court, standing at the front of the room to discuss company updates. They also performed a sort of folksy corporate comedy routine, which "internal communications" staff scripted in advance. Early on Stapleton botched some technical point about Gmail in a script. One founder looked up during the stagecraft and asked, "Who wrote this?" Stapleton thought she would be fired.

She wasn't. And she quickly learned to adapt to the founders' nerd-comic sensibilities, writing jokes about how many pizzas a group of programmers could consume. TGIF's grandest invention was Dory, a computer system (named after the Pixar character) for submitting audience questions and voting on them. Those with the most votes were asked. Dory reflected Google's deepest held beliefs in data, efficiency, the wisdom of crowds. Dory felt like how a true democracy should feel and, for many there, quite different from what it felt like under George W. Bush.

Google's leaders openly chafed at Bush's Patriot Act and began flaunting the company's liberal California bona fides more often. The summer Stapleton joined, Google made the largest-ever corporate purchase of solar panels for its campus. As Google expanded, TGIF became a place for the company to recite and reinforce its values. During one TGIF that fall Al Gore dialed in on the very day he had been awarded the Nobel Peace Prize for his environmental work. "I heard that you won something today," Brin mused. "We all feel grateful to you." Googlers erupted in applause.

Stapleton's job was to reflect Google's specialness back onto itself. On top of scriptwriting, she ghostwrote executive emails touting Google's unique culture and sent an email every Friday ahead of TGIF—engrossing, weird missives one colleague remembered as "postmodern poetry." Affable and witty, Stapleton soon became a Google mascot. She appeared on "memegen," an internal company messaging board, where Googlers wrote lines like "I want whatever drugs Claire is smoking." She earned a nickname: the Bard of Google.

"This is PR, which I would think is kind of a joke, but at google it's life-or-death serious," the Bard explained to her friend over email. "They really believe that the fate of the company lies in the way the world digests it, and the golden glow that exists right now is understood as finite."

• • •

Not far from Charlie's Café, Nicole Wong worked in a row of desks with Google's other lawyers. Wong, a seasoned attorney who rarely got frazzled, read the email from Thailand's Ministry of Information in late 2006 and assumed it was a fake. It came, strangely, from a Yahoo email address. But Wong quickly confirmed its authenticity and read it again: Thailand had listed twenty YouTube videos that insulted its king, a criminal offense under the nation's lèse-majesté law. Until they came down, the email read, YouTube would go dark in the entire country. Before she replied, Wong reached for a phone. Google didn't have an office in Thailand, but it had contracted with a "scout," a person studying the feasibility of opening one there. It would be very bad if Thailand held that individual responsible for the offense. Wong dialed Bangkok in the dead of the city's night.

As soon as the scout picked up, Wong ordered, "You need to leave the country."

Thailand's email, which came shortly after the YouTube acquisition, would be the first of many jarring reminders for Google of what it had bought: a free-for-all website accessible in countries that did not look particularly fondly on free expression. When a demand or threat came from a nation-state, it usually landed on Wong's desk. After working in First Amendment law, Wong joined Google in 2004 and quickly ascended to deputy general counsel. Colleagues dubbed her "the Decider," a play on President Bush's nickname for himself. (They made her a T-shirt with a big Superwoman *D*.) Much of Google's political identity was formed in opposition to the Machiavellian moralism of the Bush-Cheney era. Google coined a company creed in its early years—"Don't be evil"—a corporate slogan for a company that hated slogans. It was meant to combat concerns that Google would do nefarious things with intimate details it had gathered from users' internet searches. In practice the motto stood for Google's steadfast belief that the internet was inherently a force for good. In 2006 the company brought search to mainland China, justifying requirements to censor results for queries like "Tiananmen Square" with an argument that the World Wide Web, even missing some pieces, would loosen China's autocratic grip.

YouTube began working on a version of its site for China under similar restrictions. Steve Chen, the co-founder born in Taiwan, voted for it. ("If we need to get into Thailand, then respect the royalty," he argued. "If we get into China, follow the rules.") But operational complications and opposition from other colleagues buried the project.

Overall people at Google saw themselves as proud objectors to government censors, when feasible. For Wong, the Decider, this was relatively easy at first. Search sent people elsewhere on the web, and Google could plausibly distance itself from sites that it only indexed and linked to. That grew trickier in 2003 when Google bought Blogger, a software tool that made web diaries a cinch and turned Google into the owner of a mountain of online content. Still, Blogger was manageable. Lawyers could parse written text quickly, and people in one country usually wrote and read entries in one language. Thai bloggers blogged in Thai, Greeks in Greek. So Google's lawyers developed a system to track legal risks according to nationalities.

Then YouTube mucked it all up. The sprawling, Babelian video site made internet governance nearly impossible, particularly as it expanded across the globe. Suddenly Greek soccer fans could make a video mocking the founder of modern Turkey to taunt their hostile neighbors. Which is exactly what happened in March 2007. Turkey had no problem understanding that Greek video. A Turkish judge had ordered internet providers to block YouTube in the country after the clip surfaced, and officials there, seemingly unaware that Google owned the site, didn't inform Google. (Later, though, outraged Turks protested outside Google's Istanbul office.) Wong spent hours on phone calls parsing Turkish law and sat at her home one night with sixty-seven different Turkish videos open on her computer, deciding.

To resolve the Thailand issue, Wong had proposed a technique that removed videos insulting the king from Thailand but nowhere else. Thai officials accepted this proposal. A similar proposal didn't take in Turkey, where YouTube remained blacklisted. Wong wanted to avoid imposing Google's norms on other countries, but she and her colleagues did discuss lines in the sand when they might refuse to comply with national restrictions—like if India, which then outlawed homosexuality, ever demanded LGBTQ videos come down. "What is the mandate? It's 'Be everywhere, get arrested nowhere and thrive in as many places as possible,'" Wong told *The New York Times* a year after the Turkey episode.

Tim Wu, a Columbia law professor, offered a different formulation on Google's rising power as primary gatekeeper and moderator of speech around the world. "To love Google, you have to be a little bit of a monarchist," he told the newspaper. "You have to have faith in the way people traditionally felt about the king."

• • •

Within YouTube, however, things felt less like a monarchy than a raucous parliament.

Before Google, Micah Schaffer, the former hacktivist, had tried writing standards for YouTube's moderation SQUAD balancing free speech with respectability. This required a familiarity with the web's wide, weird vastness. An early manual was seventy pages long and included a section named "Shock and Disgust," with tips for identifying videos for restrictions (drug paraphernalia; animal genitals; adults wearing diapers as fetish) and those for deletion (video playlists of kids in swimsuits or doing gymnastics that looked strung together for sexual purposes). "Use your judgment!" read one document above video stills of a woman holding a banana to her mouth suggestively (reject) and another eating a corn dog normally (fine). Videos intended to "maliciously spread hate against a protected group" were to come down; those touting "bigoted views" as commentary (for example, Andrew Dice Clay or Ann Coulter) should be marked as "Racy."

Locating and drawing such lines was never easy or unanimous. Moderators once found an account of a man ranting in his room about kosher food. He was preaching a fringe, anti-Semitic conspiracy accusing rabbis of enriching themselves from a "kosher tax," but YouTube staff didn't think the man directly slandered Jews. No one was sure what to do. Julie Mora-Blanco, an early moderator, found videos in her queue designed as mini-documentaries on the wisdom of phrenology. In others antiabortion activists posted footage showing only gruesome aborted fetuses. Schaffer then created a loose principle that such videos needed enough "educational, documentary or scientific" value to stay up.

Staff argued and debated. While unpolished, the moderation division felt agile and responsive. On top of watching YouTube, the SQUAD scanned recesses of the web for trends. They once were able to cut off a troll attack while reading 4chan, the toxic message board, which planned to flood YouTube with porn. Elsewhere on the web young girls posted "thinspiration" or "thinspo," alarming photos and videos praising anorexia. One of Schaffer's first policies was to "age-gate" this material, restricting it to posters aged eighteen and up and deleting videos from users marked under that age. (Some staff noticed that these deletions typically occurred without explanation to the uploaders.)

Google brought unheard-of resources to YouTube's SQUAD. Screeners were hired in Europe and Asia. Therapists were brought in. But YouTube remained wary of its buttoned-up parent. Once, a fleet of Google execs walked past the desks of YouTube moderators as they were examining some particularly confounding footage—a trend, out of Japan, of women using octopuses in sexual maneuvers—and two YouTube managers hurriedly hid the screens, nervous that the suits would freak out.

Still, Google openly embraced parts of YouTube. Its publicity team loved YouTube's ability to expose the rich and powerful, which made the site seem more legitimate. Michael DeKort, a Lockheed Martin engineer, spurred an investigation into the weapons maker after posting an exposé of its malfeasance on YouTube in August 2006. That same month the Senate campaign of a Virginia Republican was undone after his on-camera mockery of an Indian American (the "macaca moment") spread from YouTube to cable shows. But Google also brought orders and demands that, for some on the YouTube SQUAD, came off as prudish, classist, or just ludicrous. There were dead-serious requests for YouTube to remove any video that "glorified" illegal activity, including those showing graffiti, flamethrowers, and driving above the speed limit. Google salesmen wanted videos scrubbed that offended certain advertisers. Once, an entire staff meeting was devoted to addressing "booty-shaking videos."

Requests varied by geography. Brits were okay with sex but berated YouTube for hooliganism. The British culture secretary demanded the site place warning labels on clips with foul language. *Panorama*, a BBC program, aired an episode called "Children's Fight Club" in which the TV presenter expressed shock over YouTube videos showing bullying and violent assaults, noting how the company only screened footage once it was flagged. The BBC reporter grilled Rachel Whetstone, Google's policy director and a well-connected Brit, who returned to California and leaned on YouTube to clean up its videos. YouTube often tried pushing back. Schaffer would check in with his boss, "Hey, I'm going to defend the graffiti videos. Is that cool?"

Hurley replied, "Yeah, go for it."

Soon after joining Google, the SQUAD was placed under a new executive, Tom Pickett, an exacting operator who once served as a U.S. Navy TOPGUN pilot. Pickett attempted to bring order to YouTube's often improvisational moderation, a professionalism some respected. But others chafed at Google's mechanical processes. Under Google, moderators' job performances were graded on the "review gap"—how quickly it took them to judge a flagged video. Another metric: the "reversal rate"—how often one moderator's decision was overturned by another who viewed the same footage. "Getting it right was not part of the equation," recalled Mora-Blanco. These felt less and less like professional judgments than robotic decisions.

• • •

The Key of Awesome: "Crush On Obama." June 13, 2007. 3:19.

> Bass and drum thump slowly. A woman appears, all curves, tight shirt, and lip gloss. "Hey, B. It's me. If you're there, pick up. I was just watching you on C-SPAN." She sings. It's a music video spoof of America's latest raging political crush, the junior senator from Illinois, Barack Obama.

Politicians had always chased media fads to reach voters and signal hipness. A generation before, Bill Clinton played the sax on *The Arsenio Hall Show*. In 2008 the fad was YouTube. Ahead of the presidential elections, six candidates announced their campaigns on the site. None generated as much buzzy infatuation on YouTube as Obama, subject of viral comic hits like the one above. (After his election, several Googlers would join his White House.) Steve Grove, YouTube's politics manager, loved all this buzz. He invited Obama and other candidates to a small soundstage at Google for interviews, reading out questions YouTubers had submitted. By June 2007, Grove had arranged to take this format to TV. CNN had gone all in as the tech-savvy network in the twenty-four-hour circus, with producers tracking candidates in the Election Express bus, an enormous vehicle stuffed with a satellite dish and computers.

That July, CNN's bus parked outside the Citadel, a genteel military school in Charleston, South Carolina, and the location for the Democratic debates. CNN would host and YouTube would air the event online and present video questions from YouTubers—democratic participation with a Google touch. (CNN's producer, David Bohrman, a TV veteran, initially hoped YouTube would split the cost, but the techies told him, *Sorry, YouTube doesn't make money.*)

Grove sent CNN more than three thousand questions posted in videos, which the network had to whittle down. CNN's Anderson Cooper opened the debate standing on a sleek, futuristic stage. "What you're about to see is—well, it's untried," the anchor told the audience. Cooper then played some rejected questions. A YouTuber dressed in a chicken outfit. A vlogger who asked for candidate thoughts on a cyborg Arnold Schwarzenegger as a nuclear deterrent. Grove sat in the front row at the Citadel with Hurley, Chen, and their boss, Eric Schmidt. Two summers before, Hurley was frantically trying to erase pirated CNN clips from his fledgling site. Now his logo was displayed next to CNN's in front of a national audience. He and Chen got to meet Obama and Hillary Clinton backstage. Still facing the Viacom lawsuit, YouTube was now showing it could be a place for serious world events, not just frivolities. Once the debate ended, Schmidt, a consummate political schmoozer, leaned across to YouTube's founders. "You've arrived, gentlemen," he said.

That elation, though, quickly dissolved once YouTube staff headed home. As they sat on the runway, Chen collapsed to the plane's floor in a seizure.

He woke up, confused and aching, in a hospital bed. Not until he made it back to a neurological center in San Francisco did he get a clear diagnosis: he had suffered an aneurysm. Chen later chalked this up to a lifestyle of putting in eighty- or hundred-hour workweeks on little sleep and plenty of drinking. Doctors gave him Dilantin, a heavy-duty seizure medication, and he took time off to recover. He returned to YouTube a few times for meetings and events until the fall of 2008, when another seizure hit. After that Chen, the mischievous programmer who had worked so diligently to keep YouTube alive through its infancy, stepped away more or less for good.

YouTube had gone from a silly clip site to a stage for important world affairs, at an alarming speed. And handling all that fell to Chad Hurley—at least for a brief moment.

CHAPTER 7

Pedal to the Metal

"Time to turn off your television. And *turn* your *computers* on!"

Katy Perry sashayed down steps on the gaudy stage set Google built, wearing a pink-and-black sparkly number, commanding the audience to *turn on.* She walked straight into a hippie YouTuber holding a placard that read FREE HUGS. Hurley stood backstage, unamused.

He had shot down ideas for this sort of corporate pageantry before. In the same spirit he had rejected proposals to turn YouTube's office into a production studio, arguing that this would give an unfair edge to certain users over others. But now, in November 2008, he had been outvoted. "YouTube Live" proceeded—a stiff promotional event, streamed online and before a live audience from the San Francisco waterfront, where the company featured Perry and a handful of viral YouTubers, handed a "Visionary Award" to the queen of Jordan, and doled out free handheld video cameras.

Much had changed in 2008. The year began with Hurley stepping in stride with Google's first edict for YouTube: grow the audience. He did so relentlessly: by 2008, the site had nearly double the video traffic of its closest competitor, MySpace, then a dying star. Some chalked up Hurley's success to "luck," in having created a workable video site just when Google and the world wanted one. But others credited Hurley's instincts and sensibility for maintaining a service that normal people, even the less technically adept, found easy to use. David King, a YouTube manager, recalled Hurley dismissing proposals for features as "too geeky" or not intuitive enough. "You know, he's just a pretty normal guy," said King. At a Hollywood *GQ* party after Google's acquisition, Hurley was spotted drinking past 2:00 a.m. with one of the lesser guys from *Jackass*. Once, on a Google business trip, Hurley asked to stop for a cheeseburger—a sin for Silicon Valley's health-conscious elite. Staff found him easily approachable, especially compared with Google's wooden executives. "His lodestar in developing YouTube," said King, "was that he had a real sense of what mainstream is. And that turns out to be quite useful."

Hurley also lacked the world-conquering narcissism of other tech founders. *Time*'s Person of the Year cover right after the Google purchase showed the video player Hurley designed emblazoned not with his face but with the word "You." He didn't seem to mind. He did most of Google's publicity bidding—the *Oprah* appearance, a globe-trotting tour to inaugurate new YouTube offices and localized versions of its site. In one frantic sprint Hurley blitzed through Europe. A night in Berlin, then Moscow, then Paris. YouTube.de, YouTube.ru, YouTube.fr. He stood next to the queen when she started a Buckingham Palace account (and drank at a pub with the crown princes the night before). In Tokyo he and Chen stumbled upon YouTube Café, an eatery serving fare under Hurley's exact logo. *Awesome*. At a reception Google hosted for the Republican convention, Henry Kissinger approached Bob Boorstin, a former Clinton aide whom Google had hired for its D.C. operation, and requested an introduction to the golden boy. "Now, tell me about this YouTube thing," Nixon's consigliere asked Hurley, who graciously obliged.

But Hurley's normal guy vibes didn't always play well at Google, where some degree of world-conquering narcissism was required. Former employees noted his lack of attention to managerial details. Gillette, the first office manager, couldn't recall ever having a proper one-on-one meeting with him. Google brought in a management coach, but YouTube was by then already too mired in its own corporate morass. After costly expansions into new countries and months of struggle to make an ad model work, the company was hemorrhaging money. (It was on track, by the following year, to lose roughly $500 million, twice what it brought in.) Managerially, its arms were tied in knots. YouTube engineers reported up the company chain to executives at Google in Mountain View, who wrote their job reviews and budgets. Critical components flowed from Mountain View, too: an algorithm YouTube once used to pilot a new formula for serving related videos sat on the server of one Google programmer; when he was out sick, YouTube was helpless. Its sales staff also reported up to Google. Neither Google sales nor engineering considered YouTube, the dogs-on-skateboards site, a top priority. Hurley managed the third leg, product, out of San Bruno, but even he reported up to yet another Google executive, who reported up to Schmidt. One director brought in around that time described YouTube as a "three-headed hydra monster"—one that Google felt Hurley was not up to handling alone.

Also, Google moved his goalposts.

In February, Eric Schmidt set a new edict for YouTube: devise a working business plan. Schmidt called this his "highest priority" for Google that year, but the edict caught Hurley by surprise. "You didn't tell us to work on it," he protested to his CEO.

"Well, times have changed," Schmidt replied.

By March 2008, Google's stock price had fallen 40 percent from the autumn. GOODBYE, GOOGLE, ran a headline in *Forbes*. Critics knocked the company as a "one-trick pony" (search) burdened with expensive side projects. YouTube, in particular, looked like deadweight. Sure, it captured the attention of youth, but so did a rapidly rising company called Facebook.

This new order—make money—would lead to a drastic overhaul of YouTube's operations in the coming years. Those inside saw it coming when a new face from Google started showing up in San Bruno: Salar Kamangar, Google's "secret president" and inventor of the search ads auction. He worked in the same office as Hurley as a "facilitator" or "co-CEO," depending on whom you asked, an arrangement most didn't quite understand.

They did surmise that he had been dispatched there to whip YouTube into profitable shape. A new marketing director joined, too, and began planning "YouTube Live," the splashy November event. A sprinkling of the site's viral hitmakers were invited: Obama Girl; Tay Zonday, the "Chocolate Rain" singer; the teens behind Smosh; and "Will It Blend?," an account from a Utah manufacturer whose founder stuffed various objects into his company's blenders on camera. Freddie Wong, the film school student, was asked to re-create his *Guitar Hero* clip. He stood backstage absorbing the spectacle, a little amused. No one there knew who he was.

Others at YouTube had joined Hurley in objecting to the promotional stunt. The event "went against our DNA," recalled Ricardo Reyes, YouTube's communication chief. "It was produced. And we weren't in the business of producing content." At least not yet.

• • •

As YouTube grew, staff started seeing problems with another part of the service: its comment section. Michele Flannery always warned YouTubers about it, particularly the women who posted on the site.

Flannery managed music for YouTube's coolhunter team. A former local radio director, she loved finding hidden, esoteric gems on the site, digging through it as she would a dusty record crate. Always prowling for unconventional artists, she unearthed gifted ukulele players and singular indie rockers. "Make it really personal and intimate," she advised musicians posting on YouTube, "like you're sitting in your bedroom." Before featuring them on the home page, Flannery also warned musicians about the inevitable onslaught of attention and insults.

Including a section for comments initially helped YouTube dispatch its rivals; an ability to interact beneath videos and with video makers fostered devoted followings and kept people glued to the site. But the darker side was always there. Chatter below some videos devolved into juvenile battles or a parade of spam. "We knew it was a quagmire," confessed Hong Qu, an early YouTube designer. Once, before Google's acquisition, Hurley and Chen tasked Qu and a few colleagues with fixing comments. Digg, a web aggregation site, was popular at the time, and someone proposed lifting Digg's feedback system: one pixelated thumb pointed up, another pointed down. With those icons viewers could vote on comments, surfacing the most up-thumbed comments to the top. Qu worried this might elevate the juvenile bullies flooding the site. "If you ever read the Federalist Papers, James Madison warns about mob rule!" he pleaded in a meeting. *Blinks.* Thumbs it was. They had to devise a solution fast, and besides, one engineer noted, a weighted comment section meant the computer systems had more data to parse, always a plus.

A few years later, YouTube would swap out its five-star rating system for the two thumbs, like and dislike, beneath every single video.

Soon Flannery saw the mob turn on the musicians she discovered. "The comments were insanely awful," she remembered, "especially if you were featuring a woman or a person of color, someone who didn't fit the standards." Many comments on videos from DeStorm Power, a Black musician in Brooklyn, were racist slurs or invective like "Go back to Africa." Power shook off the "keyboard killers," as he called them. "It was just sad," he recalled. The company did give its broadcasters an ability to filter comments, but they had to select each term to filter. YouTubers like iJustine passed "bad-word lists" around to one another. Commenters, Flannery determined, were playing a crass sport, pushing limits of acceptability to see how far they could go. Eventually, she stopped looking at the comments section altogether and told YouTubers to do the same.

Hunter Walk thought comments could be fixed. Walk, a tall, loquacious YouTube manager, had a textbook Google résumé—consultancy, Stanford MBA, rose-colored internet glasses. Before Google he had worked on

Second Life, the shade-too-early virtual reality game. He came over to YouTube shortly after the acquisition and was the "first person to be full-on Google," said Chris Zacharias, a former YouTube developer. While Walk's go-getter, corporate style rubbed some YouTube natives the wrong way, he earned the trust of Hurley, becoming his top deputy managing the look and feel of the site, the *product*. Colleagues found Walk sometimes "annoyingly disciplined" about growth, as one put it, shooting down ideas that weren't immediately about spawning more video views, like a spin-off feature to record karaoke performances. But he had an evident affection for YouTube's idiosyncratic culture, and he shared Hurley's distaste for disrupting it with ads. Under Schmidt's new orders, YouTube began slapping more commercials on the site and hiring more "monetization" engineers. Walk once greeted one tartly, "What are you doing to ruin my user experience today?"

Walk had assigned two staffers to mitigate the disastrous state of YouTube comments. Then, suddenly, they were reassigned. YouTube's pivot to sales at Schmidt's command came just before the housing bubble burst and everything tanked. Google did not suffer like the banks, but it had to freeze hiring for the first time ever and trim elaborate perks born during its rapid global explosion (all-expenses retreats in China, Reese's Peanut Butter Cups imported into India). Walk's two staffers were moved to work on YouTube monetization. Later, looking back, Walk would call this decision "YouTube's original sin."

• • •

Those working in YouTube's moneymaking unit, even with all this sudden attention and resources, still felt existential dread. In the fall of 2008, Google recruited a new manager to oversee the look and feel of YouTube ads. Shishir Mehrotra had come to Google from Microsoft for a secretive project to make televisions more interactive. Quickly, though, he discovered the project was a dud. (Page and Brin, Google's co-founders, disliked the service, in part because they found television to be vapid, a waste of time.) Higher-ups at Google nudged Mehrotra to an opening at YouTube, surely not a dud.

In one of his first meetings, Mehrotra sat with Schmidt and Google's new chief financial officer, a Canadian telecom executive named Patrick Pichette. The CFO put three charts on the table. The first showed the money YouTube was shedding—hundreds of millions of dollars a quarter. Second chart: how much YouTube lost every time someone watched a video. The third: YouTube's views over time, a line pointing skyward. The CFO didn't even mention its gigantic, looming Viacom lawsuit, but his message was clear: YouTube was a smash consumer hit en route to bankruptcy. "This is the worst business on the planet," the CFO said. He asked if Google should look into selling YouTube or shutting it down.

Such questions, Mehrotra would learn, were recurring at Google, only half in jest. Nearly every quarter its financial chief would eye a growing list of the company's internet assets and science-fair projects—bottomless free web video, free global maps, and later self-driving cars and computerized eyeglasses—and wonder aloud how long they should remain funded. Google rarely pulled plugs abruptly. Instead, it usually let fading projects slowly die on their vines; eventually, the Google Video searchable TV database did just that.

YouTube, however, had a turn of luck that helped it avoid that fate.

Viacom's billion-dollar lawsuit had dragged on and on for more than three years, with growing bitterness. (After Philippe Dauman, Viacom's CEO, berated Google as pirates during a 2008 conference, Schmidt approached *The New Yorker*'s Ken Auletta and growled, "Everything Philippe said is a lie. And you can quote me!") Google took the fight to court. Copyright law found a website at fault if the website had "actual knowledge" of infringing material it hosted. So Google lawyers created a presentation designed to show how flimsy this idea was. They played three videos: a modified reel from Stephen Colbert's Comedy Central show; another from Fox

News of Bill Clinton jabbing his finger at an anchor; and a third, a lo-fi, zany clip of the YouTube star Brookers. Then the lawyers asked: *Which clip should stay on YouTube, and which was pirated?* Everyone always got it wrong. Fox had permitted Clinton's segment to live on YouTube, and the Colbert clip was okay under "fair use," a copyright clause that permitted borrowed footage for scholarship, criticism, or satire. It was the Brookers one, produced under her NBC deal, that needed to come down. The point of this exercise was to show that people couldn't readily identify who owned a copyright, had license to use it, or violated it by watching videos. So how could YouTube?

But YouTube's real legal shiv came from uncovering the inconsistencies: Viacom had okayed several of the very same clips its lawsuit listed as stolen. At the offices of YouTube's outside counsel, Micah Schaffer pored over spreadsheets of relevant videos as if he were on a forensics expedition. His brother was in the comic troupe behind "Lazy Sunday," and he knew *SNL*'s marketing unit didn't mind promotional boosts from the clip's virality on YouTube. Similar cases came up in Google's dig for dirt. CBS, a Viacom network, had sent a takedown notice for a clip with Katie Couric posted by TXCANY—an account, Google discovered, run by a CBS marketing agent. Viacom employees had gone to Kinko's to upload clips from computers that couldn't be linked back to Viacom. The company had demanded a reel from Al Gore's *Inconvenient Truth* come down, but Paramount Classics, Viacom's studio, had sent a conflicting email: "Clip is okay."

"Oh my God," Schaffer marveled aloud. "These people are fucking idiots."

Schaffer and several other early YouTube recruits were deposed in downtown San Francisco. Company lawyers presided over training sessions beforehand. (*"I don't recall." Good.*) On Google's internal web forum staff were awarded badges, like Boy Scouts, for feats such as fixing a bug in search. There was also a badge for being deposed. "Among other things,"

Hurley said in his deposition, "we realized that we were regularly making mistakes and taking down videos that actually were authorized and had been uploaded by the content owner." On June 24, 2010, a Manhattan district judge ruled that YouTube was protected under safe harbor provisions of the DMCA. It was a summary judgment, not a definitive one, which meant Viacom could swat it back to the courts. (It did, dragging the case out for another three years.) But YouTube decided to take a victory lap. Zahavah Levine, its lawyer, wrote a blog post on the ruling titled "Broadcast Yourself." Another brief post on the company's site declared "YouTube Wins Case Against Viacom."

At Google headquarters staff celebrated the following day at its TGIF. Free beer flowed. Before Page and Brin came out to speak, a screen played a clip of Jon Stewart, the *Daily Show* star and Viacom employee. A lawyer for Google's outside counsel grabbed Schaffer. *Pack your bag. We're going to Vegas.* Several YouTube employees flew out to bask in victory. Heather Gillette, YouTube's first copyright manager whose vigilance about the practice annoyed engineers, wasn't invited on the junket. But she did get an apology email. "[It] was all the hard work you did," one engineer wrote, "that won us the Viacom case in such dramatic and convincing fashion."

• • •

YouTube's victory over Viacom and its hunt for money were covered closely in the press as signs of a Web 2.0 child growing up. Around that time, a less public but still significant change occurred. It began with car videos.

Sadia Harper, the early coolhunter who passed on pint-sized Justin Bieber, had taken over curating automotive videos. She liked cars. People liked watching cars on YouTube—racing cars, Humvees climbing walls, detailed tutorials on engines. Periodically, Harper would slot interesting ones onto YouTube's home page. One day, a programmer approached her at her desk and explained that engineers had developed an algorithm designed to pick home-page videos using viewing data designed to get optimal clicks. They wanted to test it on a trial category. *How about cars?*

The coder loaded a sample page of videos the algorithm had selected. *Enter. Refresh.* The reloaded page filled with "revving" videos—footage shot inside luxury vehicles where cameras lingered on the foot or lower half of the driver, usually a woman in heels, pumping the accelerator. Often, leather was involved.

Harper had seen those sorts of videos and intentionally ignored them. "That's a fetish," she protested. "That's not what we're about."

In YouTube's early years, its algorithms for recommendations—the set of instructions telling computers how to behave—were fairly simple. The primary ingredient was "co-visitation": when someone clicked on a video, the page's right flank, its "related videos" section, filled with clips that other viewers who clicked on that same video watched. *People who like this also like that.* Experiments, though, could go awry, sometimes showing YouTube too much of a glimpse into the internet's dark mirror. In the old San Mateo office, programmers once tweaked the system and saw video clicks shoot up, only to discover the related sections were now brimming with "boobs and asses basically," as one engineer put it. Programmers went back to the drawing board, adding more filters for decency into the code. Computing formulas for predicting particular outcomes—*this person will probably click on this related video,* for millions of people a day—need clear signals and minimal noise. The clearest signal from YouTube was what a viewer watched before, but others mattered, too (how long someone watched, what time of day, which country). More video views brought more signals. Google brought more computing horsepower, coding proficiency, and loads of signal readers. YouTube's algorithms improved. Early on they couldn't detect a butt from a peach, leaving that to humans, but now they developed skin-detection algorithms to remove obscene stuff automatically.

Related videos started clocking more clicks. The home-page formula looked ready for prime time.

At the same time, as YouTube's third year at Google closed, its coolhunters looked less and less relevant. Harper and her tight-knit team were still able to assemble YouTube.com and experiment with its canvas. They set up guest edits, handing over the site to delightful weirdos like Rob Zombie and Wes Craven for Halloween. Michele Flannery, the music curator, thought of the home page as a virtual town square, a place where viewers could congregate to discover creative works from "everyday people," as Hurley first described the site. But YouTube's turn toward profitability chipped away at that purpose and exposed an emerging double standard. When Lady Gaga's hit "Telephone" debuted, Google salespeople wanted the music video, a raunchy, slick featurette set in a women's prison, to premiere as a paid promotion on YouTube's home page. Some YouTube staff protested that a similarly raunchy video from amateur uploaders would be "age-gated" and prohibited from being featured. Lady Gaga won.

A sales leader once asked Harper to put an advertiser's video on the home page. When Harper declined, citing the video's quality, the salesman pointed to the assorted clips sitting on YouTube.com and asked, "Is *that* stuff any better?"

Some YouTube higher-ups had always had misgivings about the curation team. In court against Viacom, YouTube argued that besides removing smut, it did not proactively scout videos on its site, and yet here was a team dedicated to doing just that. As YouTube expanded globally, costs of replicating a curatorial team for each country felt too high, particularly when software could do such tasks much more cheaply. Besides, the signals indicated that the home page wasn't working; most people were going there to search, not linger or click on videos. A few staffers thought the coolhunters operated as kingmakers, picking video stars in a hidden, undemocratic way.

But the most damning case against the team was that it lacked a way to measure itself. At Google, everything was measured. Harper, who had studied mathematics, put together a data analysis trying to quantify the impact of their virtual town square. It was not enough.

A strategic shift was under way at Google. The company had determined its video site should present media tailored just so for every viewer, the way social networks did. A product manager, Brian Glick, began meeting with the editorial team to discuss ways the home page could be more "relevant" to viewers, primarily with more algorithmic decisions. Mark Day, the comedy editor, had an epiphany in one of these gatherings.

Oh, wait a minute, Brian, Day thought. *Your job is to eliminate* my *job.*

Shortly after, the coolhunters were disbanded. Most members were reassigned to work in marketing roles helping brand-name companies sell on the site. YouTube's machines would now pick the videos.

Part II

CHAPTER 8

The Diamond Factory

Danny Zappin, inmate number 08036-032, had plenty of time to kill. He spent it painting portraits of other prisoners and playing softball. Like everyone inside, he thought about how he wound up there. How Hollywood had spat him out.

Just a few years earlier he had been near its inner ring. He moved to Los Angeles in 1998 after quitting film school—a waste of time, he decided. Wiry and pugnacious, with a shock of red hair and a cocksure strut, Zappin got a bit part as a gangster in Spike Lee's *Summer of Sam*. But other roles didn't come, and besides, Zappin really wanted to create. He joined two other hopefuls behind an online studio, Certified Renegade American Product (or CrapTV.com). They pitched a late-night pilot to Comedy Central called *The Hot Show*. ("The ideas are pretty out there," a network brass boasted to *Variety*.) But the network never aired it, and funding for the website shriveled. So did Zappin's bank account.

He didn't have connections to get any further in showbiz. Desperate, he tapped a connection he did have: a drug dealer, who offered to put him on a plane as a mule to Ohio, where Zappin had grown up. It seemed like easy money, and it was, until cops busted him in the Cincinnati airport. Zappin was sentenced to the Nellis Prison Camp, an air force encampment in the Nevada desert.

So there he was, painting, hitting softballs, thinking. Zappin also spent his time plotting his next stab at Tinseltown. In small prison notebooks, he sketched out visions of a bolder CrapTV, this time in an industry free from studio heads and network brass.

Once released from prison in 2005, Zappin landed in a halfway house with slightly more freedom and internet access. He discovered MySpace and this incredible new website YouTube. Zappin became obsessed. He used cash from a job parking cars to buy cameras and then upload, upload, upload.

• • •

Sleight0fHand: "Danny Diamond Gay Bar." [date unknown]. 1:51.

> Camera angles up at a ginger tough in a track jacket, smoking a cigarette in slow motion, menacing stare. The soundtrack hits—a chintzy garage-rock chorus, "I wanna take you to a gay bar, gay bar, gay bar." The tough, now shirtless with baggy khakis, dancing like a drunk frat boy. Same menacing stare.

On YouTube Zappin invented a persona, Danny Diamond, a thuggish, absurdist performer who goaded others to best him in dance-offs. During home confinement he met Lisa Donovan, an aspiring comedian from Scarsdale, New York, with zero shame in front of a camera. They began dating. Lisa and her brother Ben, another actor, joined YouTube and published reams of videos with Zappin, mostly on Lisa's account, LisaNova, where she parodied reality TV and fellow YouTubers as a caricature of a stone-cold diva.

Zappin became LisaNova's de facto manager and shameless promoter. When YouTube held a comedy competition, he emailed its staff incessantly demanding she be crowned. YouTube's home page reserved slots for videos with the most comments and likes. So Zappin had a friend write a software bot that generated ten different comments and flooded LisaNova's account every time it posted. The maneuver technically broke YouTube's rules, but he was never caught.

To a certain set of fans during YouTube's first years, Danny Diamond and LisaNova *were* YouTube. Old media began calling. Chris Williams, a former Yahoo executive, had just sold his teen-focused web studio, Take180, to Disney and hired Lisa Donovan to promote his site on her YouTube page. Once live, the clip crossed a million views its first day, blowing past Williams's expectations. *I've got to change everything about the business plan,* he told himself. He called Disney higher-ups and proposed moving his studio's material from his site to YouTube. Disney's lawyers, wary of piracy, shot him down. By then Zappin and the Donovan siblings were discussing their own studio. Zappin wanted to model it after United Artists, the production company Charlie Chaplin co-created a century earlier to give artists control over films; back then, a studio head lamented that "the lunatics have taken over the asylum."

Zappin's lunatics would soon take over YouTube and fuel its astounding commercial growth. He nicknamed his operation the Diamond Factory and invited other YouTube stars to join as fellow studio heads, promising them control over their production. They were media rebels who took pride in transgression. One Diamond Factory regular, who posted regular videos in blackface drag, liked to call their oeuvre the "anti-Disney," a label that certainly fit Zappin and his pals. At least for the few years before Disney itself came calling.

• • •

Up in San Bruno, Chad Hurley had grown tired and bored. YouTube, his baby, had proven its bona fides as a global phenomenon and a permanent fixture of the culture. In 2010 the pint-sized singer a record producer discovered there, Justin Bieber, went double platinum. The Viacom lawsuit was over; the bastards lost. And still, every week, Hurley was summoned down to Google's office to get chewed out.

YouTube had an ad hoc board composed of Google brass, charged with enforcing the new mandate to make money. Every Wednesday, Hurley drove down to meet them with Kamangar, his new co-chief. Of utmost concern for the board was YouTube's "sell-through" rate—the percentage of available slots for online ads sold to marketers. On Google search, this metric made sense. Imagine a law firm that specializes in divorce. It might buy an ad on TV or a billboard so that people will come to recognize its name. But it will *really* want to buy an ad that appears atop Google when people search for "best divorce attorney." Google had great sell-through rates; law firms didn't see much use in advertising on YouTube.

Until the Viacom case ended, YouTube was nervous about running ads widely for fear of copyright missteps. Standoffs with record labels prevented YouTube from placing ads widely on music videos, a huge part of its site. Before the acquisition a Google exec had scrawled on an internal document an estimate that YouTube would have a sell-through rate of around 75 percent by 2010. It was nowhere close. By 2009 less than 5 percent of videos were eligible for ads, and YouTube had found sponsors for only 3 percent *of those videos.* Sometimes Google brass showed an open disdain for its business model. When one YouTube ads manager presented during a TGIF, Sergey Brin, a billionaire made rich from Google search ads, cut him off mid-speech—the interruption was a joke about YouTube's annoying preroll commercials.

Another time, Google almost blew up YouTube's entire model. One reason YouTube had struggled to recruit TV networks was that networks wanted to use their own video players. So Kamangar and Google higher-ups voted to morph the site to allow them to do so; YouTube would show its own videos as well as links to clips from Hulu, CNN, and so on, something that looked like Google search. Patrick Walker, YouTube's European chief, learned about this vote when he woke up in London. *We absolutely should not do this,* he thought. He had recruited the BBC to the site and believed he was gaining traction by pitching Europe's media moguls on YouTube as *the* new media destination. And now Google wanted to turn it into a "giant link engine." Walker phoned Hurley and sent a long, harsh email to Google brass calling it "the wrong decision." Google undid its vote.

Skirmishes like this were expected, but all the board trips, minutiae, and metrics exhausted Hurley. A colleague later described Hurley's workday as a *Dilbert* cartoon. Co-workers noticed he had checked out. Four years after the acquisition, when his Google stock turned into tradable shares, Hurley left, announcing plans to focus on Hlaska, a menswear line he had helped start before YouTube. It was a quiet end of an era, the departure of the last YouTube inventor who, despite his managerial missteps, tended to prioritize people's experience of the site above commercial success, a trait quickly fading away in Silicon Valley. Hurley ran YouTube like someone who used YouTube and his exit cleared the way for managers to run YouTube Google's way, with spreadsheets and algorithms.

Hurley's exit also left the *Dilbert* dilemma for Shishir Mehrotra. Mehrotra had come to YouTube with Microsoft's buttoned-up corporate swagger, which turned off some of the more laid-back YouTube rebels. Compact and voluble, Mehrotra was the son of two computer scientists and fiercely competitive in everything. He liked playing poker and formed his first company at age twenty-one. He joined YouTube at twenty-nine but carried himself as a seasoned hand—"the oldest youngest guy," per one co-worker. He loved complex technical problems and corporate metrics, which he enjoyed reciting during meetings.

When Mehrotra met with YouTube's ad hoc board, they proposed a fix for its sell-through problem: *Why don't you reduce the number of ad slots?* Again, on Google, this strategy made sense. Think back on the divorce law firm desperate for clients. Search thrived on scarcity: if Google showed fewer ads, law firms would pay *more* for ones they could get.

That wouldn't work on YouTube. Unlike Google search, YouTube didn't have scarce, valuable real estate to offer, and advertisers sure weren't banging on its doors. Mehrotra ran the math. It made more sense to do the opposite of what his board recommended; to improve YouTube sales, he should make available *more ads*, not fewer. Mehrotra went to Dean Gilbert, a grizzled media veteran at Google, and asked for advice.

"Look, if you do what they say, and it doesn't work, they will deny ever having said it," Gilbert told him. "If you *don't do* what they say, and it works, they will take credit for it. Just make sure the damn thing works."

The damn thing worked. YouTube increased the percentage of videos eligible for ads to 10 percent, and its sales began to climb. "Revenue solves all problems," Gilbert observed.

At around that time YouTube also began exploring ways for its advertisers to exploit the new hotness: product placement. The sexy hamburger company had done it first. Years earlier Carl's Jr. had run a TV spot with Paris Hilton washing a car, flouncing erotically while scarfing a burger. Now the fast-food joint was preparing another splashy ad campaign to sell its new $6 sandwich to young men and had set aside millions for TV. With a small amount of cash left over, an ad agency proposed spending it on little-known web stars. Google lined up nine YouTubers to promote the burger for a few thousand dollars. iJustine and Smosh posted videos. Another teenage sensation, nigahiga, made a one-minute clip wherein he rubbed the burger on his car, armpit, mouth, and nipples. Contractually, the YouTubers only had to utter the slogan "How do you eat your burger?"

George Strompolos, the skinny YouTube manager, did his own math when these videos debuted in June 2009. Those nine clips raked in more than eleven million views, outstripping the audience for the expensive commercial; even YouTubers who weren't paid made videos about the burger.

TV's advertising model was broken, he concluded, and this was the gold standard of a new one.

• • •

Danny Zappin imagined a different gold standard. That summer he sat in the backyard of his new Diamond Factory headquarters, in Venice, lounging on a multicolored hammock and holding a BlackBerry. On the phone was a suit from Sanyo, a Japanese electronics firm, which was weighing a sponsorship deal with the YouTuber. The Sanyo adman brought up the Carl's Jr. campaign and deemed it "dumb." Clumsy viral marketing, he went on, an obvious pitch that young audiences wouldn't swallow.

Zappin laughed and said, "It's kind of like, who's coming up with these brilliant ideas, you know?"

He started pacing around the backyard where a bright blue surfboard hung above him, roped to the deck siding. Zappin was pitching Sanyo on a more authentic promotion for its handheld digital cameras. "It will have all our big stars in it," Zappin said, dragging from his cigarette. "We're just not going to call it a sellout." Zappin would get LisaNova and other Diamond Factory YouTubers to upload clips inviting fans to enter a contest: make a commercial for Sanyo on YouTube and you might win a free camera. Zappin guaranteed at least ten million views with the promotion. Sanyo would pay him $60,000.

The suit sounded unconvinced.

In the backyard a colleague started filming. Zappin, once he noticed, put his phone on speaker mode and hammed it up for the camera. Advertising agencies wasted money and didn't understand YouTube, he declared. "We know who's going to get views and who's not," he said. "And that's why agencies should get out of this business and let us deal with it. Because we will not fail. We do not fail. It's not in our DNA. We know what we're doing."

Zappin was embellishing, as he often did. LisaNova had made a sponsored clip back in March, but the Diamond Factory had never done anything like this. The YouTubers had run up serious credit card debt funding their dreams. Still, they had timing on their side. The bursting housing bubble had sent the economy in a free fall. Companies were desperate for consumers; people were desperate for work. After YouTube began paying broadcasters in its partner program, those who viewed the site as an amusing hobby started to see it as a possible lifeline. Zappin, a savvy hustler, knew this. He took YouTubers under his wing, promising them fame, movie deals, merchandise lines. He was particularly fond of Shay Carl, a bearded, bearish YouTuber in Idaho who had left a job setting countertops to post frequent, zany home videos of his family (YouTube handle: Shaytards) that earned impressive traffic. Real estate opened up next door, and Zappin offered it to Carl and envisioned filling the house with YouTubers, a decade before social media influencers piled into mansions regularly. Zappin put money down on the house before Carl agreed. Many other moves were impulsive. After Philip DeFranco, a popular YouTuber (handle: sxephil), shared that YouTube had given him a $100,000 grant to upgrade film equipment, Zappin begged YouTube for the same grant; once he got it, he went on a spending binge. As a business partner, he had recruited Scott Katz, a former movie lawyer, who read the contract YouTube sent and discovered there wasn't an up-front payment, as they had thought, and that the money would come only sixty days after videos were uploaded. "He definitely leaps before he thinks," Katz recalled.

But Carl did move in, and Google's money did arrive. Zappin gave his venture a proper name, Maker Studios, and went hunting for staff, flying to New York to court Ezra Cooperstein, a director at the agency behind the Carl's Jr. ads. "Dude, you've got to come join us," Zappin urged as they walked around Midtown.

"We're building the United Artists for YouTube." Cooperstein joined. So did Jeben Berg, a YouTube employee who ran its comedy contest. Berg soon noticed that Danny Zappin operated like a benevolent Fagin, luring promising YouTubers to Los Angeles and finding them shelter. "Come here," he'd say. "I'll take care of you." Others came aboard with vague roles; staff dubbed them the FOD, friends of Danny. Zappin ran his enterprise from his gut, freewheeling and unscientific, the polar opposite of Google.

Newcomers from TV world were caught off guard. Mickey Meyer, a film school graduate who had worked on *The West Wing*, joined the Maker Studios to handle production. He walked into the makeshift studio and froze. "Is there a lighting kit?" he asked. "Where can I set things up?" He was met with blank stares. Zappin's crew had only cheap lights from Home Depot and a fabric-store green screen. Meyer swore he'd been told he would be paid $2,000 a week. Nope, $2,000 a *month*.

But still, something clicked. In his previous job Meyer had worked on a commercial for a gimmicky reality-TV offshoot and had to listen to ten different people on set bark directions about every minor detail. At Maker they just shot film, posted it on YouTube, and then shot some more. A creative energy pulsed through everything. *It's infectious,* he thought. He stayed.

Maker made its first big splash on YouTube in August with "ZOMBIES TAKE OVER YOUTUBE!!!!!!" Eleven of the biggest YouTubers crowded into the four-minute video, which they shot on a lot in North Hollywood, recruiting fans to don makeup as undead extras. Maker's primary channel, called The Station, became a YouTube hit machine. Maker headquarters at 419 Grand Boulevard became YouTube's cultural epicenter, where celebrities and hangers-on came through to shoot videos or just be seen. *The Hollywood Reporter* called it the digital era's Haight-Ashbury, though more hard liquor was consumed than LSD. One web executive recalled Zappin producing a celebratory bottle of tequila during a 2:00 p.m. meeting. Later, when he sold off the headquarters' apartment, the hustler who defined YouTube's Hollywood scene blamed it on the liability of "too many parties."

Katz, the lawyer, compared Zappin's cowboy style to a split personality.

At times he was cool and collected, but when frustrated, he would scream and badger people into compliance. He became his YouTube persona. Whenever this happened, Katz tried to break the tension. "Hey, everyone, look! It's Danny Diamond here," Katz would sing. Sometimes it worked; sometimes it didn't. The Collective, a talent agency that discovered the squeaky-voiced Fred, had gone on to sign more YouTubers and noticed Maker's ascendance. Dan Weinstein, an executive at the Collective, met with Zappin multiple times to hash out details of a merger. Finally, Weinstein made an offer. Zappin found it too low.

Normally, in business talks, an unwelcome offer is met with a counteroffer or a polite no. Zappin took a different route. He responded over email with a YouTube link. Weinstein clicked to see the video. There was Danny Diamond, face smashed into the camera, rapping a Lil Wayne song. "So I pick the world up. And I'ma drop it on your fuckin' head."

CHAPTER 9

Nerdfighters

Not far away, in downtown Los Angeles, Freddie Wong, the *Guitar Hero* star, began his YouTube glory days in an apartment with three other aspiring filmmakers. They split rent ($375 each) and ate mostly from a taco truck down the block. During the winter they rubbed their hands over computers that whirred hot from rendering visual graphics. Wong's parents worried about his living quarters, an open loft. Friends who visited just voiced confusion. "Wait, you guys don't have *walls*?"

Wong worked on special effects for Epic Pictures, a low-budget, straight-to-DVD production shop. He posted often on YouTube, a playpen for his own filmmaking, though he didn't see it as a viable professional path—not until he met DeStorm Power, the Brooklyn YouTuber, who convinced him otherwise. They had both been recruited to Vancouver during the Winter Olympics as "brand ambassadors," cheap, digital-savvy promoters for companies that didn't want to pay for celebrities. Power was in YouTube's partner program, and he explained its math to Wong. *This* many views translated to *this* much money from Google. Neither knew the full potential, but they had heard that Michael Buckley, a popular YouTuber (handle: What the Buck?), had bought a house with his intake.

Wong didn't feel the need to make millions, but he did want to make his own movies. He called Brandon Laatsch, his college roommate and collaborator, and made the case for going YouTube full time. "We could totally do this," Wong enthused.

freddiew: "The Rocket Jump." September 1, 2010. 01:35.

> A video game brought to life. Wong and his allies, draped in camo fatigues, are locked in a gunfight amid rubble. A shooter perched on a turret takes them out one by one, bloodily. Only Wong is left standing. A cartoon thought bubble appears over his head. *What if he could launch himself airborne above the shooter, like in a video game?* And so he does, bounding up to a dramatic soundtrack and victory.

The set was an abandoned lot near a friend's place; locals called it "the graveyard." Wong and his friends hauled in cameras, sandbags, a green screen, a trampoline, airsoft guns, and an empty military-grade rocket launcher. After filming for five hours, they drove back to hunch over laptops for color correction and editing. Wong found a musical composer, settled on a rate ($500), and sent over a time-coded reel. A week of work to produce ninety-five seconds of footage, a mini-movie. In his old dorm room Wong had struck one-hit viral gold, but these days YouTube felt like a place where dedicated posting could bring in dedicated audiences. YouTube's algorithms and money could reward the time and energy sunk into their mini-movies—at least for now.

So Wong's crew made one a week: real-life re-creations of video games, featuring pyrotechnic explosions and even a skydiving stunt. Wong concocted an idea for a "YouTube road trip"—a drive around the country to visit fans and make videos they requested. He called the only person he knew at YouTube, George Strompolos, and made the pitch.

"That's cool, man," Strompolos replied. "What do you need?"

Strompolos won approval to offer an advance on advertising credits (calculated at $60,842), which Wong used to buy new cameras, lights, and an

RV. A fan in Mississippi emailed to say he could blast a watermelon off Wong's head with a sniper rifle, so Wong went there and filmed that (with special effects). In Valencia, the YouTubers invented "Roller-Coaster Day" at an amusement park; while there, a young fan handed Wong and Laatsch a portrait of them as anime characters, which she had spent three days painting. Wong gleefully showed it off to the camera. Studios pitched him on making cheap straight-to-DVD movies, but he turned them down. On YouTube, Wong had total control. He could emulate indie auteurs, only with a much more immediate sense of the zeitgeist. Tarantino didn't let fans shoot at his head or hand him painted anime characters on-screen.

And unlike Hollywood, YouTube provided data. YouTubers could see in real time what viewers watched, how long they lingered, what made them click. "Hollywood is a world of voodoo and old traditions. *We released this action movie on this weekend because last year it did well,*" Wong said. "Silicon Valley is a world of reinventing the wheel." Wong was just not prepared for how fast it would do so.

• • •

A few months before starting his road trip, Wong stepped into a basement of the Century Plaza for the largest gathering of YouTubers ever. Across the street loomed CAA's massive headquarters, the Death Star, Hollywood officialdom. In the hotel the YouTubers read the conference program:

> *The Century Plaza has twice been the venue for the Grammy Awards, and six American Presidents have slept there. Indeed, The Century Plaza has seen a lot of famous people and spectacular events, but it's never seen anything like VidCon.*

Hank Green had written this greeting after he and his brother, John, decided to throw VidCon without any funding or guarantees of success. By then Hank and John were YouTube elders. Three years earlier they had launched vlogbrothers, a YouTube account where the siblings, then in their late twenties, spoke only via "textless communication." It was part novel internet diary, part performance art. Hank was a zippy, prolific blogger, and John, a young-adult novelist, already had a sizable teen fan base—his fourth book, *The Fault in Our Stars*, released in 2012, would go atmospheric. The Greens were witty, good-looking but not obnoxiously so, and unabashedly earnest. They talked about animal facts, nonsense, Neil Gaiman, the absurdities of capitalism. They were proud nerds and internet natives; on his acoustic guitar Hank had scrawled, "THIS MACHINE Pwns n00bs."

A loyal viewership flowered. The brothers called their fans Nerdfighters and began to interact with them directly. "Basically, we just get together," Hank explained in one video, "and have a good time and fight against worldsuck."

"What's worldsuck?" his brother teed up.

"It's hard to quantify exactly," Hank replied. "It's, you know, the amount of suck in the world." Their brand appealed to the awkward, bookish, pop-culture carnivores drawn to early YouTube. In 2007, three days before the final Harry Potter novel was published, Hank uploaded a song he had written pining for the book's release. YouTube's coolhunters put it on the front page. Nothing produced more fervor in those days, online or off, as Harry Potter fandom. A fan gathering two years later invited the Green brothers as guests. A lightbulb went off. The Greens hired the same organizers to create an event for YouTubers.

They invited every prominent YouTuber they could and sold tickets for $100 each ($210 for "industry" participants). About fourteen hundred people came—maybe half fans, half YouTubers. The line between the two was blurry. For many YouTube regulars VidCon was their first chance to meet fellow members of their subculture. Hank Green fawned over his favorite YouTubers; one attendee was starstruck spotting iJustine. Beatlemania shrieks erupted when a fan was selected to touch the hair of Shane Dawson, a puckish YouTuber. "I just decided that making videos by myself in my bedroom wasn't creepy," Dawson said on the hotel stage.

Wong led a seminar session on "not-so-special effects." YouTubers and teenage fans played Calvinball, the invent-your-own-rules game from *Calvin and Hobbes.* A band performed with a male accordion player in a thong. People crowded the stage holding film cameras. Danny Diamond danced. Group hugs were given backstage. "It had such a pure and precious feeling," recalled Laura Chernikoff, a vlogbrothers fan hired to work the event. For many, VidCon brought their vision of YouTube to life—a site that "felt like a bunch of creative weirdos in a room together," Hank Green said later. "It was, by far, the most magical thing I'll ever be a part of."

Two dozen YouTube employees bought tickets to VidCon, but it wasn't their event. Some who came couldn't help but notice a stark contrast to their day jobs. Andy Stack had just joined YouTube to manage its partner program. In July he trekked down to Los Angeles to meet with a major studio, which was negotiating to place some of its entertainment library on the site. The meeting started at 4:00 p.m. and dragged on—debates about copyright and ads, about what worked, didn't work, should work. No one looked as if they wanted to be there. Afterward, Stack, exhausted, drove to the Century Plaza hotel to see the VidCon crew. He met YouTubers he adored. Everyone wanted to be there. It felt electric. *Wow,* these *are the people I want to help win,* he told himself.

• • •

VidCon had one noticeable absence. In 2010 the reigning king of YouTube was Ray William Johnson. He was no Nerdfighter.

Johnson was a solo mad scientist, with thick black hair gelled into spikes, chiseled features, and bravado to match that of Danny Diamond. After starting his account in 2009, Johnson quickly surmised two simple truths of the website: people came to see viral videos, particularly from regular people; and people liked to laugh, particularly *at* regular people. He settled on a shtick of mashing viral amateur clips on YouTube with color commentary. He was the Bob Saget of YouTube's funniest home videos, only crasser and more high-octane. (One of YouTube's most common search terms back then was "funny.") Johnson had surmised something else: if his videos *about* viral videos appeared when viewers searched *for* viral videos, or in the related section, this would pay huge dividends. YouTube's algorithm loved his shtick, helping him set record growth in subscribers.

Johnson also avoided the DMCA sledgehammer by wielding "fair use" so his videos stayed up and made money, often getting more traffic than the originals he mashed together. A report claimed his annual intake from video ads made him YouTube's first millionaire. "I'm just a regular guy with an entertaining hobby that happens to make a friggin' million dollars without leaving my apartment," he said on a podcast. "Am I supposed to apologize for that? If you're jealous, just do what I'm doing, and do it better." He started a production company called Equals Three (written out as =3). Dan Weinstein, the talent manager, took Johnson to meetings with MTV Networks for a potential show, but the network wanted a cut of his earnings for years to come. Johnson made friggin' more on YouTube without leaving his apartment, so he turned them down.

Making money on YouTube, however, could be complicated. YouTube sent payments and 1099 forms, but left YouTubers to sort out the taxes on their own. In 2010, Johnson met one of YouTube's earliest dedicated moneymen, David Sievers, a twenty-two-year-old Nebraskan training to be an accountant. Sievers handled taxes for a friend in Nebraska, Dan Brown, who had achieved sudden YouTube fame filming his skills solving Rubik's Cubes. YouTube fame was a small world, and Sievers quickly added more marquee YouTuber clients, including the mad scientist Johnson.

Later that year Johnson and his new taxman traveled together to Los Angeles to pitch a web series. Sievers, who had grown up thumping a Bible, had never been to the city; Johnson bought him his first margarita there. On the trip, the pair met a scrappy redhead named Danny Zappin. Once Sievers returned home, Zappin sent him a long, unvarnished letter outlining his life story—the drug charges, the Hollywood rejection, the whole United Artists of YouTube thing—and recruiting him to join Maker Studios. Sievers took the job and moved west, bringing Ray William Johnson and his friggin' million dollars to YouTube's ragtag Hollywood crew.

When Sievers arrived in Los Angeles, he was shown Maker's office—no desks, lousy Wi-Fi—and its rooftop bar, which overlooked the ocean and a bong-ripping weed festival on Venice Beach. *Infectious.*

Maker was on the prowl for even more stars it could add to its roster. Besides accounting, Sievers's new bosses wondered, did any part of YouTube particularly interest him? He admitted he liked watching YouTubers play video games.

• • •

While the bongs ripped in Venice, a huge shift was under way in media. Google saw it in the data. Starting in 2009, TV viewership began to slip, falling for a quarter or two and then dropping off a cliff. By 2011, American TV ownership declined after twenty years of upswing. A small decline, but perceptible; the internet could claim its piece. That year, for the first time in YouTube's life, *American Idol* no longer topped the TV ratings charts. Just two years before, the world had watched Susan Boyle, a demure Scot, belt out "I Dreamed a Dream" on the British version of the show for free on YouTube. Fame didn't need *American Idol,* with its catty judges or prime-time slot. America's top-rated show became *Sunday Night Football,* which didn't live online.

YouTubers saw another shift that Google didn't fully grasp. Like cable,

YouTube's abundance meant its programs didn't have to satisfy the muddled masses as broadcast TV had to; YouTube could expand in infinite interesting directions. But YouTubers were not just amassing niche fandoms. They were building and fostering *communities*, ones with powerful bonds. They could be relatable, influential personalities, like Oprah or reality-TV stars, but were much more intimate and accessible. Freddie Wong formed bonds over a shared creative canvas, the Green brothers over shared intellectual curiosity—forms of interaction that barely looked like TV. YouTubers could use their stage to provide meaning for a complicated world, or a complicated internet. Ray William Johnson did this with his viral mashups. YouTube's coolhunters once helped people sift through the site, but they were gone. YouTube was abundant, chaotic, overwhelming to navigate and absorb. People wanted to find someone who could take them along for the ride.

CHAPTER 10

Kitesurfing TV

Felix Kjellberg opened his YouTube account in 2006 and summarily forgot his password. Thus prevented from uploading, he watched. Four years later, as a student at Chalmers University of Technology in his native Sweden, he finally posted. He recorded at his dorm-room desk with huge black headphones atop a mop of dirty blond hair. He picked an absurdist handle, PewDiePie—rhymed with "cutie pie," sounded like a toy gun—and uploaded footage of himself playing video games. Viewers arrived, and Kjellberg took to the craft with the zeal of a convert.

PewDiePie: "Call of Duty: Black Ops: Wager Match: Gun Game." December 16, 2010. 3:12.

"Today, we're going to celebrate because I got one hundred subscribers." We see Kjellberg's virtual gun but only hear his voice, fluent English accented in Swedish and internet-ese. He skillfully takes out enemy combatants. "Some of you have been asking me," he narrates, "'Do you want to be famous?'" Another kill. On-screen: *Humiliation. Demoted enemy!* "No, I don't want to be famous," he goes on. "The only reason why I do this is because it's fun and you guys are the biggest part of that." The game play ends. "I guess I'll talk to you guys tomorrow."

• • •

YouTube's next seismic shift started as a total fluke unearthed inside a company with a silly name that would launch PewDiePie's career. Machinima had begun in 2000 as a web forum dedicated to the field of using video game techniques on film; its name was a portmanteau of "machine" and "cinema." Over the decade its forum morphed into a production studio, churning out material for and about gamers and making the bet, still risky, of placing all its material on YouTube for free. Machinima's chairman likened YouTube to the third wave of mass media, after radio and TV, a wave primed for a gold rush. Machinima hired young, cheap (mostly male) recruits from gaming and media to do the necessary digging for gold.

Luke Stepleton joined the company after stints with *American Idol* and Mark Burnett, the producer who put Donald Trump on TV. At Machinima, Stepleton's new boss sat him down at a desk in its Burbank office to watch their YouTube output, ordering him, "Tell us what we're missing." Stepleton believed Machinima was missing the feeling of playing games at your friend's house. Advances in graphics and consoles had turned video games from clunky, pixelated dioramas into elaborate virtual worlds, and in doing so had spurred a gigantic subculture. Gamers weren't just buying new titles; they were tuning in online to watch other rabid gamers play, for tips on the craft or just for laughs. Regular Joes on YouTube, with handles like Hutch and Blame Truth, filmed themselves playing *Call of Duty* and *Gears of War*, narrating with color commentary and off-color jokes. These lo-fi streamers got way more views than the slickly produced stuff that Machinima made.

And they weren't even using the same YouTube! On-screen, all YouTube videos played the same, but behind the curtains YouTube had given select partners a different version of its site: one with customized software that enabled users to track their video performance and claim copyright infringements if accounts lifted their material. Machinima had this custom version. One day, at its cramped offices, Stepleton watched a colleague toy around with its pixels and drop-down buttons. His colleague opened a new YouTube account and tried an experiment. *What if Machinima's special, "partner" account could merge with a normal account?* Turned out, it could. The normal account suddenly became a YouTube "partner" too, which meant it could host ads on videos. *Eureka.*

With this loophole Stepleton realized that Machinima didn't have to produce smash YouTube hits or recruit hot YouTubers to make its videos. It could just bring existing hitmakers like Hutch and Blame Truth under its own umbrella, linked with its special account, and count their success as its own. "I could get you to a hundred million views in a day," Stepleton told his boss.

People at YouTube weren't aware that media companies could aggregate other YouTube accounts. They certainly hadn't designed the software with that goal. By the time they realized the possibility, Machinima had already begun recruiting scores of YouTubers, giving them admission into YouTube's ad program and forming the site's dominant business model over the coming half a decade. Machinima's young staff perused YouTube, searching for anyone who uploaded gaming footage, then offered these YouTubers skeletal contracts to post under its custom software. Machinima would handle any copyright claims that came up in exchange for a cut of earnings. Soon, Machinima was signing a thousand YouTubers a month.

One haul around 2011 included a Swedish kid with the handle PewDiePie.

• • •

Machinima's discovery happened to fit neatly into the new strategy Salar Kamangar was setting for YouTube: a big, calculated swing at TV. YouTube's new chief, Google employee number nine, was "cookie-cutter Google," according to a former colleague. He liked computers, logic, and chess. He was a "systems thinker," in Silicon Valley lingo. (Danny Zappin called him a "genius robot.") He was introverted, sometimes cagey. Once he scribbled something on a notepad during a meeting with a media executive, who asked what the note said. "I'm not telling you," Kamangar replied. Many at Google knew him merely as a very veteran, very wealthy corporate legend. Like other Google brass, he didn't excel in the softer skills. Once, before a call with a prominent business partner, Kamangar asked a deputy, "What's the emotion I need to convey?" But he could schmooze when needed. He was fit and lean, with dark eyebrows and a toothy grin, and *Gawker* wondered if he was Silicon Valley's "most eligible bachelor" while printing gossip that he once dated Ivanka Trump.

At the end of 2010, when Kamangar fully took over YouTube, the company celebrated. That year it had crossed $1 billion in revenue. Bonuses were doled out, and, per custom, there was a trip to Las Vegas. But trouble lurked. For the first time, the hours people spent watching YouTube had begun to plateau. Most visitors were now catching a viral video on breaks at work or school and then leaving the site. Rabid fans like the Nerdfighters still devoured footage, but average viewers clocked only five minutes a day. Those hours that TV had been losing seemed to be slipping over to the DVD-by-mail company, Netflix, which had recently started streaming shows on demand over the internet, posing an alarming threat.

YouTube, once again, embarked on an effort to make its site more *premium*. Internally, they framed the challenge differently: *How do we stop being dogs on skateboards?* Such cheesy hits had served YouTube's initial ascent swimmingly; when the co-founder Chad Hurley met the queen, he had played her a viral laughing baby. But for an advertising business, viral hits were less useful. Marketers purchased most commercials months in advance, cutting handshake deals in the spring over TV programming slated for the fall. YouTube couldn't offer any of those guarantees. Who knew when or where the next laughing baby would appear?

So YouTube would refashion its site as more predictable and familiar for viewers and advertisers, to be more like TV. To do so Kamangar borrowed a concept from cable. YouTubers shouldn't have accounts, a Web 2.0 term.

They should have *channels.* Channels could appeal to even the narrowest of interest groups, a gift of YouTube's economics. Cable networks needed an audience big enough to justify the costs of porting a television signal from a transponder to a satellite and creating 24/7 programming. On cable, producers green-lit shows, dictating how much money they got and how long they ran. YouTube didn't need that; it had infinite airtime, a freeway without traffic lights. The smallest niches could live on YouTube at virtually no cost. To explain this model, Kamangar cited his own hobbies. He liked to go kitesurfing, sometimes enjoying the luxury sport with his work pal Larry Page. "On cable, there is no kitesurfing channel, no skiing channel, no piano channel," Kamangar said in an interview. "All these topics I care about suddenly have a home." His deputy Mehrotra put it more pithily. "Online video will do to cable what cable did to broadcast," he told an audience of advertisers. Cable had splintered the old three networks into hundreds, creating neat audience segments advertisers could use to market more effectively—outdoor sports fanatics, home shoppers, right-wing news watchers, and so on. YouTube would do all that again, and the smart money should get on board.

YouTube took Machinima's fluke and ran with it. Cable networks like ESPN had ESPN2, ESPN Deportes, and so on. If YouTubers had individual channels, then companies that sponsored many channels were networks. Ergo: "multichannel networks," or MCNs. A cottage industry sprang up overnight.

Down in Los Angeles, Danny Zappin swiveled to this emerging model. He already had a dream team of YouTube stars in Maker's orbit. Now, as an MCN, he could add more stars and plug them into a bigger business. (Others at Maker later argued that the studio was the first to devise the multichannel networks idea.)

Some at YouTube didn't know what to make of these new satellite businesses. Later they would cause immense headaches, but back then MCNs filled a pressing void. Google had no direct experience in dealing with on-camera talent, performers who could be unpredictable, irrational, petulant. Even the standard trappings of stardom were unfamiliar. Dan Weinstein, a manager who represented Freddie Wong, once rang YouTube staff on behalf of another client, the Annoying Orange, a viral YouTuber who filmed talking, acid-tongued fruit. "Wait, who are you?" a YouTube director asked.

"You know how George Clooney has a manager?" Weinstein explained. "So does the Annoying Orange. I'm him. You've got to talk to me to get to him."

Kamangar and the Google crew knew 0s and 1s—categories, not people. Multichannel networks provided a convenient solution to their creator-management problem. YouTube swiftly leaned on other partners to adapt this model. Sarah Penna ran a management company in Los Angeles that had been working with YouTubers to create business plans and branch into new fields, and YouTube had flown her up to San Bruno for events with advertisers. One day YouTube informed her that perks like those might end unless she turned her firm into a multichannel network. YouTube gave the same directive to Weinstein's company, which later rebranded as Studio71, an MCN. These new entities could deal with the Torso of YouTube's beast, the vast sea of marketable amateur videos.

Now Kamangar needed someone to revive the Head.

• • •

Robert Kyncl was bred to do two things: ski and win. Raised in Czechoslovakia, he attended a state-run boarding school as a teen to train as an Olympic skier. Instead, his country toppled communism, and Kyncl traveled to school in the United States on a scholarship and channeled his fierce competitiveness into show business. He started in a talent agency mail room, climbing the rungs to HBO and then Netflix, where he cut licensing deals with studios and networks. While Netflix wasn't yet a serious

Hollywood player, Kyncl learned the world of managers, talent, and studio politics—at least he knew it better than Googlers. Svelte and angular, Kyncl had a square jaw, a staccato baritone, and conspicuous swagger.

YouTube hired Kyncl in 2010 during a major renovation period. With the Viacom lawsuit over, YouTube saw another chance to get old media on board. Google had dispatched Dean Gilbert, an executive on its TV hardware project, to YouTube to run its content division (the Head) and operations (the Torso and Long Tail). Gilbert, a hard-boiled cable veteran, didn't buy the Silicon Valley precept that all content was created equal. "Not all pixels are alike," he told his staff. He recruited fellow media alums to separate YouTube's diamonds from its rough, forming categories on the site that TV advertisers might recognize. He convinced YouTube to buy Next New Networks, a studio from TV veterans that ran several hit YouTube channels, including the one behind "Obama Girl," and had written a "playbook" for aspiring YouTubers. (Google brass loved the playbook—finally, written code.) Gilbert pushed his team to court big-ticket sports leagues—while the boss liked the kitesurfing channel, YouTube still wanted to host sports most people watched. (Google brass didn't help here. American pastimes weren't very Googley; one staffer remembered Larry Page confusing the NFL with soccer during a meeting. When YouTube's team managed to get Page to sit down with NFL commissioner Roger Goodell over streaming rights, the meeting leaked to the press—a tactic, Google suspected, the NFL used to get better terms with TV providers—and Page, irate, called off the talks.)

Under this renovation plan, Kyncl's job was to woo Hollywood. Before he joined, YouTube staff would fly down to Los Angeles, pitch bigwigs, fly back up, and hear nothing. Part of their problem was that Google, like other tech companies that parachuted into Hollywood, asked talent agencies to sign onerous insurance and vendor agreements that studios didn't require. Kyncl opened YouTube's first Los Angeles office and, according to one prominent agent, quickly nixed those agreements, making the agencies more willing to put their big TV and movie stars on YouTube.

Inside YouTube, Kyncl arrived "like a sledgehammer," one former colleague recalled. He made it clear those working for him should join him in Los Angeles and he immediately butted heads with other newcomers over turf. When he met the leaders of Next New Networks over lunch in Manhattan, where the studio YouTube acquired was based, the sledgehammer gave his blunt assessment. "I would have never bought you," Kyncl said. "What a stupid acquisition." It started a years-long rivalry between Kyncl and the studio team in New York over strategy and resources. None of Kyncl's rivals, though, outlasted him at the company.

But the East Coast newcomers did settle one debate inside YouTube: what to call its population. Initially, everyone on its site was a "user"—people who made videos and people who watched. Once stars emerged, the company tried other labels. "YouTubers" was imprecise; some were filmmakers, some makeup artists, some just weirdos with a webcam. "Partners" felt too corporate. At Next New Networks there was the "audience" and "creators," a catchall term to capture everything that went into making web media. At YouTube, "creators" stuck.

Some studio staff noticed another oddity about their adopted company. Very few people at YouTube actually spent time watching YouTube.

• • •

In Hollywood, Kyncl's first move was to replicate the Netflix playbook. He spent months negotiating with media companies to place their material on a new pay-per-view service YouTube would launch. Some TV staples, like *SNL* and late-night talk shows, had begun putting segments online, and it seemed others would follow if the money was there. But marquee shows like *Mad Men* were held up in knotty syndication deals, and big studios were still skeptical of YouTube: when Kyncl and Gilbert arrived at Disney to meet an executive, they faced only a row of lawyers. Google had also tried acquiring Netflix and, when that failed, held talks to host Netflix's library entirely on YouTube. (Back then, Netflix frequently crashed; YouTube no longer did.) Both talks fizzled, and so did Kyncl's pay-per-view project.

Kyncl switched gears once he met the former teen heartthrob Brian Robbins.

Trim, tanned, Hollywood in the flesh, Robbins hadn't aged much from his old TV star days, when he wooed Reagan-era teens in a mullet and leather jacket on ABC's *Head of the Class*. After outgrowing the mullet, he moved behind the camera to direct and produce, making some hits (*Varsity Blues*) and some clunkers. Shortly before connecting with Kyncl, Robbins got a call from his agent, who asked him to meet a squealy-voiced YouTube sensation called Fred Figglehorn.

"Is my career over?" Robbins replied. "You're introducing me to a YouTuber?"

Talent managers wanted Robbins to put Fred in movies, but Robbins was deeply skeptical. He remained so until a family vacation in Miami Beach, where, lounging at the Fontainebleau luxury hotel, he watched his two preteen sons ignore their suite's big-screen TV and stay glued to YouTube on their computers. Robbins asked the boys, "Do you guys know who Fred is?" Instantly, they imitated Fred's squeal and recited lines of dialogue verbatim, just as their dad could with his favorite movies. Robbins prodded his focus group further. "Would you guys want to see a Fred movie?"

"Tonight?!"

Robbins made it. He spent a million producing *Fred: The Movie* and pocketed $3 million. Upon its release, Robbins watched the movie trend worldwide on Twitter, the internet generation's Siskel and Ebert. Robbins devised a plan: put a million dollars into short videos made for teens, with decent production values, and other Freds would emerge, spinning more gold. He pitched this shorts idea to Nickelodeon, but the network declined, so he developed the concept himself, gave it a bubblegum name,

AwesomenessTV, and crafted a c-suite-flavored presentation full of high-definition reels. Robbins showed these to Kyncl at YouTube's Hollywood office, carting in his own big computer screen to capture the aura.

Kyncl loved the concept—a seasoned Hollywood hotshot with a network that would live on YouTube. Kyncl had underlings adapt Robbins's presentation to explain the Head's big gambit. "To have credibility in this town," Kyncl told staff, "you have to put your money where your mouth is." Under his plan, YouTube would fund honchos like Robbins to make original, hi-def material for the site, as a way to generate reliable programming that advertisers would line up to sponsor. Google's cash would come as advances on ad credits, like the sixty grand Freddie Wong had gotten for his road trip—only much bigger. Kyncl proposed doling out up to $5 million apiece to twenty channels.

To find these worthy recipients, Kyncl recruited veterans from digital start-ups and movie studios for a "covert Black Ops team," as one member described it, designed to keep plans shielded from rivals. The Black Ops team began with a list of two hundred popular YouTube categories—horse-riding fanatics, gearheads, fashion enthusiasts—and whittled it down to twenty. Scanning their internal figures for trends, they discovered how coarse Google's data mining could be: a chart of YouTube traffic patterns showed a huge concentration of "military" titles. *Huh?* It turned out that machines were categorizing YouTubers playing *Call of Duty* as "military."

At times Kyncl asked his team to find inspiration from Hudson News, the standard airport magazine seller with a glossy mag for every interest. Kyncl's team heard more than five hundred pitches, cycling through them in thirty-minute rounds every day. Even lions from publishing like Rupert Murdoch paraded in. (The Murdochs, old Google foes, knew Kyncl socially.) Kyncl pitched media wunderkinds like Shane Smith, a founder of *Vice*, the once-punk magazine refashioning itself as punkish web TV. Smith once met Kyncl and Patrick Walker at a suite at the Aria hotel in Las Vegas. Five minutes into the meeting, Smith cut in. "You guys like to gamble?" he asked. The Google men said sure. Smith clapped his hands, conjuring a private table with $5,000 stacks, and suggested chatting over a game of blackjack. *Vice* received a YouTube grant.

After furious debate YouTube decided to structure these grants as anti-Hollywood: YouTube would hold rights on the material for only a year and take no equity. Kyncl approved a slate of ideas designed to draft celebrities into becoming YouTubers: a comedy channel for Shaquille O'Neal, a dance one for Madonna, skateboarding for Tony Hawk. Google sold ad space in advance and projected high viewership. Advertisers flooded in, eager to sponsor A-listers. Ahead of the debut Kyncl gave an interview with *The New Yorker* in Google's Manhattan office, where the company had named certain conference rooms after TV hits. Kyncl sat in the "Cosby Show" room and offered some customary swagger. "It's certainly not going to be easy," he said. "But, as a friend who just landed a job at one of the networks said, 'At least you guys are swinging for the fences.'"

Within YouTube, Kyncl used another metaphor for his pet project. Its code name was Lighthouse. YouTube had lovely shores, but Hollywood needed a guiding light to find them. One manager had a blunter description for Kyncl's plan: "an attempt to shoehorn a mini-Netflix into YouTube." To some at the company, funding select channels felt like preferential treatment. And back during the Viacom brawl, YouTube had chafed at creating its own programs for fear of legal backlash. Now staff started asking a vexing question: *Are we a media company or a tech company?*

• • •

Media companies were, of course, chatty. In October, as YouTube finalized its slate of funded channels, word leaked out. When a *Wall Street Journal* reporter started sniffing around, executives at YouTube summoned content personnel and lawyers into a conference room in San Bruno. They were worried that word of the grants going public might sabotage unfinished deals. Staff were told they had twenty-four hours to sign contracts, or they might not happen. "This is a red alert!" someone barked. Everyone sprinted to the phones.

Around that time a prominent YouTuber aware of the program advised Kyncl's team that it might look bad if they gave all the money to celebrities and old media and none to YouTubers. Another scramble commenced with staff prowling for YouTubers to fund. Machinima and Maker Studios got money. A representative from YouTube called Hank Green, the Nerdfighter vlogger, and quickly explained the Black Ops project. "If you could have a proposal for us by Monday, I might be able to get it in front of them," he was told. It was Friday. Green spent his weekend assembling a PowerPoint presentation on two YouTube educational programs he called SciShow and CrashCourse. He typed "Budget" on his last slide and invented a sum. YouTube gave it to him.

For many the Black Ops project was a colossal bungle. One hundred channels were funded at first, and YouTube took another shot at the program the next year. Kyncl "spent like a drunken sailor," quipped one executive involved. Once the videos premiered, no one wanted to watch celebrities like Madonna and Shaq. Detractors blamed Kyncl for an obsession with Hollywood at the expense of the site's scrappy creators. But defenders described the grants as a stimulus program for YouTube, an injection of commercial activity that raised its profile in Hollywood and on Madison Avenue. Brian Robbins credited Kyncl's "brilliance" for this maneuver: "He got the advertising community awake and legitimized what was not a legitimate platform." Everyone agreed that the few funded channels that *did* thrive came from existing YouTubers like the Green brothers. YouTubers understood YouTube, often better than the company. "We discovered our superpower was creators," recalled Ivana Kirkbride, an executive who worked on funded channels. "YouTube couldn't be what it wasn't."

That lesson took several years to stick, however.

For the first funded channels program, Kyncl's team had recruited marquee directors to make *WIGS*, a scripted web series featuring movie stars like Julia Stiles. But these highbrow channels didn't focus on specific topics or themes popular on the site, which meant that YouTube's recommendation system, the main driver of views, ignored the videos. A barebones vlogging channel from Mexico, Werevertumorro, which also received funding, bested the Hollywood stars in traffic. A frustrated Kyncl, on a trip to San Bruno, once asked YouTube managers why *WIGS* wasn't appearing more often in search results. *Why did YouTube's home page look as if it were built for fifteen-year-old boys?*

The managers shrugged. YouTube's algorithms set the search results and home page, just following the clicks.

CHAPTER 11

See It Now

Television reached maturity on November 18, 1951, when Edward R. Murrow broadcast the Korean War into American living rooms. Americans knew Murrow. A decade earlier he had narrated Hitler's terror and D-day from rooftops in London over the radio. *See It Now*, his evening telecast on CBS, debuted in 1951 with a program on the ugliness of America's latest war, focusing on its young soldiers. "They may need some blood," Murrow said, looking into the camera. "Can you spare a pint?" Until then, television was mostly known for jesters like Milton Berle and Jack Benny. After *See It Now* debuted, a fellow newsman told Murrow he couldn't remember a half hour "that has made the belly of the cathode tube so powerfully absorbing every second of the way."

YouTube's Murrow moment came on June 19, 2009, when it broadcast the revolution unfolding in Iran.

Iranians had begun posting grainy cell-phone footage of street protests and the regime's ferocious response. A trickle became a torrent. Iran's Green Movement was the biggest news story in the world, but cable networks, beset with cutbacks and restricted from entering Iran, struggled to cover it. Ricardo Reyes, YouTube's communications chief, received a call from a perplexed CNN producer. "You guys are getting more stuff than we are," the producer said. "How did you get it?"

Reyes explained how YouTube worked and invited the network to visit.

On June 19, Wolf Blitzer's nightly show aired a segment from YouTube's headquarters. A cameraman set up in front of a desktop monitor, zooming in from San Bruno to document the revolution for millions. Reyes pointed a CNN reporter to a Google map of Tehran showing digital displays of each uploaded clip, many of which showed disturbing, graphic violence. Normally, those would come down, Victoria Grand, YouTube's new policy director, a Google import, told the reporter. "YouTube is not a shock site." But, she continued, they made an exemption for videos with a "clear documentary purpose"—media the world should see.

"So, Wolf," the reporter concluded, tossing it back to Blitzer in the studio, "really a watershed moment for YouTube."

The next day, a Saturday, Iranian security forces shot and killed Neda Agha-Soltan, a student in Tehran. A shaky video of her death—a screaming throng forming around her small frame, blood pouring from her eyes—landed across the globe on the screen of Julie Mora-Blanco, an early YouTube moderator who had been promoted to deal with knotty screening scenarios. It was grim and getting lots of attention. Three years earlier YouTube staff had debated how to handle uploaded footage of Saddam Hussein's hanging, opting to keep up some clips and remove others as too violent, but the company still lacked exact protocols for such events. Mora-Blanco knew her decision could be globally significant. She roped in colleagues, who consulted with a lawyer to apply a "newsworthiness" exemption for the video. A coder designed a small graphic content warning label at the start of the clip, which stayed up. Footage of that gruesome death would come to define Iran's political upheaval and the world's view of it. This was raw, wrenching content, and it could be seen only on YouTube.

Yet the company struggled to adjust to its newfound responsibility. Neither its machines nor its moderators were experts on the Middle East. Errors occurred. Once, brutal videos posted from Cairo showed police officers torturing a woman and heaving the corpse of another in a trash bag. *Too violent.* YouTube removed the videos and the account that posted them.

Human rights groups quickly informed Google's lawyers that the account belonged to a prominent Egyptian activist documenting police abuse. YouTube put the videos back up.

Unsurprisingly, Iran's regime eventually blocked YouTube in the country. Hackers built a work-around—a computer networking tool that enabled Iranians to send footage to an email address, which then automatically uploaded to a YouTube account. But the work-around had a hitch: anyone could use it. Someone started sending copyrighted videos to the email address. YouTube had a firm rule: three copyright strikes and the offending channel was automatically removed. So that account and its thousands of videos documenting Iran's upheaval disappeared. Some YouTube staff suspected Iranian government agents sabotaged the account but had no proof. They were powerless to fix it either way.

• • •

Mark Little watched the Green Revolution erupt on YouTube from his home in Dublin. For two decades Little had been working in Irish radio, reporting on wars in Iraq and Afghanistan. Budget cuts kept him from going to Iran, but he saw the raw on-the-ground footage on YouTube and was transfixed by its potential. *Who needed foreign correspondents anymore?*

Social media was exploding into a global phenomenon. In 2009, Twitter catapulted from a techie niche site into a mass spectacle; Facebook introduced its "Like" button and nearly tripled its user base to 350 million. Little joined a handful of news industry veterans smitten with social media's power—how it replaced the immediacy of the news wire, the authority of the anchor, the emotional weight of eyewitness accounts. This struck him again weeks after Iran's initial protests, when Michael Jackson died. Little was attending a wedding in Ireland when younger attendees shared the news about the King of Pop. "How do you know?" he asked one. "Twitter," came the reply. Wedding guests danced to "Thriller" in tribute ten minutes before newspapers confirmed Jackson's death.

For Little, this was sobering. Anyone on the internet could report news, and people would buy it. *If everybody can tell a story*, he wondered, *whom do you trust?* Little formed Storyful, a digital newsroom designed to answer this question. It would filter social media output from places like Iran by verifying its authenticity and significance. He recruited a small staff and pitched news outlets on using Storyful as a foreign desk for the digital age. Soon Little connected with Steve Grove, the young politics manager at YouTube. After his editorial team was disbanded, Grove had doubled down on forming partnerships with outside groups to market YouTube as a destination for serious media.

Then a street vendor in Tunis set himself ablaze, starting the Arab Spring. In Egypt, where protesters swarmed the streets, 100,000 YouTube videos were posted in a short span, a huge jump in the country's usage. Grove called David Clinch, a CNN veteran who managed Storyful's U.S. operation, and asked if Storyful could help curate Arab Spring footage pouring in. Clinch hung up and immediately called Little in Ireland, relaying the request.

"Are we being paid?" Little asked.

Clinch froze. "I don't know."

YouTube did pay, around $10,000 a month, making it clear the company wanted editorial hands sifting through content from the Middle East. But YouTube wouldn't do this themselves; it hired engineers, not journalists, and the Viacom suit had persuaded YouTube to shun any perception of editorial control. Google, however, viewed the Arab Spring as a silver-platter chance for its video unit to shake its dogs-on-skateboard reputation. At the second VidCon in 2011, Kamangar gave a speech to that effect. Some of his staff were nervous beforehand; their boss was smart but could be awkward, and he was trying to balance a push for prestige and profit at the Head while keeping the free-spirited creator class in the Torso happy. On the VidCon stage, he cited his site's mind-blowing stats: every minute, forty-eight new hours of video were uploaded; every day, three billion videos were watched. Media had tried for a century to involve "everyday people," he said, "but it hasn't been until today, with YouTube, that everyday people actually *are* the media. You guys don't just watch news; you make news." He then invited onstage the Annoying Orange, the YouTube comic.

Meanwhile, news kept pouring onto YouTube from anonymous, incredible accounts across the Arab world. Storyful's job was to verify this footage so YouTube could spotlight it for millions to see.

كتيبة المقداد بن عمرو تدمر دبابة في داريا باستخدام قاذف بي

March 5, 2013. 2:43.

> A soldier sits on a green chair, a keffiyeh covering all but his eyes. Soon, we are up, trailing three men carrying rocket grenade launchers over rubble. *Blasts.* We are running now with the men, taking cover in a crumbling structure. The camera operator sprints to a small, blown-out window. *Explosion.* The camera falls in a blur. Then back up, in focus, a tank smolders, billowing fire and clouds of black smoke. "Allāhu akbar! Allāhu akbar!"

The video's Arabic title claimed its footage showed Syrian rebels in Darayya, a Damascus suburb, taking out a government tank. Little's team watched it in their small Dublin office and began their forensics. First, they looked for identifying landmarks—a water tower, an angular building in the background. They scoured Google Maps and photo-sharing websites for water towers in Darayya. They confirmed the explosion's time and location using Facebook posts and satellite imagery. Then they found another video, shot with a GoPro from a different angle of the same exploding tank. *Verified.* Eventually, protesters using YouTube picked up on these cues and began their videos by framing the camera on morning newspapers (confirming the date) before panning up to a recognizable minaret (confirming the place). Little marveled. Storyful had become the new newsreel. "The archive of now," he called it. "The internet of awesome." Edward Murrow and Wolf Blitzer were no longer writing first drafts of history. Unknown YouTube accounts from Darayya were.

YouTube seemed like a willing partner on this expedition. Although, in hindsight, signs of discord were there from the outset. When Storyful first laid out its process to staff at YouTube, Clinch heard a Google coder scoff at the proposal to have expert hands curate footage. *We have an algorithm for that.*

• • •

Google was also dragged into the Arab Spring in a way it didn't expect. Soon after protests erupted in Egypt, Wael Ghonim, a Cairo native working for Google in Dubai, disappeared. Unbeknownst to his company, Ghonim had started a popular Facebook page under a pseudonym, spurring Egyptians to take to the streets. As protests grew, the computer nerd boarded a flight to Cairo and went dark after landing. "We don't know where he is," Patrick Walker grimly told staff at a meeting in Munich. Google ran notices on its home page in Egypt about Ghonim's disappearance and hired private security officers to search for him. While his Facebook posts were secret, Ghonim's support for the protests was not, and some at Google raised concerns about intervening, worried that advocating for an activist employee might be seen as an American company supporting a government overthrow. Googlers in the region already felt like they were in a precarious position. Around this time, the company selected an executive in Israel to oversee its policy operation across the Middle East but asked staff not to make that appointment public. Some working there felt that disclosure would contribute to existing perceptions in the Arab world that Google personnel were spies.

Eleven days after disappearing, Ghonim was released from a Cairo prison. He gave an interview on Egyptian TV, blurting between sobs, "I am not a hero. I was only using the keyboard." Still, at a Google event welcoming him back in London, CEO Eric Schmidt called Ghonim a "hero." According to company doctrine, people *could* be heroic simply using computer keyboards.

Besides geopolitics, Ghonim's ordeal underscored a very different concern for Google. Ghonim inspired and mobilized so many—and had terrified an authoritarian regime—with posts on *Facebook*. The social network, already a looming threat to Google's central internet seat, now emerged as the new global public square, *the* place where revolutions unfolded. Storyful, as it tracked the Arab Spring, had to shift much of its verification work from YouTube to Facebook, because so many vital posts were appearing primarily on the social network.

Meanwhile, the euphoric idealism of the Arab Spring started to sour. YouTube registered this in 2012. That July an account named "sam bacile" uploaded a fourteen-minute trailer to a film, *Innocence of Muslims*. Sam Bacile was one of several aliases for an Egyptian Coptic Christian, living in California, with a long rap sheet of fraud and drug charges. His trailer was terrible—regional-theater-level acting in a sword-and-sandal plot, portraying the Prophet Muhammad as a pedophile and a brute. Few noticed; YouTube certainly didn't—until September, when an Islamophobic blogger wrote about the trailer in Arabic, circulating it more widely.

A tinderbox erupted. Tens of thousands protested over what seemed like an ugly depiction of the Prophet in an American movie. Egyptian newscasts played the YouTube clip, prompting calls for marches on the U.S. embassy. Mobs set cinemas aflame in Pakistan, where six people died in the unrest. Google staff visiting Islamabad awoke to armored vehicles carrying U.S. diplomats to their hotel. One Googler remembered the greeting: "Our embassy is under attack, and it's because of your stupid video."

That same month the French magazine *Charlie Hebdo* printed satirical cartoons of Muhammad, and terrorists murdered a U.S. diplomat in Benghazi. The Arab Spring had veered off course, sparking violent conflicts over free expression, Western imperialism, and dogma. YouTube was caught in the vortex. The company had rushed to expand across the globe, pushing citizens to broadcast in every language and nation they could, without putting enough staff in these countries to watch videos or deal with politics on the ground. Inaccurate reports swirled that the YouTube movie trailer caused the mayhem in Benghazi, adding to the confusion. "All hell broke loose," one YouTube publicist recalled.

Bob Boorstin, Google's lead operator in D.C., received the first call. Obama's State Department wanted YouTube to take the video trailer down. YouTube leaders joined tense conference calls with Google lawyers and political officials to decide the next move. Some voted to pull the clip. They could republish it later once the storms calmed. *People had died.* But lawyers dug in on a free-speech argument; Kamangar concurred. Google prided itself on not caving to demands from governments, even its own. Nicole Wong, Google's Decider lawyer, argued on technical grounds: the video didn't violate company rules against hate speech. There was no question that the trailer was asinine; it personally offended everyone on the calls. But, Wong pointed out, it didn't explicitly call for harm. *Why would YouTube outlaw criticism of religion?*

The free-speech wing won. YouTube blocked the video in Egypt and Libya but kept it up elsewhere, including Pakistan, where YouTube didn't operate a local version of its site. (Pakistan then blocked YouTube.) YouTube released a statement affirming that the trailer was "clearly within our guidelines."

YouTube's guidelines, however, were never ironclad. Publicly, the company presented them as definitive principles, but it privately waffled on decisions all the time. Like a year later, when the singer Robin Thicke posted a video for his song "Blurred Lines," which showed models cavorting around topless. YouTube took it down—female nipples violated guidelines—but some staff argued the pop track clearly fit within an "artistic" exemption the company had created for videos. During a meeting on the subject, a top YouTube lawyer, Lance Kavanaugh, called in while driving, so a colleague had to awkwardly describe the entire video to him so he could decide whether the seminude models cat walking were artistic or SG ("sexually gratifying"). One executive voted for a takedown by arguing, "My wife wouldn't watch this," according to a person in the room. But YouTube reinstated the video. Afterward, Victoria Grand, YouTube's policy chief, used the term "Blurred Lines" in presentations to describe YouTube's tightrope moderation process.

While the company relied on pop stardom to gauge "artistic" merit, it applied a more laissez-faire philosophy for videos that were politically or culturally sensitive. Grand frequently cited YouTube's decision on *Innocence of Muslims* as a cornerstone company value—one that would stay entrenched for years. She explained these values in a 2014 speech, in which she called the internet a "marketplace of ideas."

> *Of course, defending access to information means colliding head-on with governments and others who seek censorship of ideas they find dangerous, content they deem offensive, attitudes and speakers they consider threatening.*

Later, when *Charlie Hebdo*'s offices were attacked, Google purchased copies of the magazine to display in its offices, a gesture of free-speech solidarity.

• • •

Innocence of Muslims clearly revealed YouTube's power to provoke. As Middle East unrest continued, Mark Little noticed another disturbing power. His forensics started identifying fakes.

At first Storyful debunked innocent tricks, like viral images of a shark swimming on a flooded highway during Hurricane Irene. Then hoaxes started appearing in the Arab Spring footage. A video purported to show a flowing river from July, when the river had actually run dry. Another clip showed rebel forces burying a Syrian soldier alive, but the grave looked suspiciously shallow. Storyful staff sensed something off and reached out to fellow sleuths, who discovered dubbed audio. Alarmingly, it looked as if the Syrian government had manipulated footage to wage information war.

But by then Little had a more pressing concern: YouTube was no longer sure what to do with its Arab Spring curators. The company preferred that Storyful make meaningful money, but who wanted to put ads in front of Syrian rebels? "It's starting to put people off," Little was told. "If YouTube becomes the death channel, then kids are going to stop watching." Priorities kept shifting. YouTube said it wanted to spotlight world events, but it seemed the company had concluded propping up news programming was too heavy a lift. Steve Grove, Little's initial patron at YouTube, had moved on to the next big thing at Google, a social network called Google Plus. Little's subsequent patrons tried to arrange an acquisition, with YouTube buying Storyful, but it didn't pan out. Eventually, a YouTube manager suggested Little shift his model a bit. *Why don't you create a multichannel network?*

From outside, where he sat, YouTube seemed a little chaotic. Little didn't know that from the inside it had begun to feel that way, too.

CHAPTER 12

Will It Make the Boat Go Faster?

Larry Page, dressed in a black T-shirt and jeans, whispered into a microphone. He was nearing forty, with hair that had more salt than pepper, and his voice had turned hoarse, a little like Kermit the Frog's, signs of a vocal cord affliction. But Google's co-founder was smiling, giddy even. His company had grown so quickly that it had to move its weekly meeting, TGIF, to a sizable amphitheater on campus. Page presided over the meeting on a spare, elevated stage facing rows of employees, rapt in attention.

He had reclaimed his CEO job. Eric Schmidt had stepped aside to become chairman in early 2011. ("Day-to-day adult supervision no longer needed!" Schmidt tweeted.) As returning CEO, Page had to concern himself not with financial ruin—Google posted a record $29 billion in sales in 2010—but rather with the death of innovation. Namely, becoming Microsoft. Microsoft had once ruled technology but had grown bloated and bureaucratic, missed two big consumer trends—the internet and mobile phones—and turned into a cautionary tale, at least for the moment. With Facebook and iPhones ascending, Page worried that Google might be headed down the same path. He read books on successful, innovative CEOs. He visited Steve Jobs, Apple's ailing founder, who told him Google's strategy was "all over the map." Google had to "put more wood behind fewer arrows," is how Page put it to underlings. He nixed dozens of projects and invested in priorities, like his prized mobile unit, Android. Once ruled by committee, Google would now be ruled more like Apple. Page's aphorisms, Larry-isms, were enshrined. Everyone should be "uncomfortably excited," should possess "a healthy disregard for the impossible." The top Larry-ism: "10x." As in, *that thing you're working on, why not make it ten times bigger?* When Google's new web browser, Chrome, failed to meet its usage goals, Page called for higher ones. When roboticists making self-driving cars readied vehicles that could handle contained settings like a college campus, Page ordered the vehicles to work on all roads. *10x!*

Inside YouTube, Page's 10x decree prompted a dramatic overhaul of its business goals and operations. In a few short months, YouTube would lay the groundwork for an expansion in size and economic activity beyond what anyone imagined. Those changes would convert YouTube into a tremendous commercial success, but they would also end up creating the kind of perverse incentives that soon mired the company in controversy.

All that began with trying to follow a Larry-ism: having a "healthy disregard for the impossible." Whenever Page spoke to his staff as returning CEO, he reminded them of Google's ambition to expand into market opportunities few thought possible. The robot cars sat inside an experimental lab called Google X, or "the moonshot factory." Another experiment was launched to research human longevity—or, if you prefer, to solve death. In New York a cadre of designers and admen formed Google's Creative Lab, a workshop designed to capture the sort of artistic sensibility in technology that won Apple so much praise.

And that New York workshop was where Claire Stapleton, Bard of Google, scripter of TGIF routines, was heading. During her last TGIF before moving, as Page addressed his troops, Stapleton stood by a wall offstage in a white blouse, expecting to go unnoticed. Page went off script, asking in his Kermit-voice onstage, "Can we get a round of applause for

Claire Stapleton and her incredible email writing?" Stapleton turned deep red, covering her face in embarrassment.

"She's apparently a little bit shy," Page said as applause died down. "She likes to express herself more through computer means."

Years later, after she moved to YouTube, Stapleton would look back on this memory fondly, as one of the last bright spots for her time at Google, before things started veering off course.

• • •

YouTube's radical transformation under Page's return started innocently enough: with the office slide.

It debuted on a summer Monday in 2011. Bright red, huge, ridiculous—the industrial playground equipment installed inside YouTube's new office accommodated three people at a time and sloped down from its third floor. YouTube had purchased this larger property across the street from its first San Bruno home as its staff swelled. An old Gap headquarters, the three-story building welcomed visitors with a spacious atrium exposing floors above, encased with tall windows that made it look like a mall. The slide was part of Kamangar's plan to make YouTube's work space more Googley. Other Google offices had massage rooms and private chefs, special perks to make them cozy and desirable. At its new digs, YouTube had a rock-climbing wall and gigantic TV panels (playing YouTube). Conference rooms were renamed after viral hits. On the Monday the slide was introduced, everyone at YouTube—from programmers to the new on-site cooks in chef's whites—crowded around the banisters to watch Kamangar and two colleagues inaugurate their new perk. The threesome held hands and slid down, alarmingly fast, the boss taking an awkward stumble at the end.

Shishir Mehrotra, Kamangar's deputy, couldn't stand the slide. Not for its Googliness, but for spoiling his management plans. Mehrotra, a metrics fiend, wanted desks at YouTube to have wheels. Every few months YouTube shuffled its seating chart, obeying a Google dictum that moving around personnel sparked ingenuity. This was a logistical nightmare. If people's desks could roll, Mehrotra believed, the reshuffling could be managed with less disruption. YouTube's realty division rejected the idea. They were too concerned that a wheeled desk might accidentally careen down the slide.

The desks were stuck, just like YouTube.

As 2011 ended, Mehrotra and other leaders were deeply concerned about stasis. Google's brightest minds wanted to work at its "moonshot factory" or on other cool stuff. By contrast, YouTube seemed unfocused, Google's JV team. It had all sorts of competing projects and interests—Head and Torso, a secret new music service (code name: Nirvana), the no-longer-secret Black Ops effort. Old-timers felt YouTube had lost some of its original character and charm. "We kind of became the Walmart of video," recalled Chris Zacharias, a developer who left in 2010. The Arab Spring had energized Google's Davos contingent, not YouTube's rank and file: the death channel thing could be tiring. Besides, most were preoccupied with less prestigious issues than geopolitics. Like cleavage.

As YouTube grew, so did the number of uploaders trying to game its system. Danny Diamond wasn't the only one pumping the site with fake comments. Others surmised that adding " . . . " to a video title could drive views, given how YouTube displayed preview thumbnail images on its home page (for example, "Check Out the New Amazing Justin Bieb . . ." *Click.* Not actually Bieber). In 2010, Apple's iPad premiered, a spiffy tablet that people could flip to suit their computing needs. Digital pirates learned they could evade YouTube's copyright filters with the iPad by posting movies upside down. YouTube coders had to patch that hole. So began an unending game of Whac-A-Mole against crafty creators trying to flaunt or bend the rules. One executive likened them to "velociraptors at the fence."

One group was particularly troublesome. Back then, YouTube let only creators in its partner program pick their thumbnails, but other YouTubers figured out which video frames were automatically pulled for the images. A group of women exploited this perfectly. They posted videos as responses to popular clips, staring at the camera with low-cut shirts, making sure a salacious shot made it in the thumbnail frame. They then rode the viral video coattails. These "reply girls" swamped the site that summer. Fifteen-year-old boys could not not click; YouTube's algorithmic system could not not promote. Most viewers who clicked immediately left the video, yet YouTube counted a view as soon as a video began. (In some ways, these women behaved like the savvy "growth hackers" celebrated across Silicon Valley, but unlike coders these women on YouTube received death threats for their persistence.) YouTube wrote code specifically to stymie these reply girls. Still the phenomenon contributed to a broader sense internally that rewarding videos for clicks was not working. Views were a bad metric.

And Google hated bad metrics. Each year every Google division had to draft annual business goals. As 2012 neared, YouTube's directors gathered to write theirs, armed with laptops and bottled water, settling into a conference room, far from the slide.

YouTube was then run by a small group of men. Kamangar, the chief, was most reserved, one of the few souls who could commune with Page. His lieutenants were livelier, more assertive, and argumentative. For some there at the time, YouTube's culture leaned machismo. *Work hard, play hard.* Mehrotra was legendary for his competitive poker hobby, often staying up playing online and drinking scotch until 3:00 or 4:00 a.m., only to arrive at work a few hours later, bouncing energetically from one underling's desk to another. Robert Kyncl, who was rising in the ranks, could keep up with rowdy Hollywood types. Dean Gilbert was known as a table pounder. (During one shouty meeting about livestreaming the Olympics, Mehrotra stood up, walked around a conference table directly to Gilbert, like a baseball umpire, and barked, "I can yell, too!") At one infamous gathering in Tokyo, reveling YouTube staff lost wallets and keys and almost came to blows. Someone was robbed at a bar but stayed there drinking anyway. There was little talk of children or family lives. One male manager went through a divorce in this era but didn't tell colleagues, because he picked up from the regular job evaluations ("calibration meetings") that it might show weakness. "We were masters of the universe," the manager recalled.

Kyncl typically flew up from Hollywood midweek to meet with the other chiefs and debate. Their job was to write the Google gospel for YouTube: the "objectives and key results," the holy OKRs. Page had adopted this management framework early at Google, and it permeated everything. Simply put, OKRs tracked *what* success should be and *how* to get there. Objectives had to be inspirational, key results needed timelines and numbers. Each Google employee had OKRs, which were listed on an internal directory by their name. Promotions, prestige, and bonuses (sometimes, seven-figure ones) were all tied to hitting them. Page used OKRs to wield power when he wasn't CEO and advocated for aggressive objectives when he returned to the role. A constant one was improving the speed of search results. Google should "make the web as fast as flipping through a magazine," he told employees. Page rarely visited YouTube HQ—one person there while he was CEO remembered two trips—but during his rare appearances Page stuck to one edict: make videos load faster. Staff recalled Page stopping meetings to complain about YouTube's buffering speed, which he once called, to YouTube's embarrassment, "the biggest problem Google-wide."

So, gathered in San Bruno, YouTube's masters of the universe plotted OKRs that could appease Page, solve their "reply girl" clickbait problem, and undo their stasis. They plotted ways to achieve the "hypergrowth" so prized in Silicon Valley.

The YouTube leaders rallied around three narratives for inspiration. One came from an exec in Europe who had just read a new inspirational book that told the story of Great Britain's Olympic rowing team, which had stroked its way to gold in 2000 with a simple training mantra: "Will it make the boat go faster?" Every decision—where a rower sat, what workout regimen he used, what breakfast he ate—circled back to that question. A member of the crew team had converted this mantra into a business bestseller. "Will it make the boat go faster?" the Olympian wrote. "Then do it!" The book brimmed with advice for setting "crazy goals" and fostering "cold clinical minds," using corporate-speak like "high performance conversations" and "bouncebackability." C-suite catnip.

The next narrative they loved involved soda. Years earlier the Coca-Cola Company plotted its own shake-up as executives worried about the stasis developing in the tug-of-war with Pepsi. Coke decided it should aim not for soda market share but for "percentage of the stomach." The soft drink company then bought water and juice brands, reinventing itself. "What's our stomach?" YouTube honchos asked in their meetings.

The third inspiration came from Kyncl. The former alpine skier hated indecision and felt that HQ's winding course was hindering his team. "You can't keep doing this to us—not telling us what matters," he told colleagues. He often invoked his old company, Netflix, where its CEO once set a crazy goal of twenty million digital subscribers, a laughably rosy sum at first. Netflix had hit it that January.

The Coke analogy was simple. YouTube had outgunned digital video also-rans, like Dailymotion and blip.tv. Its stomach was clearly television, a $450 billion market. Like Netflix, YouTube could set crazy goals; its executives, in fact, used an unblinking machismo term popular in tech: "big, hairy, audacious goals" (BHAGs, for short). The boat thing, though, that was harder. Mehrotra pulled up YouTube's metrics dashboard, called RASTA, a listing of each statistic the company tracked. Views, subscriptions, unique users, new users, daily active users, click-through rates, session time, visit time, likes, dislikes, comments—the list was endless. To really focus, to just make their boat go faster, they felt as if they had to pick one goal above others.

• • •

Cristos Goodrow had sent his email right as YouTube's leaders convened. For maximum attention, he sent it to all of them and gave it a compelling subject line: "Watch time, and only watch time."

Goodrow came to YouTube after two decades programming software for companies across Silicon Valley and for Google search. At YouTube he ran search and "discovery," the company's term for its systems that surfaced videos like buried treasure. A mathematician with a marine's cropped hair and build, Goodrow immediately sensed YouTube's gridlock and lack of clear marching orders. In his prognosis he proposed rewiring YouTube's machines to favor just one outcome: how long people stayed with videos. "All other things being equal, our goal is to increase watch time," he wrote in the email. Goodrow liked to debate this idea with fellow programmers, arguing over distinctions between the world's two biggest search engines, Google and YouTube. He asked colleagues to type "how to tie a bow tie" into YouTube and imagine two results: one video is a minute long, a quick tutorial on bow tying. Another lasts ten minutes, mixing instructions with jokes and maybe songs. "Which do you prefer?" Goodrow asked.

"The first one, of course," a Google colleague replied.

Silly Google. Goodrow preferred the second video. This defied Google's logic: when web surfers typed into its search box, Google measured success by how quickly it sent them to another website (if possible, one that paid Google to run search ads). But on YouTube, Goodrow argued, the longer people lingered, the happier they must be, logically. "It's a virtuous circle," he wrote later in a management book about OKRs. "Our job was to keep people engaged and hanging out with us." Goodrow turned his email into a long, running document on Google's internal message board, a go-getter initiative that Google rewarded. (An ex-colleague teased Goodrow as a "straight-A student.")

To YouTube brass, Goodrow's manifesto seemed reasonable. They, too, had the bow-tie debate and had agreed that rewarding the amount of time spent on videos was better than rewarding clicks. They debated other metrics, including one for the percentage of creators who earned at least six figures annually and another for ad sales. But a consensus formed. Google's rising nemesis, Facebook, flourished not just by racking up accounts but by keeping people *engaged*. TV certainly did. So YouTube would keep people engaged by promoting videos that racked up the most watch time.

Will it make the boat go faster? Then do it!

Now they just needed a big, hairy, audacious goal. Often they spoke of aspiring to something "sticky"—a goal that could keep viewers around and serve as a lasting corporate strategy. They ran calculations. The internet's top destination, Facebook—Pepsi to their Coke—captured about 200 million hours every month of its users' time, according to their math. *Very* sticky. TV, the big stomach, claimed four to five hours of the average American's day, depending on who was doing the counting. YouTube was then clocking around 100 million hours of viewed footage every day. So, 100 million. *10x.*

Mehrotra left the conference room and beelined to a data scientist who worked for him. "What would it mean to hit a billion?" he asked. "When could we reasonably do that?"

• • •

Mehrotra announced the new OKR the following year at YouTube's annual leadership summit in Los Angeles: YouTube would work to get one billion hours of watch time every day within four years. "Look, I know what you're all thinking," Mehrotra began. *Impossible.* Once trained as an academic, Mehrotra addressed his audience like an excitable TED talker. He walked them through the math. A billion daily hours would be five times Facebook's traffic, the largest on the internet, he explained. Then he added the punch line: "It will still only represent 20 percent of television. *This* is our percentage of the stomach."

Over the coming months, each team at YouTube gradually learned of this audacious goal designed to put YouTube on par with the other 10x moonshots of Larry Page's Google. Ad sellers would now be judged on how much revenue they brought in per hour of watching, gauging those numbers next to cable's. The computer networking unit estimated that hitting a billion daily hours meant YouTube would eat up nearly twice the bandwidth of the entire web. To model that, the unit plotted trajectories on a chart, which they called their "Break the Internet Graph." Goodrow's coders began rewiring YouTube's search and recommendation system to promote videos that generated the best watch time, not the most views. *Only videos that made their boat go faster.*

One team was not so enthusiastic. Bing Chen had recently joined YouTube to work with its most popular creators and received an email in early 2012 from Mehrotra inviting him to a room in San Bruno nicknamed the product den. Before setting its audacious goal, YouTube wanted to test prioritizing watch time in its algorithm. Inside the den, Chen was told he had five days to prepare an explanation of the planned change for creators. Chen left stunned. He knew many YouTubers had learned to thrive, even earn livings, on a system that rewarded views, not video length. This could undo all of their efforts overnight.

He and his colleagues struggled to piece together a blog post that would break the news. When they completed it, they decided to leave their post unsigned. They knew it would not go over well.

CHAPTER 13

Let's Play

Blog post, March 9, 2012: "Changes to Related and Recommended Videos," by the YouTube Team:

The last time you went channel surfing, did you enjoy (or remember) the 20 TV shows you flipped through, or just the shows you watched all the way through? Would you recommend the 20 you surfed through to a friend, or the ones you actually watched? To make the videos you watch on YouTube more enjoyable, memorable, and sharable, we're updating our Related and Recommended videos to better serve videos that keep viewers entertained.

Trevor O'Brien was sitting in one of YouTube's glass-paneled conference rooms when he saw a colleague frantically jumping up and down outside. It was a manager who worked with paid creators, mouthing his name. *Trev-orrr. Trev-orrr!* When O'Brien walked out to see what the problem was, the manager told him, "Your algorithm has a bug."

O'Brien was responsible for YouTube's search product and neatly fit Google's product manager mold—a fast talker with software know-how and better people skills than coders. Once YouTube switched its system to favor watch time, some staff were not prepared for the immediate upheaval it would cause for certain YouTubers, so they blamed a computer flaw. But the system was working as intended. Machines now surfaced videos that held eyeballs for the longest time. Viral hits were punished. The father behind "David after dentist," a cute-kid internet sensation video, contacted the company in a panic: views of the clip, which had been selected to run ads, nearly halved. It didn't help that no one had updated the"analytics" web dashboard for YouTubers to show "watch time" as a figure. They saw "views" and saw those views plummeting but had no idea why.

The company had to convince YouTubers that viewers who clicked weren't really sticking around. "Trust us," O'Brien advised colleagues to say. "Quality viewership is our focus."

Soon, he was fed to the wolves himself. He flew to Los Angeles to explain to YouTube's biggest producers why it had just tanked their businesses. O'Brien met with Demand Media, a company behind huge YouTube accounts like eHow, which churned out factory-style short videos tailor-made for searches for practical tasks ("How to Bathe Your Dog," "How to Wait Tables," "How to Get a Girlfriend in Middle School"). At one point it claimed to be YouTube's largest video supplier, but now its traffic had been decimated. To address this, Demand Media suits proposed making longer how-to videos. That wouldn't necessarily work, O'Brien explained, because the idea was to reward media that people actually *wanted* to watch. He compared his recommendation system to that of a restaurant reviewer. "We're just trying to make sure people find the best videos," he said.

In meetings with Machinima, another massive YouTube factory, O'Brien began this routine with an executive for the gaming network, who probed the YouTube manager for answers. Should Machinima change its titles? Should it make intro montages longer?

"No, no," O'Brien replied. "You're not thinking like a machine."

Machines liked videos that compelled people to keep watching and stay on YouTube. O'Brien offered a few more tips. Leaving a comment took more work than hitting "Like," so the algorithm weighed comments over likes. Stuff like that. Later, to his surprise, the executive visited San Bruno with glossy posters made for Machinima's offices with O'Brien's tips. *Whoa,* O'Brien thought, *they're hanging on every word to gain any edge they can.*

• • •

This wasn't the first time YouTubers felt betrayed by YouTube. The summer before, the company had introduced a snazzy remake of its website nicknamed Cosmic Panda. It was sold as a much-needed design overhaul—*TechCrunch,* an industry blog, had called YouTube.com "remarkably ugly"—and a concentrated attempt to tilt viewership away from only viral hits. YouTube soon added a home-page panel to highlight channels, a guide to its universe (code name: Hitchhiker), and more features that gave each viewer their own personalized feed based on their history and habits, like on Facebook.

Joe Penna was an affable Brazilian musician and stop-motion animator known on YouTube as MysteryGuitarMan. He often collaborated with Freddie Wong and his elegant, catchy short videos made Penna a familiar YouTube star. At VidCon he helped YouTube unveil Cosmic Panda, playing along to its presentation on his guitar. Everyone clapped; YouTube handed out panda swag. Then Penna returned home and watched his viewership collapse, falling some 60 percent, hurt by the redesign that drove viewers away from hitmakers like him. Months later, when YouTube moved to favor watch time, that trend worsened. Penna's management company, Big Frame, tracked stats for several YouTubers and watched them all drop practically overnight. YouTube's blog post introducing the switch had less than helpful advice:

> *How can you adapt to these changes? The same as you always have—create great videos that keep people engaged. It doesn't matter whether your videos are one minute or one hour.*

Most Penna videos were only a minute long, but he put hours or days of work into them. And people watched for that reason. "It just became this bloodbath," recalled Sarah Penna, Big Frame's co-founder and Joe Penna's wife. "It sucked the oxygen out of the room for people who were really creative." One Big Frame client took an even harder hit. At the start of 2012, DeStorm Power's music channel raked in about nine million views a month, making him one of the most prominent Black creators on YouTube. Since debuting, Power had grown more versatile, opening clips with a signature catchphrase—"What's up, world? Another day, another challenge"—and packaging himself as a singular brand, years before that practice was commonplace. YouTube's sudden algorithm change depleted his views and income and left his account riddled with bugs. For over a year Power and Sarah Penna went back and forth with YouTube staff before a company rep acknowledged there was a problem.

Glitches aside, YouTube defended its new algorithm. At one point after it was introduced, overall site views dropped around 25 percent. Because ad sales were tied to views, YouTube's overall revenue fell, too. But for the company that loss was worth the price of improving quality. Its new system did largely quash clickbait trends like "reply girls" designed to lure viewers under false pretenses. Rationally, computationally, in the aggregate, it was the best way to deliver footage people wanted. Interfering on behalf of select creators, such as Power, violated YouTube's sense of what felt fair.

But from outside, it sure looked as though YouTube had abandoned one of its few Black business partners. Power wrote an op-ed about his endeavor titled "Can I Count on YouTube?":

> *Some people think it's easy. "You just make videos and post them to YouTube, right?" Wrong. It's a full-time job and then some. . . . Unlike film or television, there are no off seasons. . . . When things are working, you're at the top of the world. However, because this has become our job, and such a huge part of our lives, when things are broken, it becomes a living hell.*

He soon stopped posting on YouTube.

Others searched for formats suited to the new era. Smosh, the most-subscribed account, started a gaming channel to complement its short skits. The talent manager Dan Weinstein was preparing a TV show for the YouTube hit "Epic Meal Time," an over-the-top, bacon-and-brawn cooking show, when its views collapsed 90 percent. The Annoying Orange, the talking-fruit channel YouTube's chief introduced onstage a year earlier, also lost views.

A few YouTubers did find something that worked: long daily talk shows, like on AM radio—a format soon to explode.

Freddie Wong learned of the algorithm change through YouTube's whisper network. Most of the site's inner workings—which levers worked and which didn't—were passed along as folk wisdom from one YouTuber to another at gatherings or parties. "It's going to be *this* now." *Really? How do you know?* No one really did. But in a few months, the main lever became clear: only watch time. When he began on YouTube, Wong carefully studied the science of virality, testing the right ingredients. Now its formula felt painfully simple. "Okay," he realized. "We're just going to shit out a bunch of minutes." YouTube's switch took creators who produced frequent, cheap footage and "put them on an escalator," recalled one talent manager who worked with Wong. Afterward, Wong noticed a growing trust deficit. YouTube released a new feature letting creators automatically schedule uploads, timed to moments audiences were primed to watch. Convenient. But no one Wong knew used the feature. They were too worried machines would screw it up.

He noticed another shift. At the first VidCon, Wong watched in admiration as dozens of aspiring YouTubers raced around the hotel with handheld cameras, filming one another. In 2012, VidCon moved to Anaheim, near Disneyland, and Wong saw swarms of attendees all start to hold cameras at arm's length, pointed back at themselves.

• • •

No YouTube genre benefited from the switch to favor watch time more than "Let's Play." That was the name given to footage of video gamers filming themselves playing popular or deeply bizarre titles, and inviting fans to watch. Novel games like *Minecraft* were exploding, with endless variations made for computers, not pricey consoles. Video games had narratives baked in, designed for devotee binge-watching.

And "Let's Play" had a rising grandmaster.

PewDiePie: "FUNNY GAMING MONTAGE!"
October 28, 2012. 11:00.

A greatest hits compilation, a clip show—only with better ratings than those on TV. Felix Kjellberg, as PewDiePie, appears in a corner of the screen, with huge headphones, a sweep of blond hair, and stubble. He narrates games unfolding on-screen, a cacophony of "fucks" and delight, imitations of *South Park*'s Cartman. Segments from more than twenty gaming sessions are spliced together. "I'm so fucking scared right now," he says, narrating a slasher-horror game, his specialty. *Scream.* Nine minutes in, Kjellberg opens fan mail. A condom. "Stay awesome, Pewds," he reads. Goofy grin.

Kjellberg was perfectly placed for YouTube's shift. He lived overseas, where YouTube wanted more creators, and he intuitively understood YouTube's appeal. He had started revealing more of himself on-screen in a new weekly series, "Fridays with PewDiePie," where he spoke directly to his audience. "My fans don't really care about professional high-end production videos," he said in an interview. "The loneliness in front of the computer screens brings us together." Perhaps more important, PewDiePie's gaming footage worked really well for search. Beneath his montage video sat a list of links and titles of each game he had played. YouTube's system devoured these words, allowing it to surface PewDiePie's clip when people searched for these games or watched similar footage.

Everyone in YouTube's growing commercial hubs took notice. Gamers brought young male audiences in droves, great for certain advertisers. At Danny Zappin's Maker Studios, the business leaders noticed that Kjellberg, blue-eyed and handsome, drew in more female viewers than other gamers.

For Maker, "Let's Play" offered a chance to overhaul its business model. The studio had recruited more YouTube stars and even some celebrities, like Snoop Dogg, to appear in videos shot inside its new office, which had a towering green screen. By early 2011 Maker had one hundred and fifty channels in its network and claimed over three hundred million YouTube views a month, rivaling entire TV networks in reach. Zappin had raised $1.5 million in venture capital. But many of Maker's channels had scripted, costly productions, and YouTube's ad intake didn't always cover expenses. Once YouTube switched to watch time, the economic tilt became impossible to ignore. Gamers filmed themselves playing and talking. Maybe they edited a bit, added some animation, but usually footage just went straight to YouTube. They didn't need production overhead or towering green screens. And fans watched for ten minutes or longer. David Sievers, the early YouTube taxman, had begun running Maker's new gaming efforts and the former church youth leader grew fond of the cursing "Let's Play" grandmaster, PewDiePie.

With YouTube's shift and new pressure to meet venture capital targets, Maker launched a raid on Machinima, the gaming network, luring YouTubers with promises of better financial terms. During the raid, Maker officials noticed that Machinima had signed Kjellberg to a flimsy "perpetuity" contract that he could escape. Once the YouTuber did, he signed with Maker and PewDiePie's viewership only skyrocketed further. Maker decided to throw him a party after he crossed three million subscribers, placing the gamer in YouTube's upper echelon. Its staff created a party flyer celebrating this achievement. They had to rip it up when PewDiePie crossed four million subscribers as they were planning, and then even more. Eventually, they settled on a party to mark six million.

Maker rented a warehouse in Culver City, hired taco trucks, and invited hundreds. Kjellberg, flattered, had few demands. "I hear in America, you can get a cake with your face on it," he told one staffer. They obliged.

• • •

Blog post, April 12, 2012: "Being a YouTube Creator Just Got Even More Rewarding," by the YouTube Team:

> *Partners are already producing videos that draw large and loyal audiences while building careers for themselves. But we know that there are many more creators out there with talents to share, and we want to empower them to achieve their goals, whatever they may be.*

It was a terribly busy spring for YouTube. A month after its algorithm change, the company made another sweeping move that would ultimately be just as consequential: it opened the floodgates and let practically everyone make money.

Five years earlier YouTube had started sharing advertising sales with a few creators, a number that had swelled to roughly thirty thousand. But the process didn't sit right with YouTube's leaders. For one, the economics weren't working. Madison Avenue marketers had warmed up to YouTube, but the service's huge reach brought in audiences overseas, from the Middle East and Southeast Asia, where digital plumbing wasn't set up to run many commercials or any at all. Some creators turned their ads off, still worried about offending viewers. The situation left YouTube with a shortage of available slots to run ads relative to the money it *could be* making. Mehrotra had faced a similar dilemma years before, during the *Dilbert* run-ins with YouTube's ad hoc board. A similar solution seemed obvious: make more room for ads.

Also, the field looked skewed. MCNs like Maker and Machinima were scooping up hundreds of channels in fell swoops, partially because this was the easiest way for many YouTubers to get paid. That made MCNs kingmakers. YouTube didn't like kingmakers. Kamangar and Mehrotra frequently touted their principle that the site should offer "a level playing field." Level fields had put creators on par with celebrities, citizens with news anchors, beauty vloggers with fashionistas. And the Arab Spring, not yet a deathly winter, justified this egalitarianism. To certain YouTubers gutted from the shifting algorithm, widening the payouts felt rather unfair. But to YouTube, which always looked at data in the aggregate, it looked like spreading wealth around.

So YouTube began allowing every uploader to apply for ads, as long as they didn't break copyright or other rules against hate speech or graphic violence too often. The number of channels running ads expanded from thirty thousand to more than three million, ballooning YouTube's unwieldy Torso and Long Tail overnight. It was, without question, the boldest experiment in mass media and internet self-governance, creating one of the largest web economies ever.

In hindsight many involved would spot the flaws in letting millions broadcast themselves with Google's backing and virtually no checks. A few people at YouTube later said they had proposed thresholds for running ads, like getting a certain number of viewership hours. Dean Gilbert, the content division chief, repeated his mantra that "not all pixels are the same," arguing that different categories of footage demanded different ad rates. But the level playing field argument won.

And in hindsight some recalled other missteps. The company didn't measure the percentage of watch time that came from videos viewers flagged as inappropriate or undesirable. While preparing to expand its ads program, staff didn't hold lengthy debates over who should have a right to "monetize." Mostly they worried that many creators would realize they couldn't get enough traction to earn livable wages. Most wouldn't. Many would still try.

CHAPTER 14

Disney Baby Pop-Up Pals Easter Eggs SURPRISE

Harry Jho read the email but did not believe it. A human from YouTube? He had long since given up on speaking to an actual human from the company. Once at an event an employee handed him a business card, a promising sign. He looked down to see an email, support@google.com, and no name.

Jho worked on Wall Street as a securities lawyer for Bank of America. With his wife, Sona, he also ran Mother Goose Club, a YouTube channel where colorful animals sang nursery rhymes. After the couple taught English in Korea in the 1990s, Harry moved into banking, and Sona, who held a Harvard graduate degree in education, produced shows for public-access stations and the local PBS. The Jhos, who were Korean American, had two young children and noticed how few faces on kids TV looked like theirs. As educators they saw television's pedagogical flaws. To learn, kids should see lips move, but Barney's mouth never did. Baby Einstein mostly showed toys. So they started Mother Goose Club, investing in a studio and hiring a diverse set of actors to don animal costumes and sing "The Itsy Bitsy Spider" and "Hickory Dickory Dock." It was like Teletubbies, only less trippy and inane. The Jhos planned to sell DVDs to parents, ginning up interest for a possible TV show. YouTube offered a convenient place to store clips, and in 2008 Harry Jho started an account there, not thinking much of it.

Two years in, though, he started checking the account's numbers after leaving work. *One thousand views.* He checked the next day. *Ten thousand.* He couldn't find many other videos for kids on YouTube. *Maybe, instead of television,* he thought, *we can be the first to do this.* At least before Nickelodeon took the field over.

It was the spring of 2011 when the human from YouTube emailed, extending an invitation to Google's Manhattan office. The staffer showed Jho designs for the site redesign and shared some tips. Finally, Jho asked the question he was itching to ask, "Why did you call *us*?"

"You might be the biggest YouTuber in New York," the staffer replied. This was news to Jho.

Harry and Sona Jho, soft-spoken professionals who wore glasses and sensible clothes, looked more like PTA parents than YouTube stars. But the iPad had debuted a year before, a handy device for frazzled parents of toddlers. Soon enough, YouTube would add an auto-play function that mechanically teed up one video after another. After the Google meeting the Jhos saw even more traffic on their site. YouTube let them into its ads program. A year later YouTube switched to prioritize watch time, and very quickly Mother Goose Club got company. It began with BluCollection, an anonymous account that only posted videos of a man's hands scooting toy figurines across a floor. The Jhos watched as these clips appeared in the sidebar next to theirs, one by one. Similar videos followed, carpeting the entire sidebar. Then they saw these videos take over YouTube.

• • •

Parents and bureaucrats have always cared what the kids are watching. In the 1970s a federation of advocates and educators who had helped put *Sesame Street* on air pushed for tighter regulation of commercial activity on children's TV, worried that kids could not distinguish programs from ads. Saturday morning cartoons were forbidden to pitch products. A 1990 law, the so-called Kidvid rules, went further, requiring broadcasters reaching children to air certain hours of educational programs and place time limits on how often commercials were aired. Networks tried bending the rules, but regulators held the threat of a license removal.

Then cyberspace arrived. An alarmist 1995 *Time* cover showed a blond boy at a keyboard, his eyes lit in horror-schlock glow, above the menacing word "Cyberporn." "When kids are plugged in," *Time* asked, "will they be exposed to the seamiest side of human sexuality?" Lawmakers governing the modern internet were so obsessed with threats of sex and violence that they ignored other concerns, such as the balance of educational content in media and the potential developmental impacts of unchecked consumerism. Privacy activists, worried about the creeping panopticon of web trackers like Google's cookies, pushed Congress to regulate children's browsing. Websites were openly inviting kids to share the sorts of personal details marketers valued. "Good citizens of the Web," read a promotional site for the movie *Batman Forever*, "help Commissioner Gordon with the Gotham Census." A minor victory for activists came in 1998 with the Children's Online Privacy Protection Act (COPPA), which prohibited websites from collecting information from minors under thirteen for use in targeted advertising. But the law gave enforcement to a different agency (FTC) from the one overseeing television (FCC) and had none of the rules concerning educational programming or commercials that TV had. Old media also had rules about talent. Certain states, like California, restricted the hours child actors could appear in TV or movies and set safeguards for their earnings. The internet didn't.

But kids were clearly heading online, and the massive kid's entertainment complex was eagerly coming with them.

YouTube had seen this migration early. Before the Google acquisition, YouTube vice president Kevin Donahue, a former Cartoon Network producer, pitched YouTube's founders on a kids' version of their site. They directed him to the company lawyer, who shot down the idea. COPPA required websites to do acrobatics to pull something like this off, and YouTube was then so thinly staffed that it needed all its legal resources for copyright issues. YouTube required uploaders to check a box stating that they were over thirteen. The site's terms of service declared it was only for people above that age, and so, on paper, it was.

Google had arrived at similar conclusions. Yahoo, its old archnemesis, once ran a kids' website (Yahooligans), and a few times a year someone at Google would propose a kid-friendly version of Google search. The idea never made it past the sticking point: *How do we decide what is kid-friendly?* A few parents on staff did spot nursery rhymes, ABCs, and toy clips clearly made for toddlers—even before BluCollection, the Jhos were not alone—and fretted about their quality. "Kind of total crap," recalled one mother there.

Any proposals for cleaning the crap had to go through Hunter Walk, the product manager who once tried to scrub the comments section. Walk embraced YouTube's youth culture cachet and knew the kids' world; he interned at Mattel during business school and once worked at a children's bookstore. Yet when colleagues pitched a kid-friendly YouTube, he said no. YouTube simply didn't have enough premium kids' material to make this anything but a lousy version of cable, he reasoned. (Viacom owned Nickelodeon, so YouTube wasn't getting its shows.) YouTube had some TV classics—the company played a *Sesame Street* clip at the second VidCon—but called these "nostalgic," not kids' content. Staff knew juvenilia like Fred were big hits but had convinced themselves that Fred's audience was mostly teens, bored with TV, and that anyone under thirteen watched with adult supervision, like the site's fine print said they must.

But enough stuff clearly made for grade-schoolers or even preschoolers was piling onto the site that some at YouTube felt they had to do something. Walk took a break from YouTube in early 2011, after a painful repetitive strain injury to his wrists, and he returned that fall with a half-baked idea: "YouTube for Good"—a patchwork initiative to improve features for activists, nonprofits, and schools. Most school districts blocked YouTube, wary of its internet free-for-all. But educational videos were blossoming: the Green brothers had their shows, and Salman Khan, a hedge fund analyst, had started uploading mathematics lessons as the Khan Academy channel, part of a Silicon Valley wave devoted to upending higher education. So Walk led an effort to give these creators a name (EduTubers), tools, and attention. He lobbied schools and politicians on YouTube's benefits for students. If schools let YouTube in, the company reasoned, more quality kids' material could flourish.

Other parts of Walk's YouTube for Good project were designed to make the site look more respectable. Coders built a tool to blur faces in videos so that protesters could upload footage with less fear of repercussions. One team worked on a feature to thread tweets and other online chatter about YouTube videos onto the site and its home page, giving it the feel of a news site. Earlier, to sell the channels idea, Walk had christened YouTube "the living room for the world" (a screw-you to TV). Now he framed it as the world's classroom and town square.

Still, as the company tried to prop up its educational and wholesome sides, it was blindsided by a strange beast born within its walls, charging hard in another direction.

• • •

The beast, as always, appeared first in the data. Some of YouTube's coolhunters, the old home-page curators, had settled into a marketing unit called YouTube Trends. Each week the unit sent around the "What's Trending" report, a roundup of the site's emerging fads. Business staff also monitored a chart of the site's top one hundred ad earners. One odd channel started landing in the trending reports *and* soaring up the earnings chart.

DisneyCollectorBR: "Giant Princess Kinder Surprise Eggs Disney Frozen Elsa Anna Minnie Mickey Play-Doh Huevos Sorpresa." March 24, 2014. 14:27.

"Hey, guys, Disney Collector here." We see two dozen toy eggs of various sizes, bearing recognizable figures on their wrappers from various children's entertainment franchises. The voice is feminine, lilting, slightly accented. We don't see her face. The camera holds tight on her hands, fingernails painted black with delicate little Disney princess portraits on top. She announces each egg methodically. "Mickey Mouuuuse." She begins to unwrap the toy eggs, first peeling back the foil casing, a soft, crisp sound. Then the chocolaty layer, a satisfying crackle. Then the tiny plastic capsule holding a toy, a treasure. Then another.

YouTube had never seen a force like DisneyCollectorBR. By that summer the channel's most popular video, a four-minute unwrapping, had 90 million views. Overall its videos were watched a whopping 2.4 billion times. *Tubefilter*, an online *Billboard* for YouTube, placed DisneyCollectorBR as the third most viewed channel behind PewDiePie and Katy Perry. Soon the channel claimed gold. A research firm estimated it raked in as much as $13 million a year from YouTube ads. The videos contained something uncanny and new, tapping into neurons in children's brains in a way that few fully understood. Certainly, no one at YouTube did. "I think she disappears in the mind of the children," one marketer told a reporter. Unboxing videos had begun years before in tech reviewer circles, with footage treating iPods and smartphones as fetish items. Now the Kinder Egg, a marginal product developed in Italy, took on totemic significance. U.S. officials banned it, citing the small toys inside as choking hazards, so YouTubers chasing DisneyCollectorBR's trend started buying these eggs on eBay, like contraband.

At Maker staff watched this fad closely and dubbed the sites "hands channels." Other YouTubers preferred "the faceless ones." Like earlier YouTube hits, these channels sought views using Google's central corridor, search. Look at the mishmash written beneath DisneyCollectorBR's video:

> *Princess egg, frozen eggs, Scooby doo, hello kitty, angry birds, sofia the first, winnie the pooh, toy story, playdoh surprise*

A keyword soup. Toy unboxing titles followed similar logic. "Choco Toys Surprise Mashems & Fashems DC Marvel Avengers Batman Hulk IRON MAN"; "Disney Baby Pop-up Pals Easter Eggs SURPRISE Mickey Goofy Donald Pluto Dumbo." These titles weren't made for the intended viewer or even their parents. These were made for algorithms—for machines to scrape and absorb. Disney, like many media giants, refused to put prized material on YouTube. So when people typed into the search bar "Frozen Elsa" or "Marvel Avengers"—Disney had bought Marvel in 2009—machines showed them the faceless ones.

One person on YouTube's Hollywood team watched the trend in dismay, thinking, *The algorithm is really effed up.* More sinister stuff prowled on YouTube's outskirts. A public relations staffer once looked up a sampling of videos viewers had flagged for YouTube's moderation unit. There was Bugs Bunny, a clip easily appealing to young kids, edited into a violent first-person shooter game. "Maybe it wouldn't traumatize kids," the staffer decided. "But it's really *weird.*" It was relatively benign compared with the weird to come.

Most faceless channels, like DisneyCollectorBR, were also anonymous. Early stars used pseudonymous handles but typically sought fame with real names or at least faces. They had managers, agents, hangers-on, Twitter profiles. To earn ad money, YouTubers had to provide the company with a legal name and an email address, but YouTube liked to keep this information separate from its staff for security reasons. YouTube faced an unprecedented situation with DisneyCollectorBR: the company knew next to nothing about its most popular channel.

A human at YouTube now called Harry Jho with a different question: "Do *you* know who they are?"

• • •

Jho had never quit his banking job even after Google's money had begun flowing in, because it never flowed steadily enough. Some months, during summers or holidays, the Jhos' channel made $700,000 from YouTube ads, a huge haul. But at other times it dropped to $150,000. How could they hire a big staff and ensure steady salaries? If YouTube were their sole income, "we would have gone crazy from the stress," Jho recalled.

A Maker representative called him once his channel took off. Jho was interested, until the second and third calls, each from a new Maker recruiter, and Jho realized they were simply dialing down a list. He went to YouTube events—VidCon and Kidscreen Summit, a children's entertainment equivalent—and asked other creators what value they saw from joining these MCNs. *Not much.* YouTube had become more welcoming, but the company didn't distinguish his programming from that of others, or at least its machines didn't. Once, the Mother Goose Club YouTube page was overtaken with promotions for a new horror film, an *Exorcist* spin-off. Right beside "Skip to My Lou" was a thumbnail of a demon-possessed girl shrieking. *We're a kid's channel,* Jho thought. *No one wants to see that.* He tried in vain to complain to YouTube. Eventually, he found a fix: if he bought ads for his *own channel* to run on YouTube pages, not only did the *Exorcist* disappear, but his traffic shot up.

By 2014, Kinder Eggs had overtaken YouTube. The formula to get into the related videos sidebar—and, thus, to get in front of kids—looked clear. The Jhos held a meeting in their Manhattan office. They looked at the columns of bright, keyword-stuffed videos from DisneyCollectorBR and its countless imitators. "It's really cheap to make these videos," Jho observed. "We could set up a room. Go buy these toys for a couple thousand bucks." They looked at the columns again.

Finally, a friend who was there in the office piped in. "This is just like porn," he said. "This is toy porn."

They dropped the proposal.

• • •

Something else happened as the faceless ones spread on YouTube: more people at the company started having children, and those with older kids heard them ask incessantly for their own screens.

Silicon Valley had long had a widespread, ironic parenting philosophy rooted in avoiding its own inventions. Steve Jobs reportedly limited his children's use of technology. Like Jobs, staff at YouTube spent hours at work reviewing code and business plans all geared to maximize time spent on YouTube and then went home and told their kids to get off YouTube. The site's addictive appeal for sponge-brained preadolescents was obvious. Some employees felt as if they were working at a cigarette company. YouTube's chiefs, loyal Googlers, attempted to measure this problem and turn it into a metric.

They invented categories: "Delicious" and "Nutritious." Much of YouTube mirrored the old scornful derision of TV as "bubble gum for the eyes"—colorful, sugary, craved. But plenty of videos were educational and healthy, too, YouTube thought. (It surely had more of those than TV did, if you added the hours up.) Delicious videos certainly improved watch time, although some staff expressed concern that this type of viewing could be fleeting. *After gorging on snacks, don't you feel guilty?* What if people felt the same way after watching hours of unboxing? What if parents, after seeing kids glued to the faceless ones, took away the candy? "We're only going to be successful long-term if people are happy at the end of their sessions," Mehrotra told staff during a meeting in San Bruno, "not just at the beginning and middle." (After Walk stepped away, Mehrotra had taken over product as well as engineering.) YouTube gauged how long people watched, but staff wanted something more qualitative to measure Nutritious videos. The YouTube for Good team created a survey to run at the end of some videos. It showed little checkboxes and asked, "What is a better use of an hour of your time?"

- Reading a book
- Going to the gym
- Watching television
- Watching YouTube

YouTube wasn't TV; it couldn't give Nutritious videos a prime-time slot. Instead, Walk proposed assigning them a "goodness score," granting educational footage from creators like Khan Academy and the Green brothers more weight in YouTube's search and discovery system. In meetings and internal correspondence YouTube referred to this as adding "broccoli" to the site (or sometimes "chocolate-covered broccoli"). Some drafted broccoli OKRs. The Torso division, which managed its ever-sprawling creator class, drew plans to get 30 percent of watch time from Nutritious videos. Coders working on YouTube search and ads all discussed the effort.

Then, in a fateful twist, these discussions petered out. No company-wide objectives and key results were set.

The same sticking point that had hobbled Google's kid search reappeared: *What exactly is Nutritious? How do we decide? Can we program quality into algorithms? Should we?* A few working on YouTube's education projects pivoted to developing a special app for kids. Walk left the company. No metrics were established. "If you can't figure out how to measure it," recalled one executive there, "you just pretend it doesn't exist."

Besides, bigger, uglier brawls were starting to demand more of everyone's time in San Bruno—brawls outside Google and within.

CHAPTER 15

The Five Families

Late on a Friday night in 2012 Salar Kamangar phoned Francisco Varela, a deputy who handled YouTube's relations with business partners. YouTube's chief was uneasy. "Are you sure about this?" he asked.

The company was preparing to make a huge gamble. Years earlier, it had struck a deal with Apple to preload YouTube's app on every iPhone (outside of China), which practically guaranteed YouTube millions of eyeballs. In exchange, Apple took a small cut of YouTube sales and, since the deal occurred under Steve Jobs, a design obsessive, Apple dictated the design of YouTube's iPhone app. (Jobs died in 2011.) YouTube staff felt that Apple didn't add new app features YouTube wanted fast enough, or sometimes ignored them entirely, and with internet usage quickly shifting to phones, YouTube worried about being at the mercy of Apple's whims. Kamangar's troops lobbied him to ask Apple to relinquish the out-of-the-box iPhone slot for control of the app, betting that consumers would download YouTube all on their own.

Varela's team had made a similar wager with YouTube's app on smart TVs, and it paid off. But iPhones brought in millions more viewers. "Salar," Varela replied when his boss asked for reassurance. "You're not going to lose a single user."

It worked. After Apple pulled the plug, YouTube's usage on iPhones barely budged, a sign of YouTube's centrality in people's lives. The following year the site would claim a billion monthly visitors. Soon shopping malls selling iPhones and smart TVs started stocking speakers and refrigerators with internet connections, video displays, and new ways to watch YouTube.

Getting YouTube distributed on a range of gadgets wasn't easy—the entire company was roped into the rushed effort to remake and market its own iPhone app, and negotiating with most electronics partners felt like pulling teeth.

By all accounts, however, those episodes were far less stressful than dealing with other parts of Google.

At the time, Google could best be described as a collection of fiefdoms running various internet utilities: search, maps, browsers, ads. Most divisions were run by alpha executives, type As, who sat on Larry Page's management council, the coveted "L Team." No fiefdom was as insular and absolute as Android. Its leader, Andy Rubin, a brilliant programmer and robotics nerd, built his Google fiefdom by giving free operating software to the legion of phone makers trying to rival Apple. (In return, Android required phone makers to preload Google apps.) Android staff ate in their own special cafeteria at Google—with custom Japanese fare, Rubin's favorite—and practiced a work culture, like Apple's, maniacally centered on one man. Tall and lean, with a bald pate and glasses, Rubin even looked like Jobs. Stories emerged of Rubin's similar tech-genius tirades. Far worse would come later.

Rubin had recruited a few staff from YouTube, including Levine, its first lawyer, to create a digital store for music and movies, a foil to Apple's iTunes. Rubin's coders also controlled YouTube's app on Android phones. Several directors at YouTube felt that *it* should run Google's music service instead—music videos, after the watch-time transition, were exploding—and that YouTube should control its own app. They pushed Kamangar, who usually avoided confrontation, to confront Rubin. When Kamangar did, deputies saw him return to San Bruno drained of energy.

By 2013, YouTube had become enmeshed in wearying fights with practically every other part of Google. Android fought over music and apps.

Networking fought over bandwidth. Search fought over engineers. Sales fought over control. Google's sales team wanted to sell ads as a package, including video, which YouTube brass didn't want to surrender. YouTube wanted to sell ads like TV did, with rebates and upfront deals. Debates could turn into loud pissing contests. The L Team, at the time, had few women. One former director described Google's leadership as a "big swinging dick room."

YouTube's absolute worst fight, though—one that sucked the life force out of the people running the company—was over Google Plus.

Once upon a time Google's social network was its absolute top priority. Facebook was growing like a weed, and every moment people spent there was one not spent on the broader internet, Google's bazaar. Starting in 2011, Page steered his entire ship toward Google Plus (shorthand: Google+). Like Facebook, Google+ invited people to share emotions and inanities, add friends, and interact online using real names. Google had a deep concern that the web—and web advertising—would gravitate to social connections and networks, leaving Google behind. Everyone's bonuses and OKRs were tied to their contribution to the growth of Google+.

Google+ descended upon YouTube as a plague. It started with redesigns—feeds and videos tailored just for you. Then YouTube coders were required to add tiny G+ widgets to the site, encouraging people to share ads, because Facebook had "shareable" ads. (Virtually no one shared YouTube ads.) Certain projects, like the YouTube for Good effort to incorporate tweets and online chatter on the home page, were shelved to make way for Google+. New orders arrived as stone tablets from on high. At one point, higher-ups held a serious conversation about porting the entirety of YouTube.com onto the Google+ video tab, effectively ending YouTube. Kamangar successfully shut this down, but the debate itself demonstrated that although he wore the title of YouTube chief, he lived under Google's thumb.

Years later, after Google+ died, its postmortems varied. Once it became clear that it wouldn't rival Facebook, Google+ lived zombielike as a "collective hallucination," recalled Hunter Walk, YouTube's former product director. Some faulted the top-down commands. Others blamed the genetics of the company, where social connections did not compute. Googlers had nicknamed Google the Borg, after the unfeeling cyborg mass from *Star Trek*. "The Borg-mind doesn't understand emotions and community," recalled Laszlo Bock, Google's HR chief at the time.

The Borg also entirely ignored the social media it already owned. Technically, much of YouTube operated like standard, one-sided broadcast media, or as a parasocial network—a psychological phenomenon where an audience finds intimacy with a performer it doesn't personally know. YouTube had jettisoned some of its Web 2.0 social features but certainly remained a place where millions interacted every day, forming deep social networks of trust and fierce allegiance. Such considerations weren't part of Google's plans. In late 2013, YouTube began requiring viewers to sign in online with Google+ accounts to write comments and placed comments from frequent Google+ "power users" higher under videos. YouTubeland's ire was immediate. Jawed Karim, YouTube's third co-founder, who disappeared when he left for grad school, returned to the site's first-ever video, his own, to type below it: "Why the fuck do I need a Google+ account to comment on a video?"

YouTube creators also needed to sign up for Google+ accounts to manage their channels. Google touted the social network's success—"300 million monthly active users!"—without mentioning how many of them were forced to join. "It was all YouTubers!" recalled the VidCon founder Hank Green. "They could not see what they had. They could only see what they didn't have." Many of YouTube's early team members viewed this neglect for the site's residents as a stinging rebuke of their culture. Some saw it as a dire turn for the wider internet, which had begun to put growth above the concerns of everyday people. (Facebook would later become a case in point.) "That's the fate of all social media," said Julie Mora-Blanco, an early YouTube moderator who went to work for Twitter. "You start with a small group, you cultivate a community, you get really big, you ditch the community. You've got a platform."

Another darker reality surfaced during the Google+ heyday. Yonatan Zunger, a Google programmer who ran the machinery for its social network, made frequent trips to San Bruno to deliver the Google+ mandates to an increasingly unreceptive audience. While working on Google+, Zunger noticed a disturbing trend: people wrote ugly posts there, either in rage or just to troll—posts that came close to violating rules on hate speech without breaking them. He saw the same thing happening on YouTube. If a video broke the rules, YouTube removed it; otherwise, all videos were on a level playing field. Footage that claimed the Holocaust was fake, for instance, flirted with YouTube's rules on speech but wasn't deemed a violation. During a discussion with YouTube's policy team, Zunger proposed adding a third tier for these debatable clips: keep them up but note internally that they were close to the line. If accounts came close too often, pull them from video recommendations. Policy staff argued that such interventions would violate YouTube's commitment to free speech and, maybe worse, threaten its liability protections. Engineers, the company's ruling caste, seconded this. Cristos Goodrow, an engineering director, later explained that the company then strove to treat any videos kept on the site equally. "Look, if it's not good enough to recommend, it just shouldn't be on YouTube," Goodrow said at the time.

Later, when catastrophe struck and YouTube had oceans more footage to handle, Goodrow and his colleagues would finally heed Zunger's advice, years too late.

• • •

As Google+ chipped away at one part of YouTube, the multichannel networks wreaked havoc on another.

George Strompolos, the YouTube manager who started its partner program, had plunged into this gold rush early. In 2010 he left Google to go solo. He moved to Los Angeles, crashed on Danny Zappin's couch for a while, and considered joining Maker Studios. Instead, he created a rival MCN called Fullscreen, setting up offices a few miles east. Compared with Hollywood types, Strompolos was staid, Googley. And unlike Zappin, who ran a renegade studio, Strompolos wanted a well-oiled machine. Max Benator, an early digital media executive, likened the pair to their industry's Steve Jobs and Bill Gates. Strompolos pitched investors on his network's "flywheel," a beloved tech term, wherein he would sign YouTube channels, give them software doodads to grow—like a pop-up square inside videos for plugging prior clips—and use money from that growth to sign more channels. He imagined subnetworks, where star creators could recruit others, like how Dr. Dre went from rapper to hip-hop mogul.

Not long after launching, Strompolos caught up with an old colleague at YouTube, Jamie Byrne, and shared that Fullscreen had signed tens of thousands of channels. Byrne had no idea. Fullscreen and the other networks had exploded right under YouTube's nose.

Fullscreen's Dr. Dre strategy, though, quickly smacked into the unwieldy, unpredictable reality of YouTube.

FPSRussia: "TOP 3 WEAPONS TO SURVIVE THE APOCALYPSE." November 10, 2012. 6:15.

"So the first weapon I want to go over is the Remington 870 12-gauge shotgun." A YouTuber named Kyle Myers stands in a field and speaks in a clumsy Russian accent. The table before him displays three massive guns. Myers settles into his zany shtick as Dimitri, a bantering arms instructor. "Most superstores, not even gun stores, carry firearms in the United States." Loyal viewers know he will soon blow something up. He does not disappoint. He aims the sniper rifle and *kaboom*.

At the start of 2013, FPSRussia was the ninth most popular channel on YouTube. The channel was the brainchild of Kyle Myers and Keith Ratliff, a firearms enthusiast with a bumper sticker on his car that read, "I ♥ Guns and Coffee." Strompolos connected with Ratliff and sold him on an ambitious project: FPSRussia would form a YouTube subnetwork with Fullscreen, focused on rugged outdoor sports like hunting. Brands like Yeti would sponsor it. Ratliff was on board, only he asked to be careful not to reveal his location and identity, because he ran a gun store in Georgia and didn't want to jeopardize that business.

It was an unusual request but seemed reasonable. Strompolos prepared paperwork for the business venture.

Then he woke up to the news that Ratliff had been shot dead.

Fullscreen's staff were understandably rattled. Sometimes they had to field distressed calls from creators, but that came with the territory, and most days were spent in the whimsical world of YouTube skits, parodies, and antics. Not this. Worse, Ratliff's murder remained unsolved. "Are we part of organized crime now?" wondered Phil Ranta, a Fullscreen executive. Internet conspiracies about the murder started to swirl, taking seed on a fertile ground for conspiracies, YouTube.

None of Fullscreen's other network deals were this jarring, but they all increasingly made less economic sense. Scale made sense. Strompolos had learned that at Google; the company worshipped software and business models that *scaled*, working the same way widely, with minimal cost and interference. The MCN recipe for scale was as follows: add as many channels, views, and ads as possible. *Repeat.* At Fullscreen scale occurred in its "boiler room." Rows of twentysomethings worked in its two-thousand-foot warehouse in Culver City, just south of Hollywood, sprawling out across long tables stacked with computer monitors. Each screen showed a spreadsheet of the top 100,000 channels on YouTube. (Eventually, the top million.) Each recruiter had to reach out to a thousand channels a day—cold emails, texts, and phone calls. At one point almost half the fifty-person staff were working as channel signers.

By then YouTube had created a simple software process that let YouTubers join Fullscreen with just three clicks. Trouble was, other MCNs also had this three-click tool. And so a war began, a battle for scale that did little to benefit the actual creators making videos.

At YouTube, Andy Stack, the payments director, noticed the brewing conflict when he saw the paperwork creators showed him. Machinima, the gaming network, sent out contracts that claimed YouTube paid $2 CPMs—a couple bucks for every thousand video views. If the creator signed with Machinima for multiple years, they would get $3 guaranteed. *Great deal, right?* (Machinima staff then measured success by the number of YouTubers quitting day jobs.) Stack clicked a few links on YouTube's internal system to discover these creators actually earned more on their own, without giving a cut to an MCN. *Bad deal.*

But most YouTubers didn't have access to concerned YouTube personnel. They had only the whisper network and MCN recruiters, who constantly upped the ante. In Hollywood, managers and agents took a standard cut for every production deal. MCNs initially did so with YouTube's ad payouts (70 percent to creators, 30 percent to networks) but ditched that pretty quickly once competition heated up. Machinima gave its guarantees. Maker tried countering those by offering gamers better ad split terms. New MCNs suddenly appeared with even sweeter deals. Eighty percent to creators! Ninety! Ninety-five! Each one wrestling for a thinning slice of YouTube's pie. Networks were often willing to lose money signing creators just to keep them away from rivals.

When Dan Weinstein tried to recruit a YouTuber to his MCN, he discovered Machinima paid the star twice the amount earned from YouTube ads. *That's crazy,* Weinstein thought. *How are they doing that?*

This business model seemed plausible so long as YouTube's ad rates kept going up, as advertisers kept coming over from TV, and MCNs remained one of the select few with access to this gusher and other business perks like endorsement deals. But when YouTube opened its ads program to the masses, MCN economics combusted. All the new videos running ads sent prices way down, and MCNs could no longer afford to throw money around wildly or justify their middleman role. Thousands of young creators who joined them, often with little or no legal guidance, suffered. Kaleb Nation, a Texas vlogger who had signed with Machinima, got a call from the network informing him that his payments would be cut down to a sixth of his current amount, despite guarantees of a fixed rate. "You can't do that," he complained. "It's in my contract!" But the fine print read that, indeed, the network could. Burned creators nicknamed the company Ma-shit-ima. Nation got out of his contract, but others couldn't.

Many at the MCNs felt it was YouTube that had the lousy terms. The company constantly rearranged the ground beneath the networks, tweaking an algorithm or upending economics without any notice. Sometimes YouTube copied them. After Fullscreen released its screen pop-up feature, YouTube released its own version six months later. Eventually, YouTube added a button on its back-end site for creators to exit an MCN with a simple click. These networks, once considered the next vanguard of media and Hollywood, found that it was impossible to build such a business on YouTube—at least by playing it straight. "It was as close to a Ponzi scheme as I will ever get in my life," recalled one MCN executive. "We were just aggregating other people's work and channels, telling them that we were going to help them make more money. A few of them did. Most didn't."

At one point Dan Weinstein convened a gathering to call a truce on the network race to the bottom. He invited four other top MCNs—"the five families," he called them, a nod to the industry's mafioso feel. The group met several times around Los Angeles, attempting to settle on a workable commission and competition model. One never stuck. Too many of the networks, Weinstein realized, were primed on investor capital and looking for exits, acquisitions that could pay investors back.

• • •

Meanwhile, the network built to give YouTubers power and control was falling apart.

Mark Suster, an investor who funded Maker Studios, had initially put money in on the condition that Danny Zappin, as an ex-con, couldn't be Maker's CEO. But Suster, a venture capitalist who preferred keeping company founders in charge, had changed his mind about the unconventional hustler and let him return to the role. *Better to have him inside the tent pissing out,* the investor thought.

The tent still got messy. When Suster emailed his renewed CEO about holding a board meeting, Zappin gave a one-line reply: "Don't tell me what to do."

Danny Diamond was no stranger to business disputes. He had had frequent heated bouts with Robert Kyncl, YouTube's Hollywood chief. Zappin's plan to let YouTubers own and operate the studio was divisive from the get-go. Several stars left Maker early on, citing disagreements over the business; another member sued Zappin; Ray William Johnson, the brash YouTube comic, exited Maker in an epic explosion, accusing the studio of demanding he hand over more money and intellectual property control. When Johnson refused, ugliness ensued. Late at night in December 2012, Zappin thumbed a text to Johnson: "You're [*sic*] lack of integrity and character are sad. Fuck you. Prepare for war . . . bitch."

Johnson went public, documenting the fight in a blog post and on Facebook, and berating Zappin's "thuggish tactics." Zappin responded in a letter to Maker staff disputing Johnson's charges and noting how formative his prison sentence had been. "I couldn't be more thrilled to be given a second chance," he wrote. But by then, that second chance was nearly up.

Zappin had split with Lisa Donovan, his girlfriend and Maker co-founder, and co-workers watched as he became increasingly emotional and erratic. The stakes were much higher now as serious money flowed into YouTube networks. During the Johnson spat, Maker was preparing to raise money from Time Warner and other investors valuing the studio at $200 million. Having Danny Diamond at the helm no longer suited the monied interests. Suster had brought in Ynon Kreiz, a hard-nosed Israeli TV executive with a reputation as a turnaround artist, and on April 16, 2013, Zappin submitted his resignation and voted with Maker's board to appoint Kreiz as CEO. Afterward, however, Zappin panicked. By his telling, he discovered that Kreiz and the board had colluded in secret with his co-founders to push him out in exchange for more shares. (Kreiz and others would deny this.) Zappin sued Maker, its board, and his ex-girlfriend, charging them with constructive fraud and a "scheme to oust" him. The lawsuit failed, and once all the creators left Maker's management, so did Zappin's vision for a United Artists of YouTube.

He also lost another prized possession amid the ordeal: soon after Maker's new leaders took over, his old YouTube channel was scrubbed from the site.

• • •

Maker's brightest star, however, was only getting bigger.

PewDiePie: "THINGS YOU DIDN'T KNOW ABOUT ME." August 9, 2013. 5:48.

Felix Kjellberg is in a well-lit close-up, all smiles. Today he is sharing twenty facts with his fans. He sold Photoshop art to buy a computer to start on YouTube; at fourteen, he won a tennis championship; he doesn't eat vegetables. "I get my nerdy side from my mom." He likes Radiohead and "rarely swears" in real life. "My girlfriend knows more about football and cars than me." He plugs his charity drive and ends with his signature maneuver, the "bro-fist," a friendly bump straight to the viewer.

Six days later PewDiePie toppled Smosh to become the most-subscribed channel on YouTube (11,915,435), an honor Kjellberg would hold for six very eventful years to come.

CHAPTER 16

Lean Back

Her fans, the *real* ones, knew Ingrid Nilsen's entire life. They knew her morning skin-care routine and her evening ritual. Her favorite concealers, eyeliners, and shampoos. They knew how she loathed public speaking and her chronic acne, how devastating her father's death was, how she just barely graduated from high school (all Cs and Ds), and how afterward, in a depressive funk, she had turned to YouTube.

Nilsen was biracial (Thai mother, white father) and never found much use in the beauty tips in magazines she read growing up in Orange County. Where she did find useful advice was on YouTube, from women who looked like her, denizens of the site known as beauty gurus. One night in 2009, at age twenty, staying up late and "in a really dark place," as Nilsen later told viewers, she started her own account, Missglamorazzi, the name inspired by a Lady Gaga song. She shot her first video, "How To: Flawless Red Lips," in her house, backlit poorly by a lamp, leaning in incredibly close to the webcam. Nilsen was pint-sized and ebullient, a natural on-screen who could ad-lib in teenspeak about makeup application, hair-care product attributes, pretty much anything.

The coolhunters had noticed the rise of makeup tutorials on YouTube, which fit neatly into the company's "how-to" bucket, and as such were videos tailor-made for search. But a much deeper appeal was at play: these YouTubers shot tutorials that anyone could follow, offering a dose of lo-fi intimacy and reassurance. "They provide a safe space to practice, fail and laugh at yourself," a fashion observer explained in *The New York Times* the year Nilsen debuted. After the watch-time transition, beauty gurus shot up in YouTube's charts. Nilsen could post ten- or fifteen-minute-long videos with relatively little editing. She expanded beyond makeup tips and skin-care regimens to clothing, food, and more personal fare ("50 Random Facts About Me!" "Zombie Pimple + Cleaning Day," "TMI Tag"). Initially the fashion industry didn't grasp the appeal. Nilsen once met with a prominent magazine's beauty editor, who proceeded to castigate Nilsen for not posting enough red-carpet-style footage. "What you're doing will never work," the editor told her.

At YouTube events and creator contests, Nilsen found herself surrounded by comics and musicians. YouTube propped up few beauty gurus, despite their rabid fandom. "It was a genre they didn't fully understand," recalled Nilsen. Eventually, YouTube's Torso team recognized the commercial potential; creators like Nilsen started doing "haul" videos, slowly unpacking each item purchased from a store. YouTube labeled them "Style," and then later, to reflect the frequent, wider-ranging posts from Nilsen and her peers, the company categorized them as "Lifestyle." YouTube's algorithm loved the category.

• • •

YouTube's algorithms were getting stronger and stronger. Mark Little, the Irish journalist behind Storyful, first saw their awesome power with the "Leanback" presentation. Viewers clicked to see footage unfurl instantly and automatically, one video after another. "No more clicking and fiddling around with your mouse," a pleasant voice explained in a promo. "All you have to do is lean back and enjoy the show." Little watched this in Google's Manhattan office, a little unnerved. In newscasts you gave people the most important stories first, informing the audience of major events before they went on with their lives. *This is totally different*, he thought. *Come in and never leave.* The more people watched certain videos, the more YouTube fed it to them. "It was creating rabbit holes not yet filled with sludge," recalled Little. "That would come later."

Back then this feature was packaged with the internet's do-gooder appeal. Jonah Peretti, a weedy hipster who ran a fledgling website called BuzzFeed, also presented in New York during the "Leanback" feature demo. BuzzFeed would soon launch a newsroom, marrying internet virality with serious reportage. Little's company, pushed to act like an MCN, adopted BuzzFeed's approach. To subsidize its serious work, curating the Arab Spring for YouTube, Storyful staff hunted for feel-good viral clips. They found the "emotional baby"—a mother's footage of her child's tearful response to her song. Call the mom, ask if ads could run on her video, split the sales, and pitch the clip to TV shows to gin up more views. Storyful called this "the mullet": business in front, party in back.

News outlets also wore the mullet. Russia Today, a TV network funded by the Kremlin, excelled at it on YouTube, mixing political coverage with tantalizing clickbait clips of cute animals, car crashes, and couples caught having public sex. (That algorithmic alchemy helped Russia Today climb YouTube's charts for years, without much worry from the company, until Russian politics became radioactive.)

The "Leanback" feature didn't last on YouTube, but the concept certainly did. More viewers moved from one recommended clip to the next, bringing in more watch time and, just as critically, more *data*. At San Bruno, YouTube staff rarely watched videos, but they watched video data constantly. In particular they paid attention to the seesaw of data on ads and viewership. More frequent ads turned people away: YouTube didn't have the established cadence of TV commercials. Ads appeared about every seven minutes of viewing time, though that figure was inconsistent (some channels had longer ad-free windows) and inefficient (thus, an irritant for

Google). Alexei Stolboushkin, a programmer working under Mehrotra, suggested that YouTube let algorithms decide.

So began one of YouTube's most beneficial business moves and a telling sign that the company often had no idea what its machines were doing. It was called Dynamic Ad Loads (code name: Dallas). Around this time, machine learning was all the rage at Google. A form of artificial intelligence, machine learning had existed for years merely as a theory, waiting for enough computing horsepower and piles of data to put it to use. Google had both. Most software until then was hard coded. (*If* this, *then* that. *If* a viewer sees this Ingrid Nilsen video, *then* show that ad.) Machine learning systems programmed themselves, detecting patterns in data to, say, recognize a face in a photo or a nipple in a video. Good ones outsmarted humans, at least at specific tasks. Oftentimes decisions these machine systems made were inexplicable, even to the humans behind them. YouTube's engineers proposed gathering enough data on viewing patterns for a model to predict an optimal rate of showing ads without scaring off viewers. Every machine learning model needed goalposts. Mehrotra drew a graph with "Watch Time" on one axis and "Revenue" on another. To strike a balance, he told his coders he was willing to take a 1 percent drop in watch time for a 2 percent increase in ads, but nothing more.

Engineers ran tests for Dallas, tweaking experiences for certain viewers without telling them. At YouTube Stats, a meeting Mehrotra held every Friday, they presented the befuddling results: the machines found a way to show more ads *and* improve watch time.

"How can it possibly be positive on both?" Mehrotra asked.

"No idea," an engineer replied. "Can we ship?" As in, *Can we use this on all videos?* Mehrotra asked them to report back the following week with an explanation for the system's behavior. They returned. Same results. More ads, more time engaged with YouTube. No idea why.

Eventually, coders deduced the machine logic. If YouTube played an ad for a viewer as soon as she visited the site, she would likely leave. But wait for her to linger, to stay ten, twenty minutes watching videos, then show an ad, and she was more tolerant of commercial interruptions. The machines had deduced a formula that kept people watching more video footage and, ultimately, more ads. That paired nicely with a novel format YouTube also unleashed around this time: "skippable ads" that viewers could fast-forward through. Advertisers only paid for these when viewers didn't skip, which meant the commercials had to be compelling (unless the viewer was lazy or was a kid who couldn't read the Skip button). YouTube salespeople loved the format: if people *were* watching Nike ads, Adidas wanted to run enticing spots too, and might even pay more to outbid Nike in YouTube's ad auction. After this, YouTube's business, one former executive recalled, became "a freight train that couldn't stop."

Stolboushkin, the lead engineer, was asked to discuss Dallas at YouTube's all-staff meeting in San Bruno. He was allotted only ten minutes, so he offered a simple metaphor, apt for the monetary bounty the algorithm unleashed. *Imagine videos were food and ads were booze*, he began onstage. Dallas was a robot sommelier, picking the right combination. On a screen the engineer displayed an image of a steak meal next to a bottle of Château Margaux poured into a crystal glass.

• • •

While YouTube refined its algorithms, its viewers were being drawn into a growing corner of the site.

Fans knew one of the site's staples simply as Stef. This balding, stocky, avuncular Canadian, with an accent hinting of Irish roots, talked about his sad childhood, about dating and marriage, about big, serious topics—he could talk about anything—speaking directly to young, disaffected men going through hard times, promising them lights at the end of their tunnels. They listened.

Stefan Molyneux, a former IT businessman, had refashioned himself in his late thirties as a grandiloquent guru. Like others who made money from computers in the dot-com boom, he wore loose polos and enjoyed the sound of his own voice. In 2005 he began Freedomain Radio, a podcast and a movement. He joined YouTube soon thereafter, posting videos with search engine boilerplate, such as "An Introduction to Philosophy," and self-help-style lectures à la Tony Robbins, many over one or two hours long. Years later, after the financial crisis, Molyneux spoke about the economic pain and anomie in its wake. "College students have a damn right to be depressed," he told viewers. "Their society is unsustainable." He delivered his lectures framed as commentary on *Harry Potter* and *Star Wars*. Some viewers were captivated equally by his worldview and by slices of personal life he shared. Caleb Cain, a college dropout in West Virginia who liked the Dead Kennedys and Michael Moore documentaries, discovered Molyneux in his YouTube sidebar and admired the domestic bliss the guru spoke of with his wife and daughter. *I want all that stuff,* Cain told himself. *If I just watch more and more, I'll be like Stef.*

Molyneux didn't start on YouTube as especially political. If the subject came up, he was a libertarian or an "anarcho-capitalist." But politics started to creep in, particularly after America elected a Black president.

Stefan Molyneux: "The Story of Your Enslavement." April 17, 2010. 13:09.

It appears to be a documentary about human nature and economics, with soft fades and archival footage. It is a diatribe. Molyneux narrates a textbook history lesson on how slavery evolved to modern society, before delivering his punch. "Nothing could be further from the reality," he tells us. Images appear of caged animals. "In your country, your tax farm"—this he spits out—"your farmer grants you certain freedoms, not because he cares about your liberties, but because he wants to increase his profits." Camera cuts to a Tea Party protest, where a poster shows President Obama above the word "Fascism." "Are you beginning to see," Molyneux asks, "the nature of the cage you were born into?"

By then Molyneux had already concerned some parents. Barbara Weed, a British councillor, grew alarmed when her son suddenly left home, leaving only a note that said, "Please do not contact me." Weed discovered he had joined others in following podcast advice from Molyneux to abandon their family of origin—to "deFOO," he called it—if they were unable to work through problems with therapy or other means. Molyneux and his wife, a therapist named Christina Papadopoulos, preached this online and at gatherings at their home. (A Canadian psychology board later reprimanded Papadopoulos for professional misconduct.) They invited listeners to donate for special courses not on YouTube, offering a $500 fan subscriber level called "Philosopher King." As early as 2008, when Weed went public with her story, newspapers used the word "cult" in their coverage of Molyneux.

YouTube had no rules in place to investigate what its creators did off its site. With so many people uploading, it could barely police its own backyard. But systems like Dallas made YouTube much more adept at raking in ad money for its broadcasters, and, back then, the company tended to give all creators equal access to its bounty.

This was well before aggrieved men online were considered an unstoppable political force, though signs were appearing. "I'm sure a few marriages broke up because of feminism," Molyneux told a Canadian reporter in 2008. "It doesn't make feminism a cult."

CHAPTER 17

The Mother of Google

Susan Wojcicki's flight was canceled. It was 2010, well before her arrival at YouTube, and she was en route to Washington, D.C., for a hobnobbing gala celebrating *Fortune* magazine's annual ranking of the "Fifty Most Powerful Women in Business." Barack Obama was speaking. Wojcicki had made *Fortune*'s list for the first time, but with work and family obligations she didn't find it was worth the hassle to book another flight.

A decade after Google started in her garage, Wojcicki wore a variation of the same dirty-blond bob haircut and possessed the same adoration for Google. She was a perfect Google operator, judicious in moving bits and hitting metrics. After Google Video folded to make room for YouTube, Wojcicki moved to Google's oil well, its ads business. She ran AdSense, the company's system that blanketed websites with digital billboards. Credited for its invention, she was given a Google "Founders' Award," a retention bonus meant to compensate staff as if they had formed their own company. Google could count the women in its c-suite on one hand. Wojcicki was one.

And yet others from this elite group were getting far more public attention. Sheryl Sandberg, a Google saleswoman, had taken the number-two job at Facebook. Sandberg, a natural politician, held regular soirees at her home in Silicon Valley, and the Facebook position earned her flattering magazine profiles and accolades. She was No. 22 on *Fortune*'s 2009 power rankings. Also on their list was Marissa Mayer, Google's product czar and one of the company's colloquial "mini-founders" along with Wojcicki. Both Mayer and Sandberg could be smooth, confident on camera and onstage. Wojcicki could occasionally be stilted. Women in the c-suite tended to be graded on a different curve, which favored media visibility and stage presence. Google's publicity team was tasked with getting Wojcicki on more stages and more visible in the press. At one point the second Google search result when people typed in her name linked to "Susan Wojcicki's big lie," a *Gawker* item accusing her of fabricating her role in inventing AdSense. "Yes," she admitted to the journalist Steve Levy about the link. "It does bug me." But Google never challenged the sanctity of its search rankings to scrub a result, even one its execs found unflattering.

Wojcicki finally made the *Fortune* list in 2010 as No. 43, twenty-seven spots below Sandberg that year. Mayer was No. 42.

After Wojcicki missed her flight, one of her public relations handlers gently advised, "You kind of have to go." Sandberg and Mayer were probably already there, networking. Wojcicki went.

• • •

Susan Diane Wojcicki was born in 1968 in Santa Clara, California, in the beating heart of Silicon Valley, into a family later canonized for being exceptional. Her father, Stanley, a particle physicist, grew up Catholic in Krakow, right by train tracks to Auschwitz. Nazis commandeered his family's apartment before he escaped to Sweden. When Susan was young, Stanford University named Stanley its physics department chair. Susan's mother, Esther, was born to Russian Jewish immigrants on New York's Lower East Side and became a vivacious journalist and educator, well known across Silicon Valley for teaching at Palo Alto High School, near Stanford. The Wojcickis had three girls in rapid succession and raised them to be industrious, clipping coupons and balancing checkbooks before they could drive. When Susan was four, she rushed up to her mother in the parking lot of her Stanford preschool. "We had marshmallows today," the child said, "and I got two." Esther Wojcicki soon learned her eldest was a participant in a psychology experiment. Walter Mischel, a Stanford professor, placed children in a room with a treat, telling them to eat it now or wait fifteen minutes for seconds. Forty years later kids who waited were physically fitter, "more cognitively and socially competitive adolescents," and less troubled adults, so concluded the fabled study. Little Susan waited longer for the marshmallow than any other preschooler.

Mischel's results came under scrutiny later, but they remained part of the lore of its early participant. "Susan is one of the most patient and logical people I know," Esther Wojcicki wrote in *How to Raise Successful People*, her book on parenting, after recounting the marshmallow episode. "Nothing fazes her. She has enormous self-control. She surrounds herself with employees that she trusts and respects. She had all these traits as a young girl, and not because she'd been born that way, but because she'd been practicing for years."

Susan studied history at Harvard and spent a year in India working as a photojournalist. Soon she veered back to the more practical, earning a master's in economics and another in business, and began a career in technology. She married Dennis Troper, a technologist who would later join her at Google. Her sister Janet became an epidemiologist, and Anne, the youngest, worked on Wall Street before returning home to become Valley royalty. Anne co-created the genetics testing firm 23andMe and married Google's Sergey Brin in an unorthodox ceremony in the Bahamas. Affable and charming, Anne easily won praise. *Vanity Fair* called her "Jennifer Aniston in Birkenstocks"; another magazine named her "the most daring CEO in America." Susan, in contrast, came off as reserved, humble, and conscientious, a "classic older sister," according to one acquaintance.

After YouTube joined Google, Wojcicki had helped orchestrate the acquisition of DoubleClick and turned Google into an ads juggernaut by placing its banners on virtually any website that wanted them. Perhaps most critically, she had the ear of Larry Page; she was "a Larry whisperer," said Kim Scott, a former Google director. "When people couldn't get him to see reason, she always could," recalled Scott. Wojcicki was even-tempered, rarely imposing, and adept at fading into backgrounds. Once, during Google Zeitgeist—an annual exclusive company event for tycoons, politicians, and celebrities eager to schmooze Google executives—an attendee spotted Wojcicki quietly taking notes, still in the hiking gear she had worn to scale a nearby mountain. No one recognized her. When it was called for, though, Wojcicki could flash her shrewdness. A former Google executive recalled a meeting with Rupert Murdoch's News Corp, the steadfast Google foe, over a business arrangement. The foe was playing hardball. "All right, we just won't pay you," Wojcicki said coolly. "How does that sound?"

Google's publicity efforts highlighted not her shrewdness but certain adaptive, anodyne traits (hardworking, nonthreatening, loyal) that played well in the male-heavy industry. A public image began to take shape. A complimentary profile from 2011, in *The Mercury News*, a San Jose newspaper, labeled Wojcicki "the most important Googler you've never heard of." Colleagues praised her there for never hogging the spotlight, while the article described her as instrumental in buying DoubleClick *and* YouTube, giving her more credit for the latter than many people involved remembered.

But the piece really focused on motherhood. Wojcicki was the first woman to give birth at Google and created its maternity leave policy, something the Google boys hadn't thought of. By 2011 she had four children, whom she chauffeured around in a Toyota Highlander as a "soccer mom," the article noted. "My kids know I'm home every night for dinner," Wojcicki told the paper. The publicity tactic seemed savvy: Wojcicki could be "dry," recalled one former business partner, "but [came] alive a bit when she talks about her role as a mother." *Fortune* dubbed her "the mother of Google." A *Forbes* article remarked upon her "self-deprecating plainness," wearing jeans and hoodies around the office, and quoted an anonymous ex-colleague who compared her to Chance the Gardener, the Peter Sellers character in the satire *Being There*. Every article—and many watercooler chats—about Wojcicki obligatorily mentioned her garage that birthed Google, her maternal nest for the boy geniuses. (Google later purchased the lot and marked its address on Google Maps as "Susan's Garage.") This had a way of diminishing Wojcicki's accomplishments and personality as an executive, and some certainly dismissed her this way. (One colleague would later deride her as "the luckiest garage owner.") Yet her early intimacy with Google's founders was important, not as a marker of nepotism, but as an explanation for her standing within the company. Larry Page and Sergey Brin became so successful, so rich, so powerful, and subjected to so much scrutiny, at so young an age, that they surrounded themselves with competent people who knew them before all that. "The founders trust Susan maybe more than anybody on the planet," said Patrick Keane, an early Google sales director. "You could never get Susan rattled, no matter how challenging the moment was."

Laszlo Bock, Google's longtime head of HR who left in 2016, compared Google to a family enterprise. Page and Brin reshuffled Google's stock in 2011 to give themselves ten times the voting power of normal shareholders. Wojcicki's husband and brother-in-law also worked for Google. Her mother was hired to consult on education programs. Google invested in Anne's company. "In a family-owned firm," Bock said, "what gets rewarded more than anything is loyalty."

• • •

By 2013, though, Susan Wojcicki found herself stuck in the thick of Google's family rivalries. She sat on Page's L Team, which had frequent arguments. And someone from her own division was gunning for her.

Sridhar Ramaswamy, a bespectacled, exacting Google engineer, had climbed up its ad unit ranks to sit on par with Wojcicki. They had disparate styles. Ramaswamy cared about details, wading into computer code documents, and knew metrics intimately. Co-workers described Wojcicki as a "big-picture" thinker, preferring to delegate details to loyal lieutenants. (One former Googler recalled bringing a sixty-slide business proposal to Wojcicki, who, five slides in, turned to a deputy, asked for his gut feeling, then ended the meeting early.) She and Ramaswamy also had a bitter dispute. Wojcicki wanted to use search queries to inform ads people saw on all those banner ads Google ran across the web; if advertisers could target consumers based on searches *and* on websites they might spend gobs more with Google. Ramaswamy opposed this for business reasons—he felt cheaper banner commercials would dent the fat margins search ads had—and because of a steadfast aversion many programmers held: seeing ads across the internet based on searches might creep people out. Page had long preached keeping search data separate from everything else. But others, like Wojcicki, felt this wasn't keeping up with trends in the ad industry, which sought ever more data.

This fight came to a head during meetings Page held at the end of 2013 to set his company's long-term direction. (Google 2.0, he called it.) Normally reserved, Ramaswamy grew animated and, in one meeting, reportedly accused Wojcicki of "hiding the truth" about her intentions. Apart from these gatherings, the two executives avoided being in the same room. Ads staff recalled attending a review meeting with Wojcicki every Tuesday, then an identical one with Ramaswamy on Thursday. One person who worked for Wojcicki said the angriest she ever became was when that person worked on a project for Ramaswamy. ("Susan and I had differences of opinions," Ramaswamy said later. "But these were differences about policy and business decisions. It was nothing personal.") Eventually, Page was forced to pick one of them to run his ads operation.

Meanwhile, Wojcicki's sister Anne and Sergey Brin were going through a tabloid-style divorce. Brin, Google's whimsical co-founder, had fallen for a twentysomething marketing staffer for Google Glass who modeled the dorky headgear Google pitched as the future of computing. The staffer also happened to be dating an executive at Google's Android division. Brin and Anne Wojcicki went public with their split in August 2013, noting, through a publicist, that they "remain good friends and partners." Anne absconded to Fiji with girlfriends to unwind with yoga. Brin went to Burning Man. But the breakup and awkward love rectangle were a recurring topic in Google's gossip mill, adding even more tension. Page, a devout family man, didn't speak with his co-founder for some time afterward. Wojcicki told some confidants that she hated how the divorce unfolded, particularly Brin's behavior during the episode.

At YouTube Google's family spats weren't as unbearable as its growing bureaucracy. In 2011, Kathleen Grace, a YouTube employee in Los Angeles, had pitched Kamangar on starting a YouTube production studio, selling him with images of gated Hollywood studios—Paramount Pictures and Universal Studios—followed by a mock-up of YouTube's space, wide open. (Slogan: "We're the place that says 'Yes.'") Kamangar approved the idea in about thirty minutes. Three years later Grace wanted approval for a far cheaper project and had to plead for approval in front of a committee of fifteen, four separate times. (The committee had formed to address concerns that YouTube had too many projects.) Meanwhile YouTube leaders felt more like shuttle diplomats, kowtowing to different Google divisions, than empowered executives. "It's forcing me to become good at something I don't want to be good at," Mehrotra once complained to Kamangar.

During 2013, YouTube staff rarely saw their chief. Kamangar came to San Bruno about once a week. He seemed increasingly unenthused about YouTube. (A director bumped into him at the San Francisco airport; Kamangar waved hello, wearing a backpack, heading off on a hiking trip. Another person recalled a critical meeting, on YouTube Kids, when the boss was idly flipping through his Facebook page.) But Kamangar was concerned about bloat. Each Monday, he attended the L Team "Google 2.0" daylong meetings, where Page liked to float ideas for Google to regain its old, agile startup magic. Kamangar spent further private time with Page brainstorming a more dramatic corporate restructuring.

For all intents and purposes YouTube's number two, Mehrotra, ran the place. Dean Gilbert, the content chief, had retired, and Mehrotra called most shots for a management staff that had grown intensely devoted to him. Kamangar began bringing his deputy to one-on-one meetings with Page, and sometimes didn't show up himself. Everyone assumed Kamangar would soon step away and hand Mehrotra the reins.

Then plans changed.

For more than a decade, Mehrotra had met regularly with Bill Campbell, a flinty business veteran everyone across Silicon Valley called "coach." Campbell had mentored Larry Page, Steve Jobs, and a host of untested technology tycoons. Page had Campbell coach YouTube's founders and often tapped him to have difficult conversations that the recalcitrant CEO avoided. In early 2014 Campbell asked Mehrotra to move up their regular chat and convene over videoconference instead. (The coach, then growing ill, would die of cancer two years later.) Mehrotra joined the call prepared with a list of knotty managerial issues to discuss.

"Actually, I have a different topic for you," Campbell began. "No one else is going to tell you this. But Larry has picked someone else to run YouTube."

Mehrotra, shocked, asked, "Who?"

"It's Susan."

• • •

That February some YouTube staff were visiting Google's Zurich office, a corporate Candy Land filled with pinball machines, a Lego room, a wine cellar, and foam baths for napping. Zurich had a plastic slide, too. YouTube's employees were out late drinking with local colleagues when they saw the flurry of messages begin. *There's a new chief. Wait, who? What!?*

Back in San Bruno most people were shocked when the news came. A Google c-suiter few had ever worked with or even met was taking charge.

Those close to Wojcicki knew she had been restless and itching for a loftier executive role, like many of her peers had. By 2014 her conflict with the ads engineer Ramaswamy felt untenable. Wojcicki held private talks to join Tesla as a chief operating officer, number two under Elon Musk, an old PayPal Mafia member. But Page wanted her to stay. He knew Kamangar wanted out. And by then, Page had begun plotting his own exit—a plan to hand Google off to a trusted deputy, Sundar Pichai. In a conversation, Laszlo Bock, Page's HR chief, suggested Page could more easily clear the way for his chosen successor by moving Wojcicki to YouTube. So, YouTube it was. Page eventually broke the news to Mehrotra, telling him, "Just keep doing what you're doing."

In the press, and among some YouTubers, news of an ads operator being appointed to run YouTube signaled that Google was preparing to wring even more revenue from the site. "Susan has a healthy disregard for the impossible," Page said in a statement to *The New York Times*. (The newspaper mistakenly published a photo of Anne, not Susan, to accompany the article on the appointment.) At YouTube HQ, Mehrotra teared up sharing the news. Wojcicki gave a brief speech and began setting up meetings with each team. When she met Cristos Goodrow, the algorithm overseer and watch-time cheerleader, he came prepared with a forty-six-slide presentation. Slide five showed the progress on reaching a billion hours of daily viewership. "By the way, we're way behind," Goodrow told his new boss. "I'm freaked out and I hope that you're at least a little freaked out." He hated missing goals.

In Silicon Valley new executives often change a company's course to leave their stamp or show authority. After joining YouTube, Wojcicki met with Google's networking staff, which invited her to do so: *they* were freaked out about the strain YouTube's hefty watch-time goal was placing on company servers and wanted to curb the plans to relieve stress on bandwidth.

There's no evidence anyone warned her about a different strain these goals placed on YouTube, how prioritizing engagement and spreading commercial funds as widely as possible paved the way for the trainwrecks to come. At the time, Wojcicki disagreed with the networking staff. "Let's just stay with the plan for now," she told them.

Part III

CHAPTER 18

Down the Tubes

Claire Stapleton, the Bard of Google, was a coolhunter now. Of sorts.

She hadn't heard the nickname, but she knew a team at YouTube had once had a more active editorial role before being shuttled into the marketing unit, where she arrived in early 2014. An HR official had suggested Stapleton apply for "curation strategy manager." She read the job description and asked herself, *Literally, what* is *this?* Still, she took the role. In her first week YouTube sent her to Paris to meet engineers scripting its recommendation system. These coders discussed the various signals that went into the algorithm, but they didn't answer her basic question about her gig. A colleague from London did, obliquely, with a link sent over company chat. Stapleton clicked. "On the Phenomenon of Bullshit Jobs," a screed from an anthropologist:

> *It's as if someone out there were making up pointless jobs just for the sake of keeping us all working. And here, precisely, lies the mystery . . . Hell is a collection of individuals who are spending the bulk of their time working on a task they don't like and are not especially good at.*

Stapleton had spent barely two years in New York with Google's Creative Lab, its spark-a-bit-of-Apple-magic unit. The Lab tried adopting Apple's famed workaholism, pushing underlings to *just grind it* on late nights and weekends. But they weren't inventing iPhones. They were doing rebrands, "mock-ups," and several projects that, like many at Google, went nowhere. She noticed that the people whom the Lab pushed hardest were not direct staff like her but contractors, outsourced hires without Google's equity or benefits. They even wore different-colored badges in the office. Back in California, on Google's communications team, Stapleton had felt challenged and dedicated to meaningful work; she considered that division, run by a female executive, "a matriarchy." Here, she just felt unmanaged. Once, angling for ways to win over her male boss, she asked a co-worker for tips. "He's a guy's guy," her co-worker said. "He just really likes working with the kind of dude he can get a beer with after work, you know?"

Nearing thirty, Stapleton had detached from the Google bubble and settled into a Brooklyn milieu of farm-to-table dinners, cashmere sweaters, and prestige television. She briefly considered leaving the company but was drawn to YouTube for its appealing cultural cachet. By then it was a gigantic ocean of culture, with new estuaries and weather patterns forming every day. ASMR videos (whispers and crackling sounds, which triggered euphoria in certain listeners) were exploding on the site. So was *mukbang*, a gluttonous eating curiosity imported from Korean TV where viewers watched people devour food; on YouTube, it was wilder and messier. Stapleton's division tracked these trends, mostly by standing on the side and taking notes. She was assigned to curate the "Spotlight" page (youtube.com/youtube), a playlist of clips the company found particularly admirable.

She dug in, writing descriptions for playlists in the same whimsical, voicey style that had won her accolades with her old Google TGIF scripts. "Keep it straight," her manager admonished. "No personality. It should sound like a computer generated the copy."

"My whole *thing* is having a personality!" she objected.

YouTube, she was told, had to operate as a neutral platform and avoid editorializing to steer clear of legal troubles. Indeed, soon after Stapleton joined, YouTube would confront a new trend on its site that hardened this cautious stance and made its earlier legal squabbles look minuscule by comparison.

After Stapleton's manager shot down her idea, she went rogue. She started a personal email newsletter, "Down the 'Tube," a list of links selected to capture some of YouTube's old magic, little refreshing droplets from its ocean. She linked to videos of cute cheetah cubs, kangaroo boxing scored to Tchaikovsky, a drummer's ingenious rendition of a cattle auctioneer. "May the streaming be with you," she signed off. She linked to more imported TV clips than homegrown YouTube stars, but, hey, the company had certainly been bending that way, too. Several executives signed up to receive her emails. Her final note of that year reflected on what her company had unleashed:

Happy hols, y'all!

> **GETTING SERIOUS FOR A SECOND . . . with over 72 billion hours of video watched on YouTube this year—mostly torpid fragments of pop culture—it IS possible we're seeing too much. *The more there was to see, the harder it was to see anything, but the harder you looked, the more you saw*—or so the old guy's saying goes.**
> **Bidding adieu to 2014 with the hope that there's a little depth left in the shallow habits of seeing stuff. That there's some flicker of meaning possible in the constant blur of motion and fractured film stills of these super benumbed times. But maybe not. *gingerly steps away from computer***

• • •

By 2014 a mantra had spread through YouTube, a corporate incantation to explain its place in the life cycle of all great businesses: *Joke, threat, obvious.*

Shishir Mehrotra had learned it from an old boss at Microsoft and passed it down to his troops. "First, you're a joke, and nobody believes in you," Mehrotra explained. "Then you're a threat, and everybody's scared of you. And then you're obvious, and everybody just assumes that what you're going to do is going to work." YouTube recognized the *joke* phase—dogs on skateboards, the laughingstock Google JV team. *Threat* had come in stages—Viacom, certainly, and then Hulu and clumsy attempts from old media to crash the internet party or ignore it. Once YouTube's boat got moving, the *obvious* sank in: YouTube, media dictated by the masses and made for dirt cheap, was the future of entertainment.

With YouTube's model established, other industries pounced. Steven Spielberg's studio, DreamWorks, maker of *Shrek* and other franchises, spent over $100 million to buy AwesomenessTV, the YouTube network for teens. AT&T spent more for the MCN Fullscreen. Disney prepared to open its checkbook too. Netflix and other online programmers started recruiting YouTube stars. Google made plans to spend nearly $1 billion for Twitch, a livestreaming site all the rage with gamers, and merge it with YouTube. But talks collapsed when Google balked at the price and potential antitrust scrutiny, and Amazon swooped in to buy Twitch instead. Facebook started tinkering with video.

Susan Wojcicki became YouTube's leader during the obvious phase. She didn't need to invent YouTube's business plan from whole cloth. She needed to give it more fuel; by 2013, YouTube only brought in $3 billion a year, still a fraction of television's overall market. The business press would chide YouTube for having a billion viewers but no profit. In Wojcicki's first major appearance on the job, she stood on a stage at Madison Square Garden for Brandcast, an annual event YouTube conceived to showcase its programming for advertisers, just as TV networks did, only YouTubier. From the stage Wojcicki announced YouTube's newest service: Google Preferred, a way for advertisers to run pricier ads on just the most popular slate of videos, the fruit of a long effort to create a premium programming shelf. This being YouTube, the company didn't handpick these programs, but rather chose them based on an algorithmic logic it did not divulge. "We're in the middle of a big revolution with video," the new CEO told her audience. Pharrell Williams performed "Happy" afterward.

Showbiz's center of gravity was shifting. Later that autumn Wojcicki appeared at a swanky *Vanity Fair* event in San Francisco on a panel with her new peers, the CEOs of HBO and the cable giant Discovery Communications. The male media honchos were in their standard Silicon Valley attire, jeans and loafers. Wojcicki wore heels and a simple silver dress. *Would cable still be around in ten years?* the panel moderator asked her. Wojcicki smiled slyly and answered, "Maybe." *The New York Times* carried a long profile of her later that year, noting that she "has quietly become one of the most powerful media executives in the world." She posed for the piece in a midnight-blue dress, visibly pregnant with her fifth child.

Her reception within YouTube was less universally enthusiastic.

Wojcicki gave her first talk at the company with her usual unpolished delivery—lots of "um"s and "like"s. A female junior staffer in the audience remembered that nonchalance as inspiring and "great to see." Most YouTube staff had pedestrian concerns at the time: the top-voted employee question for Wojcicki during her first all-staff meeting asked why YouTube's cafeteria had swapped out Greek yogurt for the regular kind. Some media veterans welcomed her arrival as a sign of Google's faith in YouTube. Although Hurley and Kamangar were sometimes introduced as YouTube "CEO," they technically weren't; Google listed them as senior vice presidents. Wojcicki took the CEO title officially, an indication of her high stature in the corporate family. "It felt like the cavalry was coming," recalled Tim Shey, a YouTube creator relations director. Others at YouTube saw Wojcicki as a bit of an interloper, one maybe too enamored with Hollywood's old guard. (Another director recalled her effusing that Brad Pitt had been among the first to congratulate her on her new role.) Mehrotra took six months off after the CEO transition to decide his next move. Wojcicki lobbied him to stay to help her get her footing, but he left and, within the year, so did most of his deputies. Some were bored with the obvious; a YouTube director, who had helped place the service on nearly every phone and gadget imaginable, threw in the towel once YouTube began bartering for space on airplane drop-down screens. *No worlds left to conquer.* Many had an allegiance to Mehrotra and felt irked when he was passed over.

Or perhaps it was the new boss's Googliness.

Early on this was a tangible corporate asset. In a book on management, Laszlo Bock, Google's longtime HR boss, wrote that the company hired for "Googley" traits: humility, conscientiousness, and "comfort with ambiguity." Being willing to plunge down a slide at work without shame—that's Googley. Being Googley had a motto: "Don't politick. Use data." During a debate before buying YouTube, one Google executive questioned if the company should be profiting from pirated material, asking over email, "Is this Googley?" But over time "being Googley" had also morphed into an epithet for someone with a slavish devotion to Google's management culture and system of pedigree, someone who would subsume emotion and eccentricity for the greater Google cause, who could be a bit of a cipher. "I couldn't tell you who she is," said one person who worked with Wojcicki for years. "She's a very blank slate."

Corporate America used "layered" as the polite term for the rather embarrassing demotion that occurs when someone is slotted above you in the professional ladder. Google ranked every staffer according to seniority and metrics, naturally. Many managers at YouTube, the JV team, were relatively junior in comparison to other Google units. Wojcicki moved into San Bruno, lining her office bookshelf with photos from her son's bar mitzvah and a box of her Google business cards accrued over the years. She also brought waves of managers from Google, vice presidents and senior directors and distinguished engineers. She layered. An entire apparatus moved in that struck some at YouTube as decidedly more corporate. *Google has infiltrated the building,* one staffer recalled thinking. Wojcicki liked trusted advisers, a management trait some admired. Though another YouTube manager, part of the exodus, billed this as a desire for "yes-people who kiss the ring." Some saw these new recruits as a sign that Wojcicki was re-creating her Google ads operation at YouTube. During her maternity leave at YouTube, she continued to dial into only one weekly business meeting, a caucus devoted to premium advertising.

While the company managed this bumpy transition, YouTube suddenly faced an entirely new nightmare.

• • •

"A Message to America," a video posted to YouTube in August, began with footage of President Obama announcing airstrikes in Iraq. Two minutes in, the frame changed: a man in an orange jumpsuit knelt in a stark desert, lit in high definition, a tiny mic clipped to his collar. Another man stood behind him, cloaked entirely in black. The man kneeling, an American, spoke a hostage recitation. Then he was gone. "Any attempt by you, Obama," the cloaked man said to the camera, "to deny the Muslims their rights of living in safety under the Islamic caliphate will result in the bloodshed of your people." The video, shot and uploaded by ISIS, ended with a shot of another jumpsuit-clad captive kneeling in a desert, a threat.

The first hostage, James Foley, a reporter abducted in Syria two years earlier, was decapitated. Two months before posting the Foley video, the militant group had declared a brutal, premodern caliphate and now waged war on the ground and on modern social media. Members of ISIS began flooding YouTube with videos of sermons against Western sins and these long, highly produced hostage horrors. The worst floods appeared during the day in Europe, forcing Google's smaller operations there to scramble. The company removed the Foley video, citing its policy against graphic violence, but had to tread more carefully on some other ISIS footage. Scores of news outlets posted clips about ISIS, YouTube didn't want to remove those. And, being YouTube, it didn't necessarily want to dictate who was a news outlet and who wasn't. The Arab Spring had shown that anyone could be.

But the ISIS influx demanded action. In Paris, where Google had a plush office in the city's center, nearly everyone there was suddenly deployed as content moderators for the week even if it wasn't their job. They set up a large spreadsheet to track every re-upload of the Foley clip and similar horrors. One person on YouTube's business team tasked with watching these ISIS videos recalled the shock of noticing how cinematic the propaganda felt. Maniacally, ISIS uploaders spliced actual news footage into clips, making them much harder for YouTube to find. The staffer had figured how to flip through still frames, less traumatizing than moving images. They grimaced and flipped—news anchor, news anchor, news anchor, desert and orange jumpsuit. *Strike. Next video.* A co-worker found them ashen-faced at their desk after seeing a parade of such videos.

• • •

For years YouTube had been scolded for giving a platform to radical Islamists. In 2010, Anthony Weiner, a Democratic congressman, wrote an angry letter demanding YouTube remove clips of Anwar al-Awlaki, a cleric known as the "bin Laden of the internet." Google replied that it had already expunged accounts tied to people on the State Department's terrorist registries "usually in under an hour." The company mostly found these requests insufferably moralizing or naive. (Some politicians, unaware of YouTube's mechanics, asked one Google official, "Why did you put that up?") Company lawyers believed that, apart from restricting posts from people on official terrorist lists, YouTube couldn't make consistent distinctions. Some videos that enraged politicians showed men running around a desert with guns. So did U.S. Army recruiting videos. To untrained eyes, these might be hard to tell apart; it certainly was for machines. And the "terrorist" label certainly looked different in Palestine or Northern Ireland. *Why wade into those battles?*

The ISIS onslaught forced YouTube to. More ISIS killings appeared online. Scotland Yard deemed downloading and disseminating this footage an arrestable offense. Headline after headline about beheadings reminded readers that YouTube was the ISIS medium of choice—it was easy to upload and video was a great vehicle for propaganda. YouTube looked like a welcoming stage for the world's public enemy number one, and the company decided its old stance of "blurred lines" and colliding head-on with governments no longer worked.

YouTube began pulling as much ISIS material offline as it could. Twitter followed suit, alarming some of Google's old allies like the EFF, a civil liberties group that cheered YouTube in its Viacom fight. Suddenly, online information gatekeepers were relying on U.S. authorities to dictate what consequential human events to document. "Most news organizations would think about this as an ethical question, not a legal one," Jillian York, a director with the EFF, told reporters. "Giving corporations the power to censor sets a dangerous precedent."

In much of Europe, though, YouTube was criticized for not censoring *enough*. Google's good standing on the Continent had already been slipping. European politicians accused the "don't be evil" company of evading taxes, ignoring privacy, and behaving like a monopolist. Jewish groups would berate YouTube for giving airtime to Holocaust deniers. In 2013 Edward Snowden's bombshell revelations included charges that the National Security Agency (NSA) had hacked into Google's data centers. Google denied any foreknowledge. ("We are outraged," the legal chief, David Drummond, told the press.) But this made life no easier for Google's European diplomats, who were pummeled as assets of Uncle Sam.

In January, five months after the Foley video, Google representatives were summoned before the European Parliament in Brussels to justify why the company hosted so many terrorism videos. (YouTube, at the time, had no designated policy staffer in Europe.) A Google manager cited a figure YouTube would deploy for years to come to explain its intractable mass. YouTube, she said, saw three hundred hours of videos uploaded every minute; reviewing them "would be like screening a phone call before it's made."

Back in San Bruno, YouTube publicists felt as if they were dealing with an atrocity every day. Staff briefed Wojcicki on its moderation process, and one deputy recalled watching as the gravity of these decisions registered on her face. YouTube relied on its exemptions for videos with "educational" or "documentary" interest, which could depict violence or extremism with a warning screen attached. (These weren't unanimous calls; one staffer recalled seeing a video instructing viewers how to commit suicide with a bag, which remained under the "educational" tag until a policy director intervened.) But YouTube's central strategy, a very Googley one, hung on counter-speech. Lawyers shared an old legal maxim: "Sunlight is the best disinfectant." When dark spots emerged online, it was better to shine light elsewhere than to blot them out. "Enforced silence is not the answer," Drummond, Google's lawyer, said in a speech on censorship. "Technology is one of the greatest tools we have to reach at-risk youth all over the world and divert them from hate and radicalization." Obama's State Department tried counter-speech, releasing a grisly YouTube video documenting ISIS brutality meant to curb its recruiting efforts. YouTube staff brainstormed ideas for recruiting creators to make other shining lights.

After the European parliamentary session, a few Google managers met privately with Gilles de Kerchove, a genteel Belgian politician who served as the European Union counterterrorism coordinator. Often in such meetings Google personnel threw out the massive numbers of reprehensible videos the company *did* remove. But they never shared how many still remained on the site, so the numbers lacked any sense of proportion. "Is it a little? Is it a lot?" de Kerchove asked. "I have no clue." Google always reassured him they were removing terrorism videos ASAP. Occasionally, de Kerchove would go home after work, search for beheading videos on YouTube, and find several easily. It seemed to him that YouTube removed gruesome footage involving Americans more readily than those showing Arabs.

During this particular meeting, one Google manager tried the sunlight approach, asking de Kerchove if EU politicians could help the company produce videos with effective counter-speech. Perhaps they could persuade YouTube stars to help? One of de Kerchove's fellow commissioners was a Swede, as was YouTube's biggest star.

The Google manager then turned to the white-haired counterterrorism official and asked sincerely, "Do you know who PewDiePie is?"

CHAPTER 19

True News

PewDiePie: "VLOG–Singapore–BROS ARE EVERYWHERE!" May 29, 2013. 6:23.

It's paparazzi footage—screaming, clamoring bodies, security guards—and the celebrity is shooting the film. Felix Kjellberg is in Singapore for the Social Star awards, and we, his reliable bros, are there with him. We see him arrive, perched on a balcony high up in this foreign land. We see him swim in the pool, dine, get fan-swamped, take it all in. "I just went down to the lobby," he tells the camera. "And there's like one hundred fans down there, and they just started screaming when I came down. It's like, WHAT? I had no idea I had this many friends in Singapore."

Without question PewDiePie remained the brightest sun in YouTube's universe. Kjellberg's fans, his "Bro Army," were now seeing more and more of his life on-screen. They met his girlfriend, a fellow YouTuber named Marzia Bisognin, who joined him in Singapore. They sent ideas for games to play, and Kjellberg became more comfortable on camera, more daring. Online gaming culture had continued to blossom, but so had a cottage industry for capitalizing on people's infatuation with internet pan flashes and memes. People went crazy for ice-bucket challenges on Facebook; sites like Reddit and BuzzFeed minted celebrities of Shiba Inu dogs and grumpy cats.

Like other YouTubers, Kjellberg leaned in; his contribution to the viral "Harlem Shake" sensation of 2013 was to perform the dance in ponytails, pink panties, and a bra. Millions watched. His bread and butter was still playing horror games, videos made at his house and on the cheap. But Kjellberg's management, Maker Studios, was about to have a windfall that would propel his fame even further.

In 2014 the Walt Disney Company had paid more than $500 million for the network, which had sprawled from Danny Zappin's Venice backyard into a media behemoth controlling some fifty-five thousand YouTube channels. The *New York Times* story on the deal mentioned PewDiePie and Maker's grasp of digital stardom—"imagine what they can do for Iron Man, Mickey and Yoda," enthused one agent—but it did not name Zappin, Lisa-Nova, or any of the original Maker crew.

Kjellberg liked certain aspects of the acquisition. While other prominent YouTubers had moved to Los Angeles, Kjellberg, more introverted, had settled in Brighton, a seaside retirement town south of London. (In internal documents, YouTube would describe him as "a 'stay at home' guy.") Whenever Kjellberg came to California, he often requested trips to Disneyland, where he received the VIP tour. But his new corporate overlord was not always awesome—like the time PewDiePie drew from his new go-to YouTube format: unpacking the internet's vast weirdness for his bros. In one clip, he and another gamer giggled their way through a pornographic fan-fiction video game depicting characters from Disney's *Frozen*. Someone at Disney noticed the clip and sent it to the chief executive, Bob Iger, who was displeased. During a late-night phone call a Maker director tried explaining Kjellberg's shtick—that he was poking fun at the perversion of internet culture—and assured Iger that his star was in fact a big *Frozen* fan. Disney recommended that Kjellberg delete the video, and he did. Eventually, this rift over taste and cultural vernacular would sever his ties to Disney for good.

Back then, though, PewDiePie was still considered a gifted performance artist, the embodiment of a new media. YouTube placed posters of him around its office. Old media attempted to grok his appeal or dismiss it. "Meet the gibberish-spouting clown who's bringing Western civilization to a screeching halt," read a *Variety* story on PewDiePie a few months after his Singapore trip. "And how does someone you've never heard of manage to get 2.6 billion views for his videos?" the magazine asked. "Blathering like a blithering idiot, apparently." Kjellberg loathed that article and held a grudge against its author for years, according to a person who worked with him. Still, he sat for more interviews after the Disney acquisition. "It's cool to have this kind of influence," he told *The Wall Street Journal*, "but at the same time it's kind of scary." The newspaper ran a photo of him wearing a flower crown with the headline YOUTUBE'S BIGGEST DRAW PLAYS GAMES, EARNS $4 MILLION A YEAR. Media's rubbernecking at just how much YouTubers earned deeply irritated Kjellberg and his peers. It felt disrespectful. To a certain extent YouTubers, so accustomed to controlling their own massively popular channels, resented the media simply because it portrayed them in a way they couldn't control.

Yet traditional media *did* treat YouTubers as novelty acts, even if they pulled in bigger audiences than TV. When the early YouTube phenom MysteryGuitarMan appeared on CNN, Sarah Penna, his manager and spouse, told producers not to ask how much money he made. *No one asked George Clooney that.* CNN still did. In 2015, Kjellberg appeared on *The Late Show with Stephen Colbert*. The Swede wore a crisp blue suit, slicked-back hair, and a genuinely nervous look. His parents had flown in for the taping. "I want to thank the internet for letting their emperor be here for the evening," the TV host began, before asking Kjellberg to explain why people watched him play video games. "I have the best job in the world," Kjellberg replied. Colbert reminded viewers that Kjellberg made an amount the prior year "that rhymes with schmeven schmillion dollars."

YouTube never acknowledged it, but this sort of attention certainly pleased the company. Just a few years earlier no one had seen it as a viable business, let alone a home for professional media. Now here was a YouTuber with a schmeven-schmillion-dollar career. Behind the scenes YouTube started a program to assign dedicated business managers to its top rank of stars, including Kjellberg. PewDiePie never made a counterterrorism video, but he did appear in a YouTube promotional spot. Starting in 2010, YouTube's coolhunters produced a year-in-review feel-good video, "YouTube Rewind," which starred a huge stable of creators. YouTubers obsessed over these clips and who appeared in them (or who didn't). In 2015, the company's tenth year, Kjellberg arrived partway through "YouTube Rewind" to give his signature bro-fist bump to the company logo.

• • •

One hundred miles west of Kjellberg's house, in a staid southern England town, David Sherratt started his YouTube journey. He was fiendishly smart, a bit of a loner, and a bit bored at school. In 2010, when he was thirteen, Sherratt fell into the absorbing world of YouTube gaming—*Minecraft, Call of Duty,* that sort of stuff. Most gamer streamers made idle chatter, though one he watched started talking about philosophy in a few videos, and YouTube began serving up more clips about that. *Click.* From philosophy Sherratt quickly drifted into the world of the "YouTube skeptics," a loose network of atheist vloggers and provocateurs. He hadn't grown up religious or with particularly strong feelings about the subject, but these vloggers were smart and edgy; it was fun to watch them rip into believers in clip after clip.

Sherratt wasn't alone. For a certain set of young, sharp, wandering minds, the YouTube skeptics thrilled. Natalie Wynn, a precocious nineteen-year-old piano student in Boston, began devouring footage of prominent atheists like Richard Dawkins and Sam Harris. Fans uploaded their lectures and debates, adding titles with YouTube's signature panache. Wynn discovered punchier YouTube vloggers, with names like Thunderf00t, who deployed montages and mockeries (a forty-eight part series called "Why do people laugh at creationists?"). Here was a loose collection of secularists, rationalists, libertarians, and weirdos—freethinkers whom YouTube elevated to the same stage as learned scholars. Their videos tackled heavy stuff, like free will and human nature, mixing academic discourse (or the pretense of it) with delightful internet trolling. Viewers debated in comments and reply videos. For Wynn, YouTube felt like taking part in seventeenth-century European café culture. Wynn and Sherratt closely tracked the debates and knew all the personalities, like café regulars.

These viewers didn't consume YouTube like normals—casually coming for a how-to lesson, song, or viral clip. Sherratt, a thin, pale teen with short brown bangs, listened to his favorite YouTubers nonstop in his room, the footage running in the background while he played video games. An iPad app let him download videos, and he listened on the way to school and during lunch and free moments. Caleb Cain, the West Virginia fan of the Canadian guru Stefan Molyneux, also watched the skeptics, and when his job let him don headphones, he would take in twelve or fourteen hours a day of video. Kevin Roose, a *New York Times* reporter who unearthed Cain's story, described these devotees as populating an "Inner YouTube," treating the site as a "prism through which all culture and information is refracted." Roose offered this eloquent descriptor:

> Imagine a genetic mutation that gave everyone born after 1995 the ability to see ultraviolet light. Imagine that these people developed an identity around UV light, started calling themselves "UVers" and became suspicious of any media product made exclusively on the visible spectrum. As an old person with normal eyes, you would experience this change as a kind of slow cognitive decline. Every day, as more and more of the world played out in UV, you would struggle to catch glimpses of it.

While some Inner YouTubers—the Nerdfighters and PewDiePie's Bro Army—bonded over shared interests or fandom, the YouTube skeptics bonded around shared ideas. Or at least a shared passion for asserting and defending those ideas, usually loudly and outrageously. Algorithms enjoyed that.

At some point, the café conversation shifted.

Sherratt remembered it as a schism in online atheism, starting in 2012, after one member dared propose addressing ills such as racism and sexism in addition to organized religion. Others disagreed, and a splinter skeptic subculture formed. But something had been lurking in the movement even before then, an undercurrent of ugly misogyny that now boiled over. As early as 2011, the Amazing Atheist, a popular skeptic vlogger, posted clips on the "failure of feminism" and went off on "cackling cunts" on daytime TV. "Stop whining, will you," Richard Dawkins wrote in response to a woman's video diary of an uncomfortable sexual encounter. Wynn watched YouTubers like the creationist spoofer Thunderf00t begin to spoof women. YouTube recommended clips to Sherratt from Sargon of Akkad, a windbag Brit who called feminism "a toxic, sick ideology." Sherratt found the YouTuber's rage amusing and cathartic. He started a YouTube channel (handle: Spinosaurus Kin) and made videos with titles like "Feminism Is Terrorism," flamboyant stuff people might watch out of curiosity or anger. An infuriated view was still a view. Once he got to college, he started to appear in British tabloids as the face of a new men's rights movement, wearing a leather jacket and a light scowl, a proud virgin steering clear of women to avoid false charges of rape.

• • •

Before YouTube, the last big moment in unregulated media had ushered in a legion of raucous, chattering talkers—a colossal force of firebrand politics that arose and grew largely in response to women's progress. It started with Rush.

In 1987 the FCC revoked the Fairness Doctrine, a rule requiring broadcasters to air both sides of a controversial issue. The following year Rush Limbaugh, a failed disc jockey, debuted a new AM radio talk show. Limbaugh brought the shock jock DJ sensibility to politics, creating an entire brand of media drenched in sexism ("feminazis" and such) and delivered to a conservative audience that felt neglected by the mainstream press. He built a rapport with fans, who listened and called in religiously. To compete with FM radio and cable news, more AM stations turned to the mass appeal of Limbaugh and his ilk; some stations justified this with a metric called "time spent listening." Rupert Murdoch described Fox News as "talk-radio with video." By 2007, 91 percent of weekday talk radio was categorized as conservative.

Perhaps that's why these talkers hadn't initially rushed to YouTube. For years Cenk Uygur felt like the only one there. The liberal, voluble radio veteran ported his new talk show, *The Young Turks*, to YouTube in 2005, mixing regular lefty jabs at the press and politicians with clickable tabloid fare. Uygur saw few conservative shock jocks on YouTube, until around the watch-time transition, when they "started popping up all over the place." Many popped up with videos mocking Uygur's show or his name; tagging footage with "The Young Turks" or "Cenk Uygur" to appear more often in searches and recommendations. Uygur often shot back. At the time, few mainstream news channels were on YouTube, giving shock jocks an open field on political topics. Uygur noticed something strange about these newcomers, how detached some seemed from reality. "When Rush lied, he would attach it to something that was feasible," Uygur recalled. "These guys are making up stuff out of whole cloth."

Stefan Molyneux, the Canadian self-help philosopher, joined in, posting a video rebutting *The Young Turks*. But Molyneux and the anti-social-justice-warrior (SJW) brigade were really waiting for a bigger spark to set things off: Gamergate.

David Sherratt watched Gamergate spread, tracking the manufactured online controversy as best as he could. From what he grasped, a feminist video game designer had received fawning coverage from an ex-lover, and then YouTube videos and web forum posts about the scandal were attacked or entirely removed. *That seemed wrong.* Details were hard to follow, but the rage was not. Some women had criticized how female characters were depicted in video games, only deepening the ire of male gamers, furthering a perception of PC, feminist culture run amok: *the social justice warriors had come to ruin video games.* The supposed central scandal of Gamergate—that a video game received biased coverage—was not true, but this did not stop Gamergate from spreading like a cancer, forcing several women in the gaming industry to go into hiding after they were harassed and received death threats. It was Rush Limbaugh's culture war moved online, magnified and untamed, like the internet itself.

Gamergate largely expanded on social networks and web backwaters like 4chan, but on YouTube the evolving internet scandal provided plenty of grist for vloggers to make long, engaging content about something newsworthy, the kind of stuff algorithms ate up. Molyneux, the guru, turned sharper, angrier. He began a series of shows called "True News," borrowing Limbaugh's proven tactic of framing himself as loyal opposition to the mainstream media. Molyneux posted frequent clips titled "The Truth About . . ." About Karl Marx. Israel and Palestine. Martin Luther King Jr. The Ferguson riots. About *Frozen* and about *Wonder Woman.* (Both movies were Trojan horses for feminist agendas.) The media, he said, enforced this by putting "the SJW thumbscrews right up the urethra, right into your balls." For his fans, who called him Stef, this message was compelling. "I was chasing truth," Cain, his loyal viewer, recalled later. "And Stef said, 'Here, look at this cave. There's knowledge down there. The truth is down there.'"

In 2013 the George Zimmerman trial began, a case that gripped the nation. Googlers staged their own "hoodie march," the nationwide protests that occurred on behalf of Trayvon Martin, the slain Florida teen. On YouTube Molyneux posted a thirty-five-minute video in his signature style: "The Truth About George Zimmerman and Trayvon Martin." He used Zimmerman's testimony, later disputed in court, to demonize the media, single Black mothers, and rap music. In other videos Molyneux spoke about racial differences in IQ, leaning on a euphemism called "race realism," a dog whistle for eugenics. He became obsessed with the refugee crisis, which he called "the burying of Europe," and joined others on YouTube's right flank in decrying a "replacement rate" from Muslim migration. "I don't know if the birth rate has gone so far down in Europe that none of these politicians give a shit about the kind of world that their children are going to have to grow up in," Molyneux raged in one video, staring straight into the camera. "But I do. I do."

• • •

Outright malice had always had residency on YouTube. In April 2007, when the company was still settling into Google, a report from the Southern Poverty Law Center documented the prevalence of neo-Nazi footage—white nationalist rock band clips and videos from David Duke, an ex–Klan leader and talk-radio regular. Prominent skeptic YouTubers, like the British ex-comic Pat Condell, began posting frequent videos about Islam as a backward, troubling faith as early as 2010. "No one thought anything of it," recalled the skeptic fan Natalie Wynn. "Of course, he's ranting about Islam. We're atheists. That's what we do." As long as they didn't break YouTube's broad hate-speech rules, they were welcome in the marketplace of ideas.

Two decades into talk radio's dominance, YouTube could reasonably have been prepared for how powerful these extreme political voices could be. But cloistered in liberal California, YouTube leaders rarely interacted with far-right figureheads, or even cultural conservatives. "No one knew how to deal with the right wing," recalled one staffer. For a long time the most unsavory characters remained on YouTube's back shelves and in its crawl spaces. Most were also on Facebook and Twitter anyway.

But YouTube, unlike anywhere else online, had started to pay its content producers. And YouTube rewarded long, engaging output. And YouTube had built a machine that could recommend these videos to just the right audiences who wanted to see them—a machine about to grow even more powerful.

CHAPTER 20

Disbelief

Matthew Mengerink came to YouTube in late 2015 after two decades of serious programming and managing software engineers, including long stints at PayPal and eBay—the exact sort of pedigree that Wojcicki prized. YouTube's deadline for its hairy goal of one billion hours of viewing a day was fast approaching. The company was on track to hit it, but just barely. Wojcicki had instituted a Googley motivational system: tiny bars that marked the progress of particular goals—red, yellow, green—next to the names of employees responsible for each one. Programmers, a naturally competitive sort, knew that much of their professional self-worth, bonuses, promotions, and so on rested on hitting that green.

When Mengerink arrived, they needed their spirits lifted, too. Wojcicki's first pick to lead her engineering efforts, Venkat Panchapakesan, a recruit from Google, had died from cancer, leaving the department shaken. Mengerink joined as a vice president to marshal YouTube's billion-hour goal and handle its bulk. The site had not broken the internet, as its old joke said it might, but it was close to doing so. Mengerink was shocked to learn how much of the internet YouTube consumed and how many videos were barely watched more than once or twice. Huge amounts of footage were marked as "private," unavailable to most viewers; YouTube offered this feature to allow videos to be shared with only select audiences, but it was also being used as free storage: businesses dumped entire days of security camera reels on YouTube's servers.

Many on Mengerink's new team had also been slogging toward another exhausting goal: transferring the site's software scaffolding, which the cofounder Steve Chen created a decade earlier, into a more stable programming language. As a coding project this was the equivalent of translating *War and Peace* from one language to another and *then* doing the same thing for every book report on *War and Peace* ever written. The effort did not invite innovation. There was a creeping, familiar concern that this sort of project welcomed sloth. "Be careful when you hire Googlers," a colleague warned Mengerink, "because YouTube is where Googlers go to retire."

One area did inspire: a novel, marvelous form of artificial intelligence sweeping across Silicon Valley. Mengerink witnessed the infatuation with this technology during a "design overview" meeting. Google held these ritually, gathering in conference rooms to go over even the smallest change in appearance to Google.com or other services. This particular meeting was about a change that had already taken place, a celebration of technical progress. On a big screen at the room's end, an employee projected YouTube.com: familar rows of videos stacked up on digital shelves, six abreast, with tiny arrows to the right inviting viewers to click for more. There were video titles and channel names and view counts and red Subscribe buttons. "Here's what the page would look like," the staffer said, "if we removed machine learning." *Click.* All that remained was the YouTube logo and thin lines separating the shelves. Without machine intelligence, YouTube was blank: a video Rapture.

• • •

One year before Mengerink joined, in March 2014, Larry Page sat onstage in Vancouver trying to explain this intelligence revolution. Google's CEO wore a turquoise T-shirt and an unzipped gray jacket, one of those effortless, fits-just-so types that certain billionaires and artists own. Some who worked for Page likened his calculating mind to that of a Vulcan, and with his straight bangs and futurist attire he did seem Spock-like. When he spoke, he held the microphone close, words straining through his partial vocal paralysis. Nearly all his hair had gone gray. Across a small white table from Page sat Charlie Rose, sporting his casual look, no tie, top button undone, still the face of reasoned American discourse. (It would be another three and a half years before reports of sexual harassment undid his reputation.)

This TED session was billed as a talk on Google's future, but Rose had to ask about the past. He brought up Edward Snowden's NSA revelations and Google's appalled reaction. Page smiled and mentioned a photo circulating at TED of Sergey Brin, his other half, grinning next to a pixelated Snowden, who dialed in via a teleconference robot. "To me, it's tremendously disappointing," Page said, his smile fading, and called the government's unauthorized snooping a threat to a "functioning democracy." He spoke with a disregard for the usual diplomatic decorum, a candor that only a founder of Google could pull off. "I'm sad," he went on, "that Google is in the position of protecting you"—Page glanced at Rose—"and our users from the government doing secret things that nobody knows about. It doesn't make any sense." Page seemed tired. Politics and stages never thrilled him. This would be one of his last interviews. Seventeen months later he would hand over Google to Sundar Pichai and recede from public life. For a moment onstage, though, he looked lively, a boyish charm familiar to those from Google's golden years. He was discussing his newest prize: DeepMind, a London company that researched artificial intelligence but did not sell any products or services. Google paid $650 million for it. DeepMind's brilliance came from its fix for "unsupervised" learning, Page whispered into the mic, and when Rose didn't immediately cotton, Page asked, "Maybe I can show the video?"

A screen behind them lit up with old arcade games. DeepMind had constructed a computer model to master these games on its own, without instructions or supervision, as old chess computers had required. Arcade classics flipped across the screen—*Enduro, River Raid, Battlezone.* "The system just sees what you see, the pixels," Page narrated. "And it's learned to play these games—*the same program*, it's learned to play these games with superhuman performance. We've not been able to do things like this with computers before." Then the screen showed *Boxing*, a prehistoric Atari arcade game, an overhead shot in phosphor green of two squiggly lines duking it out. No contest. "It figures out it can pin the opponent down. The computer is on the left," Page said, watching his boxer pulverize the other, grinning. "It's just racking up points."

Google had long been working on this kind of superhuman intelligence without making such efforts public. A covert band of coders was assigned to the company's "moonshot" lab in 2011 to develop computer systems that could, in effect, mimic human ways of thinking. The group called itself Brain. Computer systems had exhibited progress in detecting human speech and the contents of the images shown them. They could whup human opponents at chess. But these were relatively flimsy achievements; computers could not carry on a conversation like the computers from *Star Trek*. And they were *specific*. A model trained to play chess could not play checkers. A model could identify a cat in photos, if it were trained that way: look for four legs, pointy ears, whiskers, and a tail. But show it a dog, and all it spits out: *Not cat.* Computers needed a *general* intelligence. To summon this, the Brain team resurrected an idea that had been collecting dust on shelves. Starting in the 1940s, computer scientists began dreaming up models of machine intelligence based on "neural networks," layers of mathematical models that could process data—sights, sounds, and concepts—as the human brain does. In this way machines could learn without specified labels (cat, chess queen). But human brains have perhaps 100 billion neurons and trillions of synaptic connectors. Computers weren't powerful enough to replicate anything like that, so neural networks remained a dormant theory—until the internet began to prosper and computers gained formidable strength. Google called its early system for handling this approach to intelligence DistBelief, named after "distributed training," the practice of tying together clusters of machines, but it served as a token for the task's perceived difficulty. *If it worked, how unbelievable!*

The Google Brain coders, who initially worked on the same floor as Page and Brin, often discussed a 2005 neuroscience paper that studied epileptic seizure patients to see how they recognized people or objects. When shown particular faces, such as that of *Friends* actress Jennifer Aniston, a certain neuron in the brain connected to the formation of memories inexplicably fired. The same neuron fired when patients were shown photos of Aniston and the Eiffel Tower, which suggested that brains worked to make and encode associations. Google's programmers wanted to know if something similar would happen with machines. Could a neural network encode an image of a familiar object or a concept all on its own? That required showing a network a massive barrage of pictures.

Luckily, Google owned the biggest repository of video imagery ever assembled, a gigantic library of human experience. Brain researchers fed their neural network stills from YouTube—specifically, stills from cat videos. Millions were pumped into machines without any labels of their cat-ness. Google had constructed a network that, while far smaller than that of our brains, possessed a hundred times more neurons and synapses than any computer version before. It figured out how to spot a cat all on its own.

"We can learn what cats are," Page explained to Charlie Rose, two years later, in Vancouver. "That must be really important." From Google's dawning Page had fixated on artificial intelligence. In an interview back in 2002 he explained that effective web search, really giving people what they want, required understanding "everything in the world" and that this required an AI. A decade later he rightly predicted that machine learning would become all the rage. Amazon would release a speech-recognition gizmo called

Echo. Facebook's Mark Zuckerberg, who publicized his annual life-betterment goals, would spend one year inventing an AI butler. Tech companies threw around the slogan "mobile-first," signposting their fitness for the smartphone world; Google would declare itself "AI-first."

Once the arcade boxing clip concluded on the TED stage, Page caught his breath. "Imagine if this kind of intelligence were thrown at your schedule, your information needs," he told Rose. "We're really just at the beginning of that." Each arm of Page's company soon rewrote business plans and OKRs to incorporate as much AI as possible. These fruits appeared in Google search first. Type in an impossibly long question (*Where did the actress who plays Rachel's mom on* Friends *go to college?*) and there's the answer. Translate this question into French *et voilà*. Neural networks went into Google's email spam filters and ad-targeting dials and digital photo albums.

At YouTube neural networks plugged into its recommendation engine.

• • •

Think of YouTube's recommendation system as a gigantic, multiarmed sorting machine. It has one task: predict what video someone will watch next and deliver it. From YouTube's outset its computer programs strove to do this. But the Brain neural network could make predictions and sort in ways fallible humans and flimsier code could not. The network, by its nature, often behaved in ways its engineers could not immediately or fully comprehend.

By the time Mengerink came to YouTube, the Brain network had already been introduced. Viewers couldn't detect it, unless they noticed a steady uptick in desirable clips being presented to them. The network learned to show shorter clips when people watched on their phones and longer clips on YouTube's app for TVs; both decisions improved overall watch hours. The network learned to sort episodic video series automatically. It connected dots and made associations, encoding them. When people watched clips about *The Avengers,* the network detected that they would also be interested in clips about Robert Downey Jr. This may not seem so hard—blockbuster star in blockbuster movie—but imagine doing this across millions of videos, thousands of topics, dozens of tongues. Two years in, the Brain network would recommend some two hundred million different videos a day in seventy-six languages.

It also discovered formulas. "If I watch this video from a comedian, our recommendations were pretty good at saying, 'Here's another one just like it,'" one YouTube director told a reporter. "But the Google Brain model figures out other comedians who are similar but not exactly the same—even more adjacent relationships. It's able to see patterns that are less obvious." Inside Google, coders talked about this as serving up gems from a back catalog, fulfilling an important service for viewers. (Music videos were huge, after all.) Staff worked to identify "gateway videos," clips that kept people returning. "Then you're hooked," recalled Jack Poulson, an engineer who worked on the system. "I always felt kind of weird doing it." YouTube added more machine learning models to its components until, as demonstrated in the Rapture meeting, models ran the entire site.

Mengerink began having doubts. He had spent enough time around machine learning to know that much of its faults lie not in failing to think like humans but in thinking *too much* like humans. AI could be sexist, racist, and cruel, just like us. "Anything that surfaces a bias—it will mine that bias like nobody's business," he said later.

ISIS was still a headache. In 2015, U.S. law enforcement had about nine hundred active investigations against Islamic State "sympathizers" in all fifty states. Regular conversations were held at YouTube about how to handle footage preaching radical Islam. Sharper machine intelligence could certainly help: moderators struggled to parse the endless waves of footage, and if a machine could easily spot an ISIS flag halfway through a clip, that would make decisions much faster. But Mengerink discovered YouTube also hosted material that looked just as radical that preached *against* Islam. Or against Judaism or Black communities. He was Muslim and knew that most practicing his faith didn't use the phrase "jihad" unless it was *jihad al-nafs*, a term for the individual's struggle of the soul. He searched for "jihad" on YouTube. The feed filled with videos about ISIS and terrorism, predictably, but he scrolled down a bit and found clips from Tea Party acolytes and stare-you-dead-in-the-eyes vloggers like Molyneux. They were bitter and angry, albeit careful not to overuse ugly slurs or call for outright violence, the kind of invective YouTube removed. If the Brain network was set to maximize watch time, which it was, those sorts of videos might perform very well.

YouTube had begun to filter videos promoting Islamist terror, restricting certain clips by age or deleting them. In one meeting Mengerink proposed a similar treatment to bury other forms of radicalism, too, surfacing them less often in search and recommendations. "Make it stupidly difficult to find," he suggested. If someone watched a video openly critical of Black Americans, why not suggest a video that showed Black history in a positive light? There were plenty of those. He was told that such tweaks could appear as opposition to free speech, something Google wanted to avoid, and might disrupt the sanctity of search. He heard one concern repeatedly: "That's not very Googley." *Where was his data to support that argument?* Another person working at YouTube then recalled that after she protested a particular decision, co-workers accused her of not being "positive" and "Googley." Once, Mengerink's position was called "creepy."

His colleagues were not unaware of objectionable videos or particularly fond of them. They had begun discussing ideas for a "penalty box" treatment for troublesome creators like Alex Jones, a conspiracy theorist whose talk show, *InfoWars*, was gaining a massive audience on YouTube. Some people watched for amusement; others maybe bought into his ideas. Machines couldn't tell them apart. During an interview for a senior YouTube role at this time, one applicant was asked, "What do we do about Alex Jones?"

The right answer to such questions was that YouTube should minimize interference, and tweaking recommendations felt like interference. This reflected a firm belief in free speech, yes, but also an equally strong sense that the company was not equipped to impose moral judgment on its viewers' free will. "If people are watching *that*," Mengerink was told, "that's their choice."

Surveying the site, he realized where this argument's logic fell apart. "That doesn't work for kids," he said.

CHAPTER 21

A Boy and His Toy

The biggest star of YouTube's second decade was born in 2011. He appeared on YouTube a mere three years and five months later, when his mother filmed him unboxing his new toy.

Ryan ToysReview: "Kid playing with toys Lego Duplo Number Train." March 16, 2015. 15:13.

"Hi, Ryan!" "Hi, Mommy!" "What toy do you want today?" Ryan Kaji is crouched down in the Target aisle. He is adorable—chipmunk cheeks, dimples, big brown saucers for eyes. He is already pawing two red play trucks, but when his mother makes the offer, he abandons them. He has chosen his toy. We follow him out of the Target and to his home, where he spends the remaining fourteen minutes of the video unpacking the Lego train, bit by bit, practicing a count to ten, and guiding the train along the carpet.

Little Ryan's parents met in college in Texas when YouTube was in its infancy. Shion, Ryan's father, born in Japan, adored beatboxing YouTubers and bonded with his future wife, Loann, over Magic: The Gathering, a nerdy proto-internet card game. When Ryan, their firstborn, learned to watch, he fell for slightly older kids on YouTube like EvanTubeHD, a channel with a tyke who specialized in dismantling *Angry Birds* products. Ryan's parents later said that they captured videos of their son to show family living overseas.

After his YouTube debut, the Kajis uploaded more than one hundred videos of Ryan in a six-month stretch. Mostly he played with one or two toys (Thomas the Tank Engine or a Play-Doh set). He struck gold that July when Loann filmed her preschooler playing with a "GIANT Lightning McQueen Egg Surprise with 100+ Disney Cars Toys." The video borrowed ingredients from the faceless ones, those red-hot unboxing channels: a title of keyword mishmash, a plethora of toys, and a familiar franchise. Ryan's giant red egg, almost as tall as he, housed Pixar surprises and bore a Pixar movie logo. A year later the clip had more than half a billion views—unheard-of numbers. Ryan's channel then earned more than nineteen million views a day, double PewDiePie's count. That GIANT Egg Surprise video catapulted Ryan and his parents to immediate, unanticipated fame and fortune, heralding an entire generation of child YouTube stars.

"I don't know why so many people love that video," his father admitted to a reporter. "If I did, I'd make a lot more just like it."

• • •

A month before Ryan's debut YouTube made an announcement: "Today, we're introducing the YouTube Kids app, the first Google product built from the ground up with little ones in mind." With amateur kids' material growing like a weed, YouTube was trying to create order. Its new mobile app came with a selection of videos from the site, bigger bubblier buttons for smaller fingers, and a built-in timer and sound settings for parents. "Now," a company director wrote in a blog post, "parents can rest a little easier knowing that videos in the YouTube Kids app are narrowed down to content appropriate for kids." The blog made no mention of the fact that algorithms were doing the narrowing, not humans.

This app was free, with ads, like YouTube; the company believed this gave families equal access to its bounty. Also, YouTube had started embracing the commercial appeal of little ones. A 2014 document the company prepared for marketers boasted that it would take seven years to watch all videos marked as "unboxing" posted within just the prior twelve months—unboxed gadgets, skin creams, toys. In discussions with ad buyers YouTube staff carefully avoided the *k* word (kids); this was "co-viewing," a term for parents watching alongside children because, legally, that was how it happened. Google Preferred, YouTube's slate of higher-priced video placements for advertisers, had a section marked "Family and Children's Interest." The company didn't share exactly what videos were there, but *Tubefilter*, a web magazine, did some sleuthing to find a list, which included Mother Goose Club and DisneyCollectorBR, queen of the faceless ones. On the *Today* show a father with obsessed offspring called DisneyCollectorBR unboxing videos "crack for toddlers." Some suspected dopamine or mirror neurons at play—cells that fire off when we perform a task with a clear goal or see someone do it. *Find out what's in that surprise egg!* Did kids watch these compulsively out of genuine interest or because YouTube teed them up one after the other? The phenomenon was still too new to study properly, and the company shared virtually no data with outside researchers.

Harry and Sona Jho operated Mother Goose Club out of a tenth-floor office on Wall Street filled with treadmill desks for Harry's legal practice, and racks of colorful costumes and a green screen for YouTube nursery rhymes. On breaks from legal work, Harry diligently watched YouTube trends rise and fall, sometimes shifting overnight. At one point YouTube's related videos panel filled with "Finger Family" videos: cartoon hands swaying on-screen singing a "Baa, Baa Black Sheep" knockoff, each finger depicting a member of a nuclear family. Jho traced its origin to an old video from Korea, which probably went viral because its "Daddy finger" wore an absurd Hitler mustache. (Unintentionally, Jho assumed.) Kids seemed to like the song and swaying fingers, and the videos were innocent enough, if not especially educational. But something else was happening here. Once a few Finger Family videos did well, tons more flooded the site; most were animated, though sometimes people wore outfits to perform the song-and-sway ditties themselves. YouTube's machines, Jho realized, read that influx of material as a positive sign, so they heavily promoted the videos, inviting even more. Eventually, the Jhos made a Finger Family video, too.

The Jhos suddenly had a lot more company on YouTube. Families joined to bond together or grab ad money. Melissa Hunter, an extroverted operations director at a Manhattan realty firm, had to leave her job after a diagnosis of multiple sclerosis. "What do you want to do for the summer since Mommy can't leave the house?" she asked her eight-year-old. They both enjoyed watching doll-crafting videos on YouTube. *Why not try making those?* On their channel, Mommy and Gracie, they reviewed dolls, borrowing the silly, improvisational style of early YouTube. As views grew, Hunter also started an MCN for YouTubers who made videos for kids, discovering that most people running channels had limited media or business experience. Many early YouTubers were teens or twentysomethings, chasing dreams of making it in Hollywood or fashion. These were parents, with mortgages and college funds, who sometimes quit jobs to rely on YouTube's ad sales, which only seemed to be going up and up.

Little Ryan's astronomical success brought more toddlers to YouTube. And it brought corporate interests. Isaac Larian, a toy magnate running MGA Entertainment, maker of Bratz dolls, had learned about the YouTube trend from his children and ordered his company to design a toy made specifically to be unboxed. They invented L.O.L. Surprise!—candy-colored, bug-eyed dolls enclosed in opaque packaging ("like something out of an acid trip," one reporter observed). Unlike on TV, where Larian had to buy commercials months in advance, YouTube offered him instant product feedback. To promote his new dolls, he handed them out to huge kids' channels like CookieSwirlC, an anonymous toy handler. Soon the L.O.L. Surprise! line would be one of America's top-selling toys, generating more than

$4 billion in sales. Other toy makers had joined in, paying YouTubers to play with their products on-screen.

Harry Jho had expected this trend; toys were big on TV, after all. But he found another pattern more unsettling. Huge animation studios began popping up overseas, competing for YouTube's preschool audiences. Digital animation software was so cheap and easy to program that it seemed some videos weren't even made by humans. Animation factories churned out kids' videos in a gushing, unstoppable torrent. YouTube had rules against uploading too many copies of the same video, but that was harder to enforce when animators learned to tweak their offerings ever so slightly—a different look for each Finger Family member. Content mills had always pumped out cheap YouTube footage, but adult viewers usually ignored it enough to sink it to the site's bottom. Kids weren't as judicious. In 2015, the year YouTube released its Kids app, Jho watched automated torrents aimed at children begin to spread "like a virus without any antibodies."

• • •

Meanwhile, in San Bruno, YouTube shook up its strategy yet again. The company had plans to turn its three biggest categories—kids, music, and gaming—into apps. Wojcicki particularly homed in on music's potential. People watched the hell out of music videos like "Gangnam Style" and Lil Jon's "Turn Down for What." *Why should Spotify, a dinky Swedish company, win the music streaming game?* But YouTube's first attempt didn't take. YouTube Music Key offered viewers who paid $9.99 a month ad-free access to music videos on the site, but it failed to get huge sign-ups. That was mostly because YouTube struggled to pick which clips were music videos and which weren't. Robert Kyncl, the Hollywood chief, explained this epiphany with a personal anecdote: his daughter, while testing the music service, couldn't find songs from *Frozen* in the catalog. "To her, it's music," Kyncl told a reporter. So, YouTube redrew its plans: its entire site—Lil Jon, gamers, toy unboxers, shock jocks—would go into a paid service free from ads, like Netflix or HBO. The company called it YouTube Red, like the carpet, unconcerned at first about the name's resemblance to a popular porn site. (Later, the name changed to YouTube Premium.)

Kyncl had stuck around after Wojcicki took over and climbed further up the ladder to become YouTube's "chief business officer," overseeing relations with Hollywood, record labels, and creators. His *Frozen* anecdote doubled as a negotiating tactic. For YouTube Red his team had managed to persuade nearly every old media titan to move select material to YouTube. The only big holdout: Disney. In 2013, *Frozen* became the studio's top-grossing animated movie and unleashed a mammoth franchise. *Frozen* didn't live on YouTube, but its fans sure did, and practically every channel for kids, from little Ryan to the faceless ones, seized on this fandom by posting videos with Elsa figurines, Elsa dolls, Elsa cartoons, Elsa costumes, and Elsa titles and tags for the machines to read.

In Los Angeles, Maker Studios scrambled to fulfill its new duty as Walt Disney's digital heir, starting with a toy reviewer binge. Maker's executive, Chris Williams, who was on his second tour at Disney, signed five prominent unboxing channels into the studio's YouTube network. He tried recruiting DisneyCollectorBR, the unboxing queen, managing to speak to the woman behind the account, which no journalist had been able to do. Although she declined to join the network, Williams did persuade her to remove "Disney" from the start of her channel name. In February 2015 the *Daily Mail* outed the account's creator, possibly YouTube's highest-paid performer, as a former adult film actress—a reminder that YouTube stars were unvetted, and now tabloid fodder.

When Maker staff joined Disney, they were told the legendary studio planned to go all in on YouTube, leaning into the crazed fandom for *Frozen*, *Star Wars*, and sports. (Disney owned ESPN.) Disney moved gingerly, though, publishing sparse movie trailers and promos for its TV network on

YouTube. But when YouTube released its Kids app, the studio's alarms went off. A Disney lawyer called up David Sievers, the Maker staffer who had moved to oversee its YouTube integration, and demanded, "What the hell is going on?" Federal law mandated that "child-directed" content, media designed for viewers under thirteen, had to follow strict rules on TV and couldn't track viewer habits online. Yet the algorithm programming YouTube Kids had scraped up several clips from Disney TV shows the studio didn't consider "child-directed." Lawyers and Sievers set up a weekly call to review material in YouTube Kids and arranged privately with YouTube staff to remove any videos the studio didn't want in the app.

The rest of YouTube's app didn't get this white-glove curation treatment. And on the East Coast, another concerned group took notice. Josh Golin, a former Miramax film distributor, directed the nonprofit Campaign for a Commercial-Free Childhood, whose mission was exactly what its title said. Most of Golin's work had focused on television, but when YouTube Kids arrived, he dug into the internet's TV. Two months after its release he sent a letter to the FTC arguing that the YouTube app was full of "unfair and deceptive marketing" and toy videos that functioned as long, uninterrupted commercials. Most of the content on YouTube Kids, the letter said, would never legally fly on TV.

Advocates sent these sorts of stern notes all the time. Few in Silicon Valley or Washington, D.C., cared. So Golin kept digging and found more headline-grabbing material. A month later his group wrote another public letter. On YouTube Kids they had found wine-tasting videos, a power-saw tutorial, a profane *Casino* parody starring Bert and Ernie, clips with jokes about pedophilia, and another titled "One Huge Acid Trip?" The letter added a sample of app reviews, including a complaint that after a four-year-old had watched a Peppa Pig video, the YouTube Kids algorithm had recommended a pornographic cartoon called "Peppa Penis." "Who is filtering these videos?" one parent wrote.

YouTube apologized for the lapse. It had set automated filters to sort videos into the app, but people uploaded new content faster than filters could handle.

Yet by August, only three months after Golin's second letter, the dustup was largely forgotten. A reporter from *Time* came to witness YouTube's tenth anniversary celebration in an open outdoor space behind its office, where Wojcicki had instituted YouTube Fridays, a Googley weekly staff meeting. The party featured, *Time* reported, "a bouncy castle, a slushy machine, some jumbo-size board games, oceans of red candy and a DJ." Wojcicki donned a helmet to partake in Meltdown, a duck-the-spinning-propellers game played in a giant inflatable pool. (She was no stranger to embarrassment. Per tradition, Google execs arrived for work in costume if staff filled out more than 98 percent of employee surveys; Wojcicki usually went with animal onesies for the occasion.)

Out of the pool the executive told *Time* that her favorite YouTube video was a John Oliver rant from HBO on the value of mandatory paid maternity leave. For her fifth child, born eight months earlier, Wojcicki had taken only fourteen of the eighteen weeks Google offered. Wojcicki's experience with her kids was "now a business advantage," the magazine reported. "They are her first guinea pigs for many of her ideas." Kyncl described his boss as "a very regular person—mom who knows what regular problems mean for a lot of people."

• • •

YouTube's decision to launch its Kids app as an algorithmically run free-for-all had not been unanimous. Years later several employees said they voted to screen or curate the app's content. But they were overruled. The company knew preschoolers loved watching footage of trains on YouTube; unless it was a cartoon, that didn't seem like standard kid's programming, so it wouldn't go in a curated app. *But wasn't that kind of unpredictability*

YouTube's fundamental magic? Why keep kids from seeing that? Some staff pointed out that this would inevitably lead to showing children footage of train wrecks, given the popular YouTube pastime of watching disasters. One Google director who objected to YouTube's decision was given a Googley reply: *More information is better.*

Indeed, much of Google's operations were based on the slippery belief that a more informed public was necessarily a better-informed one.

And yet, buried under the trainwrecks and toy unboxing, YouTube *did* have a group as educational as any online. The company's earlier attempt to bring its service inside schools didn't take, but EduTubers, a cadre of creators with teaching backgrounds (or nerdy scholastic interests), had begun to flourish. Hank and John Green expanded their successful educational series to a new show for kids. Others unpacked science concepts with quaint animation or the whiz-bang-wow style of Discovery stations, only with more egghead detail. Many earned enough from YouTube ads to make videos full time. They were slightly older than other YouTube stars, with enough life experience to reject offers from MCNs. (PBS pitched a network concept but had few takers because YouTubers considered the station's programs rather uninspired.) EduTubers wanted to entertain and establish careers online, but they seemed driven by another conviction: to inform, to get things right. Some started a habit of listing their source material directly beneath their videos, even though YouTube didn't require or encourage that.

Many also seemed aware of a creeping threat to science taking root online: conspiracy theories that found new life and fuel on YouTube and social media—a noise that would, very soon, start drowning out reason.

YouTube always had videos with a dubious relationship to reality. "Loose Change," an early, influential 9/11 "truther" film, first went viral on Google Video before hopping to YouTube. Staff paid little attention to this material or considered ways to suppress it. "There was a sense of, 'The masses will figure out what the truth is. They'll be a self-correction,'" recalled Ricardo

Reyes, YouTube's former communications chief. One staffer once proposed turning the booming footage on UFOs and other paranormal topics into a proper category, like on TV's Syfy, where YouTube might strategically support creators involved, but the idea never panned out. Besides, defining a conspiracy felt thornier than defining "kid-friendly." So YouTube let it be.

EduTubers, though, tried doing something. Several published clips carefully debunking other corners of the site that claimed climate change was a hoax or the earth was flat or other clear falsehoods. Destin Sandlin, an aerospace engineer from Alabama who posted under the handle SmarterEveryDay, explained the approach in one video. He pointed to his signature crest, a rendition of Reepicheep, a tiny fencing mouse created by C. S. Lewis, his favorite author. Sandlin's wife had embroidered the character onto the polo shirts he wore on camera. "He'll take on a foe that will most certainly kill him," Sandlin said with a grin. "But if the foe is against truth, then he's the enemy, so he must be attacked."

A few inside YouTube pushed projects to better promote these EduTubers. Wojcicki liked to say how much she enjoyed watching them, particularly Simone Giertz, a Swedish inventor who posted a series of impish "shitty robot" clips. But the projects never had major support. And Wojcicki didn't make a concerted effort to get these educational creators in front of younger audiences or the people gobbling up conspiracies on her site—at least until she was compelled to do it.

At the time, her company had other business priorities to deal with.

CHAPTER 22

Spotlight

TheGridMonster: "VLOGMAS 2014 IS HERE!!!"
December 1, 2014. 16:28.

It's Ingrid Nilsen, Missglamorazzi, on her second channel, where she posts vlogs and personal miscellany. Her intro jingle plays on-screen: "I'll vlog every day 'til Christmas for you." Ingrid is without makeup, so her "skin can breathe," wearing, she tells us, a cozy Brandy Melville turtleneck and Target slippers, padding around her home. She must grocery shop. She places her phone behind the steering wheel, camera rolling, en route to Whole Foods. "All you have to do is sit back and enjoy the ride." Back home she unpacks her haul. We cook together. We see glimpses of her BF Chris, but he's not a YouTuber, though he's into Snapchat. As usual Ingrid has gifts to give, body lotion: we need to visit her Instagram to win them. She signs off, "I will see you guys tomorrow!"

Ingrid Nilsen had created Vlogmas, a holiday vlogging challenge, in 2011. She was good at it. She loved it, at first. Each year more vloggers joined in, testing their endurance during Vlogust, Vlogtober, Vlogmas in July, VEDA (vlog every day in April). As Nilsen aged and her real-life holiday responsibilities grew, Vlogmas became less fun. Producing regular, compelling clips started to feel like a monotonous job.

Nearing twenty-six, Nilsen was now a YouTube elder, part of a second generation of broadcasters who saw the platform less as a stepping-stone than as a full-time career. Earlier that year Nilsen had signed an endorsement deal as the first YouTuber CoverGirl "Glambassador," which required shooting videos for the drugstore brand. Beauty magazines that once dismissed her act now called nonstop. People devised a new name for creators like Nilsen: influencer.

Starting in 2014, Wojcicki launched an advertising campaign to promote such influencers to mainstream audiences. *Variety* published a survey that summer that showed American teens were more familiar with YouTubers like Smosh and PewDiePie than A-listers like Jennifer Lawrence and Johnny Depp. The survey whizzed around YouTube's office, confirming its new belief that its old strategy of courting celebrities was obsolete. YouTube had its own celebs. Wojcicki's ad campaign, dubbed Spotlight, plastered their faces on billboards, subway trains, and TV ads, starting with a trio of female "Lifestyle" YouTubers—one known for makeup, one for "haul" videos, one for cooking. The company had a new term for stars with commercial appeal: "endemic creators," a native species.

All this new promotion invited even more YouTubers to chase fame and fortune and made some of the site's veterans feel an immense, exhausting pressure to keep up.

• • •

Olga Kay had made her splash on YouTube before most "Lifestyle" stars, and she was famous in YouTubeland for hustling harder than everyone else. Kay first uploaded YouTube footage of juggling routines—she had joined a circus troop in Russia, her native country, at age fourteen—but soon gravitated to the thrilling intimacy of vlogging. She posted a medley of confessional diaries, antic skits, whatever stuck. Petite, with a heart-shaped face and boundless energy, Olga Kay became a staple of YouTube's first motley crew. She appeared in Maker Studios clips with LisaNova and juggled swords onstage at VidCon. No other YouTuber arrived at VidCon's hotel with a bulky iMac in tow; she had taught herself to edit footage and had to keep doing so during the conference, obviously.

To make money back then, YouTubers needed serious subscribers, and Kay noticed she could attract more with personal touches. Each time someone subscribed, she went to that person's YouTube page and left a comment: "Thank you for subscribing, from Russia with love." Others might see that and come to her page. *Rinse, repeat.* She posted while watching movies at home and during her day job in TV production. Hundreds an hour sometimes.

"Thank you for subscribing, FROM RUSSIA WITH LOVE:)"

Slowly, some success came. She received her first check from YouTube: fifty-four cents. *This is good,* she told herself. *That's going to be $5 tomorrow.* Some days she put in twelve hours of work making YouTube videos. By 2014 her YouTube earnings topped $100,000 for the third year in a row. It was good money, but that was before taxes, before investments in merchandise (she sold her own), before hiring an editor. And it required that she produce twenty videos a week across her channels, a new kind of juggling.

And then YouTube's second generation and all the demanding social apps arrived. Kay adapted to YouTube's watch-time change, uploading gaming and makeup videos. At YouTube events she learned that her fans were mostly teenage girls. Kay, then in her early thirties, made videos to show that femininity could be unconventional, oddball, blemished. She posted once a week until she discovered that no longer worked. To stay relevant, she needed to post daily.

Olga Kay: "I AM NOT READY!!!" January 21, 2015. 4:29.

She holds her iPhone at arm's length and films herself lying on the couch, in selfie pose. "Why am I making this video? Because I have lots of pimples and I just want to show you what it looks like on my face. There's so many of them." She points to one on the left ridge above her lip. "I am just like everybody else." Another reason for posting, she continues, is that she has shot two really awesome videos but they are not yet ready. "Well, I want to post *something* for you guys to let you know that I'm still here."

• • •

In April of 2015 YouTube invited more than a hundred of its stars to a studio space in Manhattan's fashion district for the inaugural YouTube Creator Summit, its very own video conference. There was fancy catering and a David Blaine magic performance, but no mobs of teenage fans. Still, creators felt a need to be on. One vlogger filmed the lunch, turning a camera on Ingrid Nilsen, who sat casually eating a salad in a gray hoodie. She performed on cue, lighting up with a smile and banter.

But Nilsen left the summit feeling inspired. YouTube offered production tips and truly celebrated its dedicated video weirdos, finally treating them like A-listers. YouTube, too, deemed the gathering a triumph. Inside the company, staff often discussed the two paths to creator success: the *SNL* model, where performers used YouTube as a springboard into movies and TV, and the Oprah model, where creators built an empire of loyal audiences right on YouTube. The summit was full of mini-Oprahs.

The summit also reminded YouTube of the growing competition in its rearview mirror. Nearly every YouTuber there pulled out their phones at some point to snap a selfie for Instagram. This was the hot moment for social apps: Instagram, Snapchat, and Vine, a service for six-second video loops. People only needed their phones to post on those apps, not the cameras and editing software YouTube demanded. Ambitious creators felt compelled to be on all the apps, all the time. Those apps didn't pay creators yet, minimizing YouTube's fears, but Vessel, a new video service from Hulu's founder, *did* and it had signed up marquee YouTubers like Nilsen to produce exclusive material. YouTube business staff weren't too worried about Vessel until they heard that Larry Page was, prompting a furious scramble for a strategy to snuff the rival out. (That strategy, called Platinum internally, involved giving select stars big advances on ad revenue to keep them loyal to YouTube.)

But no rival worried YouTube quite like Facebook.

In 2012 Facebook had paid $1 billion for Instagram, which now looked like a steal; the photo app captured the youth zeitgeist like nothing since YouTube. For a while Facebook was a useful adversary: YouTube clips shared on the social network often went gangbusters, like they did on MySpace back in YouTube's early days. Around 2014, YouTube managers noticed video traffic from Facebook start to sputter. Then tank. The social network had its own video player and it seemed to be favoring posts in its feed that used that over outside services like YouTube. Facebook planned to add a livestreaming feature and video functions to Instagram. Google Plus had, by then, withered to zilch. "Google had kicked its own ass in social," said one YouTube manager. "And now Facebook is coming after video."

Facebook's big assault came when it teamed with Nielsen, TV's standard bearer, for a system that let marketers run commercials on TV and Facebook using the same audience ratings system. Essentially, it was a screw-you to YouTube. This world—the wonky operations of digital advertising—was where Wojcicki had spent most of her career. She immediately put more systems in place to bundle YouTube channels like TV. She seized control of ad operations from the multichannel networks, which YouTube had let sell ads willy-nilly. No more. The MCNs, already rattled from YouTube's earlier changes, took a further bruising during its new fight with Facebook over measuring commercial messages online. The urgency was apparent. Google, recalled Maker Studios' David Sievers, was the "massive, eight-hundred-pound gorilla. They can't lose that war."

Yet in this war YouTube felt like an underdog. YouTube had claimed more of people's time: it went from racking up five minutes of average daily viewing sessions in 2011 to forty minutes a day in 2015 on phones alone. But Facebook earned far more in ad sales, a discrepancy that partially inspired one of Wojcicki's first big hairy goals as CEO: the 2020-20 plan, set in 2015, to cross $20 billion in revenue within five years.

Competition from social apps, little dopamine machines, might also have explained Wojcicki's other major order as YouTube chief. Early in her tenure she looked at the data and concluded that the site was very good at getting existing viewers to watch more, a nice steady ramp to its billion-hour goal. But it was less good at growing the overall number of daily viewers. Larry Page had a favorite Larry-ism about this: "the toothbrush test"; Google's products were only worthy of existence if people used them as often as they brushed their teeth. So Wojcicki tasked her engineering team with adjusting the recommendations algorithm to favor videos that brought in daily viewers, again and again.

All this explained the lavish treatment of the mini-Oprahs. YouTube hoped to equip its star creators to be their own media showrunners who could match TV's daily output and ad rates. To do that, YouTube wanted a tighter bond with its stars, though it had to overcome a sizable cultural gap. Ariel Bardin, a Googler Wojcicki had recruited as a YouTube executive, went on a tour to meet YouTubers early in his tenure. In Los Angeles he visited Matthew Patrick, a former musical actor whose well-known channel, the Game Theorists, mixed theatrical bombast with a scholastic knowledge of gaming lore, science, and YouTube. Patrick started with an icebreaker, "What YouTube channels do you watch?"

"*Vice*," Bardin replied.

Patrick flinched. The trendy Brooklyn media operation had raised more outside funding after receiving a YouTube grant. To most YouTubers, *Vice* typified the corporate-backed poseur stuff taking over their site, the kind of channel people with no real attachment to YouTube culture watched. "Okay, who else?" Patrick asked. The executive didn't have an answer.

At the first Creator Summit, Patrick sat near the front row as a DJ spun tracks ahead of the main act. Every creator there knew Patrick by his YouTube nom de guerre, MatPat. YouTube, the company, didn't. As the official presentation began, the company posted images of its creators on the gigantic video screens. *Smosh! iJustine! MatPat!* A photo was displayed of a different dude, not Patrick. *Ouch.*

Kyncl took the stage. "There are so many wonderful faces here that I love," he began in his deep baritone. "Creators such as—" *Pregnant pause.*

Uh, Patrick wondered, *could he not think of anyone's name?* Finally, Kyncl named Hannah Hart, a YouTuber whose face shone on the billboards YouTube plastered around the city.

• • •

Felix Kjellberg was dressed as a doll, getting whipped by a dominatrix and obeying orders from his own voice piped from a loudspeaker. It was all very surreal.

Inside YouTube, while Wojcicki worried about Facebook, Robert Kyncl had a different concern: his old company, Netflix. In 2013, the streaming service premiered *House of Cards*, an original series that became a megahit. Amazon had started producing streaming shows too, diving gracefully into the golden age of TV and prestige. *Why couldn't YouTube?* After failing to generate hits with A-listers, Kyncl pivoted to a new strategy. YouTube would finance its own shows, called Originals, available only to its paid subscribers and starring creators its audience loved. Netflix and Amazon had followed traditional media's model, letting showrunners and producers pick the story lines and stars. Sumner Redstone, Viacom's mogul, had a famous credo to defend the financial value of this model: "Content is king." At YouTube, they devised their own credo for their approach to the *House of Cards* era: *the audience is king.*

So that's why YouTube's audience got to watch Kjellberg get whipped in episode 9 of his YouTube series *Scare PewDiePie*, which debuted in early 2016.

Maker Studios produced the show, which placed Kjellberg in real-world simulations of horror games from his videos. *Scare PewDiePie* leaned heavily on reality-TV tropes; later in episode 9, Kjellberg fished through raw meat and roaches to solve puzzles à la *Fear Factor*. Kjellberg gamely went along, but his series received lackluster reviews and interest. Perhaps it was because Kjellberg played himself—a reserved, slightly awkward YouTuber—not PewDiePie, the cartoonish maniac he performed on-screen. Despite good production the series felt like stale TV. Those around Kjellberg noticed that filming it drained his energy.

As YouTube premiered more than a dozen of these Originals shows, many YouTubers noticed a central paradox. On the one hand, YouTube wanted gloss, pizzazz, *expense*; in Los Angeles, the company converted a 40,000-square-foot airplane hangar into a state-of-the-art production studio for select creators called YouTube Space. But YouTube's algorithm still wanted the opposite. It desired watch time and daily views; videos that delivered that were usually made cheap. But they rose to the top. One YouTube executive later lamented that its creator team had gone from nurturing talent and creativity to "searching for people that fit the metrics."

For many YouTubers, the site's foundational culture of independence and accessibility started to feel a bit antithetical to a mass-media advertising business. "To be both authentic and relatable becomes very hard when you're doing this professionally," Hank Green, the VidCon founder, lamented in a video called "Honest YouTube Talk Time." Olga Kay felt this tension all the time. She didn't have quite enough subscribers to get invited to star summits, though she did entertain offers from movie studios and networks. They looked for Hollywood-style production, but Kay knew

YouTube's fans and machines didn't want that; they preferred her talking at home to her camera. They demanded frequency. Kay continued posting about twenty videos a week. Friends would invite her out. "If I can't film there," she replied, "then I have to stay home and make content."

She occasionally made content at YouTube's Los Angeles studio but mostly went there for ad hoc consulting. An MCN wanted to sign her gaming network. She once brought the network's paperwork to Andy Stack, the YouTube manager.

Stack eyed her figures. "You're going to lose so much money," he said, advising her to turn down the offer. Like others at Google, Stack had concluded that MCNs, with their iffy contracts and mafioso tactics, were *no bueno*; another YouTube staffer called the networks "parasites."

Still, YouTube's shortcomings as a place for managing talent was becoming clearer. Stack oversaw the system that doled out payments to millions of creators in its Long Tail. Its daunting size struck him when Google lawyers called to alert him to new compliance measures required because these payments were becoming material to Google's daily revenue. All these YouTubers needed systems to track money flow and the site's frequent updates. Other companies might set up a call center or hire teams devoted to managing broadcasters. Not Google. "The Google way of solving problems is to throw machines at them, not people," said Stack. A team of engineers created a computer system to address needs without having to involve a phone call or a human. But machines could leave creators in a lurch. Under Wojcicki, YouTube's systems had become more aggressive at identifying videos that might offend advertisers (too much cursing or innuendo) and stripping them of ads. Yet after the system pulled a creator's income, it didn't clearly notify them or offer an explanation. In a meeting Stack pleaded with colleagues for better resources. "You know, if you get arrested, at least you get a phone call," he said.

Sometimes, the machines screwed up. On one occasion YouTube updated its skin-detection algorithm, which automatically flagged porn or sexual exploitation for removal. Someone turned a dial too far. The update was introduced (with no public announcement), and suddenly scores of bodybuilders on YouTube, a popular niche, saw their videos disappear, victims of a system that couldn't differentiate porn from Speedos. Kathleen Grace, a YouTube employee, watched this all unfold. *Oh*, she realized about the algorithms, *they can't know everything.*

Grace ran YouTube's Los Angeles studio, where creators came through not only to shoot but to unload. One female YouTuber, distraught over her falling viewership numbers, told Grace about her plans to rebound: she would spend six weeks filming more than three hundred videos, post one daily, and meanwhile create a whole separate YouTube show. The YouTuber had gone *10x*. Grace, who had produced and directed web video since YouTube's inception, stared at the creator's plans. *This is insane*, she thought. Another YouTuber broke down in tears, exhausted from posting regular videos that lasted more than ten minutes, long enough for an algorithmic boost. "It's just a treadmill of pain," Grace said. "It's not a way to be creative." Bing Chen, another YouTube employee who worked with creators, tried running his own channel with frequent postings as an experiment. Despite leading man looks, Chen resigned quickly. "It's a hundred-hour job," he told colleagues. "It's just so much work."

As a side project Stack had converted his home in the Hollywood Hills into the Alchemy House, an artists' space he imagined as a haven for creativity. He invited musicians and YouTubers. Olga Kay first arrived stunned to see Stack, who dressed conservatively at work, in shorts and blue nail polish; Stack, like many Googlers, was a Burning Man regular. In 2013, Stack reconverted his Alchemy House into a "decompression space." After each VidCon, he invited YouTubers to come enjoy drinks and panoramic views, free from crowds, cameras, and the pressing need to upload.

Perhaps Kay could have kept going on this treadmill if it were only twenty YouTube videos a week. But all the social apps, eager for slices of YouTube's business, began calling. Kay lost it with one call about Snapchat.

The messaging app had invented a format, Stories, that let people share fleeting photos or videos. A brand representative asked Kay if she could shoot five of them for a sponsored deal, ten seconds each. They would pay $7,000.

A pretty good offer, and yet her hands shook as she considered it. She felt angry and ill at the prospect of filming even fifty seconds more. She simply couldn't. *Wow,* she thought. *I am done. This is not good.* Is this what she wanted to do in her forties? Her sixties? Constantly live on camera, reinventing herself to grab people's attention? *Nope.* "I just want to have a normal life," she decided.

• • •

The company knew something was askew with its creator economy. But it had another corporate hubbub to deal with.

That August in 2015, Larry Page had shocked the world (and most Googlers) by announcing the creation of Alphabet, a new holding company that would split his empire into several stand-alone businesses: one for Google, one for self-driving cars, one for smart thermostats, and so on. YouTube seemed like a natural splinter; it already operated with a different name and office. Leaders there considered plans to become a separate Alphabet unit, detached from Google. Wojcicki wanted to keep reporting to Page, who had appointed himself Alphabet CEO, rather than his successor at Google, Sundar Pichai. But ultimately it was decided YouTube was too intertwined with Google's business and machinery to leave. So it stayed at Google.

That year Wojcicki had also brought on two new executives to shape YouTube's future. Neither had a background in media production, but they were both Google veterans. Neal Mohan had cut YouTube's first big ad deal, in its office above the pizza shop, as a director at DoubleClick, and he had stayed with Google's ad division after the DoubleClick acquisition.

He had two Stanford degrees, an NBA obsession, and the careful elocution of an accountant. Within Google he was known as a political master, someone who could "manage up." One YouTube director recalled having a tense standoff with Wojcicki during a Tuesday meeting. By Thursday, Wojcicki had changed her mind. Mohan "had done some Jedi mind trick behind the scenes," the director recalled. This paid off. A publication reported in 2011 that Google had given Mohan a $100 million bonus to counter an offer from Twitter. Some Google colleagues then ribbed Mohan as the "hundred-million-dollar man," much to his chagrin.

Wojcicki put Mohan in charge of YouTube's product, and he quickly became her top deputy. As his deputy, Mohan recruited Ariel Bardin, the executive fond of *Vice*—a fast-talking, blunt Israeli who had been at Google since 2004, most recently running its payments service. When he and Mohan arrived, they looked at figures in the creator economy and saw serious inequities. Most ad money flowed to the top one hundred creators. What if a few of them went elsewhere or quit producing? *Was the compensation system really a level playing field?* The new executives developed a plan to redraw YouTube's entire payment system based on mathematical measures of success. They called this Project Beane after Billy Beane, the unorthodox baseball general manager featured in *Moneyball*.

They had another nickname for the project: "boil the ocean," a nod to the Herculean engineering task required. YouTube had used this phrase before to describe big, successful efforts to remake its service for apps and TVs. So as 2016 began, the company prepared to solve the financial problems of its creator class with engineering, like usual. It didn't seem like there were other major problems on the horizon.

CHAPTER 23

Joke, Threat, Obvious

January 2016

A small set in the White House East Room had been designed as an imitation of Ingrid Nilsen's house with little succulents and white chrysanthemums against a powder-blue wall. Nilsen turned to Barack Obama and asked him about tampons.

Nilsen, the "Lifestyle" vlogger, was one of three YouTubers selected to interview the president on the official White House YouTube channel, part of an effort to market creators to older audiences as the influential figures they were online. The prior summer Nilsen had catapulted onto Google's radar with her video confession. "I'm gay. It feels so good to say that," she told viewers before breaking into joyful sobs. Her time with Obama was carefully choreographed—Nilsen spent about four hours being vetted by phone beforehand—but she managed to make news after she asked why menstrual products were taxed as "luxury goods." ("I suspect," the president replied, "it's because men were making the laws.") Nilsen seemed slightly nervous until a portion of their conversation that borrowed from her YouTube channel, where she asked people to share tokens of their "sense of self." Obama pulled some items from his pocket—a rosary from the Pope, a little Buddha from a monk, a lucky poker chip from an Iowa biker. "That's really touching," Nilsen cooed. "I loved it!"

Steve Grove, one of YouTube's first community managers, appeared on camera as Nilsen's interview closed. He worked for Google now and had traded his jeans-and-sneakers look for a crisp blue suit befitting a politician. "This has been a wonderful tradition that you've started," Grove said, thanking Obama. "We certainly hope your successor follows in your footsteps."

• • •

At that moment, Donald Trump's bewildering lead in the Republican presidential field looked like a punch line. His proposals, like endorsing a registry of all U.S. Muslims, sounded absurd. A note from Trump's doctor proclaimed he would "be the healthiest individual ever elected to the presidency." Publishing the testimonial on Facebook, Trump praised his "great genes" and mistakenly credited a dead doctor. Trump made great TV. He hosted *SNL* in November and appeared on Jimmy Kimmel's talk show. Clips from both went straight to YouTube.

In another corner of the site, though, Trump had begun to receive a very different treatment. Stefan Molyneux, the self-help guru, began a series in January called "The Untruth About Donald Trump." With Trump, YouTubers like Molyneux, who thrived on attacking media and other mainline institutions, had a powerful ally and great material. In Molyneux's new video series, he cataloged all the press "misrepresentations" of Trump. Each Molyneux video ran more than an hour. In the first, he correctly noted Trump's success in manipulating news cycles, before defending the candidate's views on immigration, women, and a litany of other positions. These videos did not target Trump's political opponents. "The big lesson here," Molyneux declared, staring at viewers, "don't let anyone tell you how to think or feel. Don't let me do it. Don't let anyone else do it. And in particular, don't let the mainstream media do it. They're not trying to inform you. They're trying to control you." These episodes did well on Reddit, where a ferocious force of Trump loyalists gathered. Later that year, Molyneux hosted authors the Southern Poverty Law Center described, respectively, as a "eugenicist" and an editor of a "white nationalist" publication. (Molyneux would refute this characterization of his guests.)

"If you find this information useful," Molyneux said at the end of his January video, "please like, subscribe, and share."

• • •

April 2016

Susan Wojcicki sat on a stage before rows of her creators at the Andaz hotel on Sunset Boulevard, in Los Angeles, to hear their concerns.

YouTube's boss, like her predecessor, Kamangar, could be a little awkward before groups. When she name-dropped YouTubers during staff meetings, she sounded, one former employee said, "like a dorky mom trying to be with the cool kids." Claire Stapleton likened her to Hillary Clinton—a workaholic who rarely let her professional guard down and was easily targeted as a scold. For Wojcicki, creators were a particularly tricky audience; they weren't her employees, and, while she exercised outsized power on their careers, she didn't manage them (or couldn't if she tried).

And this was a tricky situation. The Sunset Boulevard event in early April, called #YouTubeBlack, was organized after Akilah Hughes, a longtime YouTuber, delivered a damning indictment. On her channel, Akilah Obviously, she spoofed and dissected YouTube trends, pop culture, politics, and literature. (She befriended vlogbrother John Green after giving an on-screen "tipsy review" of his book.) She had propelled her YouTubing to a job at the media company Fusion. In 2015, Hughes wrote an article noting that none of the billboards or subway ads YouTube ran for its Spotlight campaign showed Black creators. Neither did any of its new Originals. That February, Black History Month, YouTube's Twitter account promoted over ten times more white creators than Black ones. And, Hughes wrote for Fusion,

YouTube's odious comments "make it harder for diverse creators to work on YouTube than on other social networks."

Her article circulated around YouTube, where staff always prided themselves on knocking down Hollywood's stale, male, mostly white gatekeepers. Marketing teams organized #YouTubeBlack, placing Wojcicki onstage with Adande Thorne, a.k.a. Swoozie, a Black YouTube animator and company favorite. Thorne cut to the chase. "When will we see a Black creator on a billboard?" he asked.

Soon, Wojcicki promised. "We can do better," she acknowledged.

In the evening these YouTubers, gathered for the first time, shouted along to Kendrick Lamar's "Alright." "YouTube isn't perfect," Hughes wrote afterward, "but at least executives seemed committed to progress."

Wojcicki was riding high off that creator affection a few weeks later, when she returned to a stage to speak to her biggest stars at YouTube's second Creator Summit. She touted YouTube's ad growth and said the company had a long-term commitment to its Originals shows. Then she opened the floor to questions.

At one point, a female creator asked about YouTube's bullying problem: a fellow YouTuber had made repeated hostile videos about her, "doxed" her (posted her personal information online), and sent waves of angry followers her way. She was scared. Another female creator grabbed the mic to echo the concern, arguing that it had become far too common. *What would YouTube do about it*? Wojcicki's guard went up. She offered sympathy but no tangible promises before moving on. A few questions later, more YouTubers tried asking about bullying again. Wojcicki had a similar reply. At the time, YouTube found it best to stay out of spats between creators and felt its existing rules balanced harm with free speech pretty well. Or YouTube just wasn't prepared for this problem.

Ingrid Nilsen, who sat in the audience, had grown more worried about online invective. No one had doxed her yet, but it seemed like only a matter of time. "YouTube just didn't have an answer," she recalled. "They knew the mess was a really big one."

• • •

Milo: "Milo Yiannopoulos denies Black Lives Matter protester special mic privileges." April 27, 2016. 3:30.

Milo Yiannopoulos, a Brit with frosted tips and a tired shtick as a gleeful troll, worked for *Breitbart News*. Here he is speaking at American University, part of his "Dangerous Faggot Tour" of college campuses, where he holds forth against feminists, SJWs, and "cuckservatives." A tense back-and-forth with a Black student makes this particular speech memorable, good YouTube fodder. Another student in a red Make America Great Again cap giggles at the spat.

Steve Bannon had taken over *Breitbart News* after a meandering career as a Hollywood financier and video game executive. He used deputies like Yiannopoulos to "activate" an army of disaffected, extremely online supporters. "They come in through Gamergate or whatever and then get turned onto politics and Trump," Bannon told the journalist Joshua Green. *Breitbart News* had shown its political heft two years earlier when its fire-breathing coverage spiked a Senate immigration reform bill, despite support from the conservative standard-bearers Fox News and Rush Limbaugh. Now Bannon, months from becoming Trump's chief strategist, played ringleader for the "alt-right," a network of internet personalities, provocateurs, and racists. Green, the journalist, aptly described this movement as "a rolling tumbleweed of wounded male id and aggression."

The tumbleweed rolled through *Breitbart News* and acidic message boards and social media. It rolled through YouTube.

David Sherratt, the British men's rights YouTuber, watched it spread. Many channels he regularly tracked went from discussing atheism and feminism to talking about Trump. Some, like the vlogger Sargon of Akkad, kept the same tone of ironic bemusement, ranting against the "Machiavellian schemer" Hillary Clinton and the "billionaire ghoul George Soros" pulling her campaign strings. At first, Sherratt thought Trump endorsements were a joke, a middle finger to elites. But then he wasn't sure. Videos about Trump certainly got views.

Newcomers were soon drawn into YouTube's alt-right orbit. Like other YouTube subcultures, they made cameos in each other's videos and posted replies and debates. They exploited search. A later study showed that a clip featuring Yiannopoulos "persistently" ranked atop YouTube search results for the term "Gamergate" in the summer of 2016. Searches for "Islam," "Syria," and "refugees" also spat back videos from alt-right YouTubers.

Bomb hurlers like Yiannopoulos began preaching on behalf of Brexit. Then these YouTubers started leaning more heavily on the ills of refugees. Sherratt had doubts. *I mean, they're fleeing war,* he thought. *Come on.* Later he would look back at this time and wonder what exactly he did believe and why.

• • •

July 2016

Wojcicki called an emergency meeting. A day before, Donald Trump formally accepted the Republican nomination. But this meeting was about something else that occurred that day—a short *Wall Street Journal* article noting the success media companies were seeing with videos on social networks, not YouTube. YouTube was too crowded and didn't have the easiest mechanisms for sharing video content, the article argued. "Something has to change at YouTube," an anonymous media executive griped, "or Facebook and Snapchat are going to own this world."

Wojcicki convened her public relations team to devise a plan to combat perceptions that Facebook made a more compelling destination for video than YouTube. A considerable amount of YouTube's time and attention during 2016 went to efforts to fend off this competitor. The fear was still real. YouTube once proposed a business arrangement with a smaller tech company with a condition attached: if Facebook ever offered to buy the smaller company, YouTube could veto the deal. When Facebook released a new commercial feature, Google publicists often called up reporters to remind them that Google had already released something similar.

• • •

Guillaume Chaslot had left YouTube unceremoniously, three years earlier, and had moved back to his native France to be near his aging father. Chaslot was short and animated, with a prominent brow and a long, sloping nose. When he tucked his long hair behind his ears, he looked a little like Timothée Chalamet. After completing a doctorate in computer science, Chaslot had applied to only one company, the one with the reputation for welcoming academic nerds. During his Google orientation in 2010, in California, more than half the new Googlers there were foreigners like him; a round of H-1B visas had just been approved. Chaslot loved that.

He was assigned to work on YouTube recommendations. Soon he began one of Google's "20 percent projects" to address a flaw he detected in the system. YouTube tended to show people the same perspective on repeat. After Trayvon Martin's death, thousands tuned in to YouTube for information, analysis, or catharsis. If they watched videos sympathetic to Martin, YouTube usually recommended footage echoing that. If they watched a video from the other side, maybe arguing that Martin's death was justified,

YouTube's recommendations played more of those. Chaslot drafted a plan to introduce balance: a digital catalog, called "Google History," which tracked videos about particular world or historical events. The French engineer intentionally noted how such a thing could improve YouTube watch time. Chaslot won praise for it from peers but couldn't find any interested YouTube managers, and he soon received a negative performance review (a "ding"). Google let him go.

He had more or less shaken off his dismissal back in France and didn't think much about YouTube. His father, a dedicated pharmacist, lived in a rural area and spent little time online, yet he once made a comment that shocked the computer scientist. "You know," the older man offered, "we need to start listening to Vladimir Putin." Never particularly political, his father began ticking off Putin's merits. Chaslot argued back but remained confused. *How does Russian propaganda get to my father in the middle of France?* he wondered. He suspected it came from his father's pub friends, who watched clips about Putin on the free internet television. Chaslot searched YouTube. There was Putin holding court on the damage immigration had done to Europe. There was Gérard Depardieu praising Putin in French on Russian state TV. These videos got better traffic than those from more respected media outlets.

Another weird encounter with his old company puzzled Chaslot. He was riding a bus in Paris and noticed a passenger beside him bent over his phone, drinking in YouTube videos. One clip, in French, was audible. A narrator droned on about a cabal planning to exterminate a quarter of the world's population. Chaslot, assuming this was a joke, leaned over to ask, "So who wants us dead?"

"There is a secret plan from the government," the passenger replied earnestly. "Hundreds of videos say so!"

Chaslot's eyes widened. At YouTube, he had worried about information echo chambers. He had no idea that these chambers could be filled with outright conspiracies.

On a hunch he started looking more closely at YouTube ahead of the U.S. presidential election. He built a tool to scrape public data on video recommendations. It was a limited sample—outsiders couldn't see what videos were fed to logged-in viewers—but was still revealing. One name jumped to the top of his list: Alex Jones, the gravel-voiced shock jock and walking media spectacle. That July, Jones crashed a Trump rally outside the Republican convention in Cleveland, parading in with an entourage before dozens of cameras to rail against globalists and the "New World Order." On his talk show, *InfoWars*, Jones had given airtime to cranks who argued the Sandy Hook school shooting was staged. Chaslot calculated that Jones's episodes were watched more than 300 million times within eighteen months. Once Chaslot collated the data, he found that Jones wasn't just near the top; he was the most recommended channel in his dataset. *This is insane*, Chaslot thought.

His old company either didn't know this was true or wasn't doing anything about it.

• • •

August 2016

"Be advertiser-friendly." "No graphic content or excessive strong language."

That's what YouTube decided messages to creators would say. To wrest more marketing dollars from TV, Wojcicki had determined YouTube needed more sanitized material. YouTube's algorithms had gotten better at automatically pulling commercials from videos that might offend advertisers, although the system couldn't always account for why it made particular decisions. When ads were removed, YouTube staff had scripted an automated email message that would go out to let the affected YouTubers know it had occurred and offer a way to contest the decision, which seemed fair and kind.

This didn't go as planned.

Philip DeFranco captured the outrage first and loudest (his video title: "YouTube Is Shutting Down My Channel and I'm Not Sure What to Do"). DeFranco, a dogged vlogger with spiky hair and an Everyman vibe, had been a YouTube mainstay for a decade and among the first broadcasters ever paid. He posted a daily news and gossip talk show that began with his customary greeting, "'Sup, you beautiful bastards?" In a video that August he confessed that he may not be able to get away with "bastards" anymore; it wasn't advertiser-friendly. When DeFranco and other YouTubers got these automatic notices, they did not see them as YouTube did, as kind invitations to contest filtering decisions. No, this felt like censorship. Many weren't aware of YouTube's ad-friendly rules or found them arbitrary and unfair. DeFranco learned ads were pulled from a dozen of his videos discussing the news. Yet clips of news commentary from big media companies, flooding onto YouTube, sure seemed to run ads without issue. Soon YouTube would introduce an icon of a tiny dollar sign next to each video inside the online dashboard that creators used. If the dollar sign lit up green, the video made money. If the dollar sign shone yellow, it didn't. Those yellow dollar icons became a symbol of injustice. A term took hold, one with religious weight for YouTubers: videos and creators without ads were "demonetized." For YouTube, the whole ordeal felt like a big miscommunication. "Feels," DeFranco countered on Twitter, "a little bit like getting stabbed in the back after 10 years."

YouTube had always tolerated complaints from creators. That was a fair price for their free videos, which brought in eyeballs, which brought in ad dollars, 45 percent of which went straight to YouTube. But complaints that summer, after months of mounting creator frustration, hit particularly hard. DeFranco's video spurred a hashtag that spread online channeling the ire: #YouTubePartyIsOver. Plenty of YouTubers joined in, including conservative vloggers and right-wing outfits that used the opportunity to accuse YouTube of silencing speech. PragerU, a conservative advocacy group backed by fracking industry magnates, charged the company with restricting its videos on the Ten Commandments and other biblical topics. YouTube convened another listening session in its New York studio, inviting representatives from PragerU and a few dozen other conservative YouTube channels.

Ahead of an ugly, caustic election, Google had been hearing charges from the right that its search results were skewed to favor Hillary Clinton. Google denied this, but the cries only grew louder. YouTube, like its parent company, had no interest in appearing as if it put a thumb on the political scales for any side.

• • •

YouTubers, though, had no qualms about picking political sides.

Paul Joseph Watson: "The Truth About Hillary's Bizarre Behavior." August 4, 2016. 5:53.

> "Weird seizures. Psychotic facial tics. Overexaggerated reactions. Coughing fits. Strange lesions on her tongue," a Brit narrates over montage footage edited to make the Democratic candidate look unhinged. "Is Hillary on the verge of a mental breakdown due to stress or are her strange outbursts linked to a medical condition?"

Paul Joseph Watson was an editor for Alex Jones's website, *InfoWars*, and a staple of alt-right YouTube. He posted frequent long clips on major news items and internet arcana, riffing on them as a humorist just asking questions. He borrowed Stefan Molyneux's tactic of promising to share secret knowledge the mainstream media obscured. (Watson's videos offered "the truth about" rape culture, ISIS, Ebola, and #Gamergate.) His Clinton video was comically sourced: one expert he cited was "pharma bro" Martin Shkreli, who was then being indicted for securities fraud. But the video took off, jumping to the top of Reddit's page devoted to Trump. The Drudge Report and the *National Enquirer,* ferocious Trump surrogates, introduced more spurious coverage of Clinton's health; Fox News's Sean Hannity devoted multiple nights of his show to the issue. YouTube, so intent on being a neutral platform, had become an effective weapon for fringe political warriors.

Four days after Watson's video, Google shared its list of most common searches about presidential candidates. The second top search query for Clinton read, "Is Hillary having health problems?" A popular Google search for her opponent: "When was Trump on *Letterman*?"

Trump loved television, but his favorite stage was Twitter, the place campaign insiders now tracked obsessively to gauge the latest spurt of political chaos. Months later Watson used Twitter to let them all in on a secret, writing, "I'm not sure the left understands the monumental ass-whupping being dished out to them on YouTube."

• • •

October 2016

YouTube had fallen behind. For a moment, things looked dire.

In July, in the Northern Hemisphere, people spent more time outdoors, away from screens. And the internal graph at YouTube charting its one-billion-hour goal showed a slip in trajectory. Cristos Goodrow, the engineer whose name appeared right beside the graph, checked it every day. On weekends, on vacation, on sick days. When autumn came, his team hunted for any incremental change to draw just a fraction more daily watch time. Once they found one, they made it. Over the year they made some 150 changes to help hit that goal set back in 2012.

Finally, one day in October, Goodrow checked the chart to discover that YouTube had cleared its goal, ahead of schedule.

That month some of his colleagues planned an event for the first week of November. YouTube invited creators and employees to its Manhattan studio space for an election night party. The hip-hop artist Common would perform. Nearly everyone arrived expecting to watch U.S. voters elect their first female president.

CHAPTER 24

The Party Is Over

On November 10, two days after the election of Donald Trump, Googlers gathered for an all-staff meeting. Sergey Brin, who had long been absent from daily Google affairs, emceed onstage. "As an immigrant and a refugee, I certainly find this election deeply offensive, and I know many of you do, too," Brin said. "It's a very stressful time and conflicts with many of our values."

Larry Page came up to stand beside him. Both billionaires had salt-and-pepper beards and held mics with foam covers that matched their shirts. They invited onstage a quartet of senior executives who were actually running Google. This included Sundar Pichai, Page's successor as CEO, a lanky, bespectacled product specialist and former consultant who was born in India and remarkably free from enemies inside Google. Employee Q&A began. Brin read off one question raising concerns that the algorithms of YouTube and social networks were polarizing people, making them "blind to what the other half of the world thinks." *What could Google do?* Pichai, in a stylish hoodie, reassured his staff that such questions were being asked at the highest ranks, but he wanted to see more "data-based and empirical work" on the topic first. Google still delivered information to the masses, he went on. "But I don't think it's reaching certain people at all," he added. More questions came in. Occasionally, Google's founders chimed in to punctuate a response, as when Brin offered, "Data suggests that boredom led to the rise of fascism and also the communist revolution." He paused, searching for the right words. "It sort of sneaks up sometimes, you know. Really bad things." Later, video footage of this meeting leaked to *Breitbart News,* held up as evidence of Google's bias against conservatives, a charge that would dog the company moving forward.

The next day YouTube staff convened for their weekly meeting in a courtyard dotted with eucalyptus trees. Live music usually played after the official presentation, once snacks and microbrews were carted in. Not this time. Instead, everyone went through a dazed and confused postmortem similar to that at Google's meeting. At one point an employee stood up to ask a question or at least make a point. The employee had analyzed data from the channels that uncritically cheered Trump, such as Alex Jones, under the guise of commentary or punditry. Bundled together, they had more watch time than legitimate news outlets on YouTube. *This is a crisis,* the staffer pleaded.

If YouTube brass agreed, they didn't say so. But a certifiable crisis came soon enough, and YouTube would have no idea what to do.

• • •

PewDiePie: "DELETING MY CHANNEL AT 50 MILLION."
December 2, 2016. 10:19.

"Can someone just stop YouTube from their self?" Kjellberg, scruffily bearded, stands in a small sound studio, where a neon sign of his bro-fist is mounted on the wall. He is practicing the budding art of kvetching about his internet home. "I feel like YouTube is a toddler playing with knives. Let's just take the knife away from that baby!" His issue, we learn, is a malfunction for his subscribers, who aren't seeing his videos. His

YouTuber friends are affected, too. YouTube clearly changed something but didn't tell anyone. And his views are down; some daily vlogs barely cross two million. "That's unheard of for me." *Jump cut.* "YouTube is trying to kill my channel." He will first. Once he crosses fifty million subscribers, he will pull the plug.

He did not. That threat ended up being a promotional gimmick for the second season of *Scare PewDiePie,* his YouTube Originals show. But the Lear rage from YouTube's king was genuine. A person who worked with him described the previous months as the "darkest" they had seen. The content grind had worn Kjellberg thin. At the start of the year, he launched his own YouTube network under Maker Studios, called Revelmode, gathering fellow YouTubers to shoot videos and run charity drives. He juggled that with *Scare PewDiePie* filming in Los Angeles and his own grueling production schedule. (Later, Kjellberg would tell fans he developed a daily whiskey-drinking habit to cope with the stress.) Maker pushed Kjellberg to extend his brand; he authored a paperback and began work on another YouTube series. Viacom called, offering him a Comedy Central show, but he turned them down, preferring to stick with YouTube. *Time* named him one of the hundred most influential people of 2016, posting a photo of him clad in a tuxedo at a red-carpet *Star Wars* premiere.

As Kjellberg floated closer to the mainstream, PewDiePie went further in the other direction. Starting in 2015, he grew bored with video games and transferred his shtick of playing patently absurd games into meta-commentary on the patent absurdity of the internet. His gaming videos, full of teenage boy humor, already toed the line of respectability. (His titles included "RUN LIKE YOU HAVE DIARRHEA," "THE GREAT HAND JOB," and many with the word "boobs.") When Kjellberg, like other YouTubers, began to feel the platform move under his feet as its systems tilted toward "ad-friendly" material and daily vlogging, he mocked that effort in the same spirit. His vlogging mixed earnest schmaltz ("ANNIVERSARY!")

with inanity ("DRINKING PISS FOR VIEWS," "I TRY POOP CANDY!" "I'M SO DONE"). Several videos he posted lamented the backwater of YouTube comments, a fair gripe. His complaint about subscriber glitches was also legitimate. YouTube, concerned about dormant and fake subscribers, had started cleaning up subscription counts but suffered a technical glitch and, the company later admitted, didn't communicate this well with creators. The sudden influx of TV networks accustomed to daily output placed YouTubers at a clear disadvantage under a system that craved daily views. (MatPat, the "theorist" YouTuber, made an animated video pointing this out that December, showing YouTubers tumbling off a treadmill while logos for TV talk shows jogged gamely ahead.)

Despite Kjellberg's mastery of the "Let's Play" format, in this era he consciously displayed a disregard for YouTube's algorithmic logic. No one was searching for poop candy or ways to drink piss. So his views dropped.

But Kjellberg stuck with his shtick, either to please his core audience or as comedic preference. He worshipped *South Park*, whose 2015 and 2016 seasons ribbed both PC culture and Trumpian bombast with a characteristic nihilism. (Trump and Clinton were portrayed respectively as a "giant douche" and a "turd sandwich.") *South Park* had a running gag about a Jewish character, which managed (debatably) to satirize cultural undercurrents of antisemitism. Online, though, this brand of comedy lost its polish and nuance. The alt-right and Breitbart army eagerly hurled insults and invective, while a brigade of online "shitposters" turned Pepe, a cartoon frog, into a hate symbol and often masked or excused their tactics as jokes. Some trolled for sheer thrills, while others were more politically calculating, a modern version of Nixon's "rat-fucking" fixers. "Like Trump's statements," the TV critic Emily Nussbaum would write, "their quasi-comical memeing and name-calling was so destabilizing, flipping between serious and silly, that it warped the boundaries of ordinary discourse."

The shitposters' cousins were the "edgelords," members of a web subculture who posted taboo topics to make some point or simply because they could.

Kjellberg embraced the edgelords, online and off. A former colleague recalled him joking in person about them being Jewish, like a kid tossing around the word "fag." On his channel PewDiePie reviewed "dank memes" and the topsy-turvy viral internet of Trump's candidacy. "YouTube at that time was a place where no one really knew where the limit was," Kjellberg later recalled. "A lot of channels were just pushing it as far as possible because there were no restrictions at the time." From the outside it was hard to tell what he actually believed.

Despite his antics, he seemed dedicated to preserving YouTube's integrity (or at least his vision of it). Many YouTubers felt the algorithm begin to put more weight on likes and comments as signs of engagement, evident in videos of boorish (mostly male) creators demanding viewers "Smash that Like button!" That December, Kjellberg spoofed this trend, flailing around his house shirtless, ranting about likes. For a brief moment while assuming this histrionic character, he threw up what looked like a Nazi salute.

Many who worked with Kjellberg insisted that he had no animosity or hateful beliefs. They described him as steadfastly loyal to his YouTube audience. (One person called him "a little spectrumy" in this monomania.) "He's a very kind person," said David Sievers, the early Maker Studios official. "Like many artists, he has an art. And like comedians who are practicing an art, not everyone gets it." In his videos Kjellberg slipped in and out of his "PewDiePie voice"—a gravelly shriek born in his gaming days. In one video, Kjellberg speculated that YouTube, the company, wanted to unseat him because he was a white man. While discussing Lilly Singh, a female creator of color whom YouTube's marketing had promoted, Kjellberg deployed the voice, imitating a conspiracy theorist. "I'm white. Can I make that comment?" he said. "But I do think that's a problem." This, Kjellberg explained in a follow-up video, was clearly an edgy joke.

The following month he tipped over the edge.

He had started a series of videos about Fiverr, an online gig-economy service that hired people to perform tasks for $5. Kjellberg wanted to see how far the service would go.

In one video he did his usual internet-commentary routine: he shared his screen with viewers and showed his reaction to its contents in real time. The screen showed a Fiverr account he had hired called the "Funny Guys," two young men from rural India. While laughing, the Funny Guys unfurled a paper scroll on-screen that read, DEATH TO ALL JEWS. Kjellberg covered his mouth in shock. Seconds ticked by. For a moment regret appeared. "I am sorry. I didn't think they would actually do it," Kjellberg said. "I don't feel too proud of this. I'm not going to lie. Like, I'm not anti-Semitic." He slipped back into the PewDiePie voice. "It was a funny meme. I didn't think it would work."

Still, he posted the video.

• • •

Later that month the Trump White House sent out a strange statement: a Holocaust Remembrance Day message, with no mention of Jews. Civil society groups criticized the obvious slight. Others wondered aloud if this was intentional, a coded message to an extreme fringe aligned with the new president. A reporter for *The Wall Street Journal* was curious about how this was being received on the far right and went to *The Daily Stormer*, an openly neo-Nazi web forum. There at the top of the website was a familiar face: blond, blue-eyed, Swedish. *The Daily Stormer* was advertising itself as the "#1 PewDiePie fan site." *What was the biggest YouTuber doing on a neo-Nazi website?*

Reporters at the newspaper dug through *The Daily Stormer* to find nine different PewDiePie clips that the site had highlighted as videos supporting its cause. This included the January video and another from a Fiverr clip Kjellberg had shown, where a man in a Jesus costume said, "Hitler did nothing wrong." Kjellberg occasionally used Hitler footage and Nazi imagery in his videos when pointing out some absurdity online. Another post on *The Daily Stormer* praised Kjellberg's haircut and clothing as coded fascist attire. The *Journal* prepared a story on the unsettling oddity of neo-Nazis endorsing a celebrity on the payrolls of Google and Disney. They tried repeatedly to reach Kjellberg for comment, and went to Disney and YouTube on Friday, February 10.

From there everything moved quickly, setting off a chain reaction that permanently changed YouTube and its biggest star's career.

That Sunday, Kjellberg released a short personal blog entry intended to bury the controversy. With the Fiverr clip, he wrote, he was "trying to show how crazy the modern world is." He admitted he had offended viewers but claimed it was unintentional, and he did not apologize. "I think of the content that I create as entertainment," the star wrote. "As laughable as it is to believe that I might actually endorse these people, to anyone unsure on my standpoint regarding hate-based groups: No, I don't support these people in any way." This wasn't enough. Disney wanted a public apology; the company had no interest in having its name in newspapers near any of this. At Maker Studios, Disney's digital arm, the weekend passed in a chaotic frenzy. It wasn't even their only PR disaster. (Another Maker star, the early vlogger Shay Carl, publicly confessed his bout with alcoholism and plans to enter rehab on the same day.) Bob Iger, Disney's chief, made it clear to Maker staff that Disney would stick with Kjellberg if the star apologized for his videos. Kjellberg refused.

Disney went to the *Journal* with its comment: the studio was ending its commercial deal with PewDiePie. The *Journal* article—Disney was dropping YouTube's king—was published late on Monday with a still frame of Kjellberg next to the DEATH TO ALL JEWS sign. Film editors for Kjellberg's production studio sat in a house in London slicing footage for his upcoming YouTube show, unaware of the storm. One editor pulled up the coverage online and, according to a person there, expressed the realization aloud: "Well, it looks like this company isn't going to exist."

YouTube initially told the newspaper that PewDiePie's videos didn't break its rules, arguing that the star was known for pushing the envelope. The company pulled ads on the video referencing Jews, but others *The Daily Stormer* praised went untouched. Videos "intended to be provocative or satirical" were in the clear while those inciting violence or hatred were not, a spokesperson explained. YouTube did not detail how it distinguished between the two. After the story was published, YouTube announced it was canceling *Scare PewDiePie* and removing Kjellberg from its premium ads tier.

The whole episode, like the media's initial circus-freak treatment of his channel, struck Kjellberg as utterly bizarre. When the story landed, on Valentine's Day, he was staying with his girlfriend at a rented cottage, where he opened Twitter to see a post from J. K. Rowling calling him a fascist. Trump's election had upended normal conventions of media and discourse and put everyone on edge. Kjellberg, a celebrity since his early twenties who professed an antipathy to politics, either hadn't read the room or hadn't heard the dog whistles in his work, perhaps because his audience was so insular. "You're in this area where everyone is on the same page," he said in a later interview. He also confessed that he was "pretty irresponsible" in handling the debacle.

At the time, though, he raged.

PewDiePie: "My Response." February 16, 2017. 11:05.

"It's almost like two generations of people arguing whether this is okay or not." Kjellberg unpacks his thoughts out loud in his studio. He does not blame Disney or YouTube. He blames the *Journal*, faulting it for seeing his satire as anything but. His Fiverr joke offended, sure, but he insists that it had been meant as a joke and so was the other stuff and "This is insane!" As the video rolls, his frustration builds. He mentions the newspaper's earlier fixation on his earnings. "Old-school media," he goes on, "does not like internet personalities because they're scared of us." He shows the recent *Journal* article on his screen, zooming in on the bylines. "I'm still making videos. Nice try, *Wall Street Journal*." He holds up his middle finger and sucks it. "Try again, motherfuckers." Then, tearing up, he thanks the "YouTube community" for its support.

Any meaningful conversations from the episode—how hate groups used, or distorted, pop culture; how two megacorporations profited from, or fueled, irresponsible satire; how jokes worked in the Trump era, or maybe didn't—were drowned out by the ensuing noise. PewDiePie loyalists (and bandwagon trolls) flooded the *Journal* reporters online with barbs, digging up information on their families. One reporter discovered that pixelated swastikas could appear in email subject lines. The newspaper had to hire private security after an employee and their family received death threats. A phalanx of YouTubers and fans, already skeptical of the mainstream press, grew even more distrustful.

Pundits wrote dozens of think pieces, but the sharpest take came on YouTube. Matthew Patrick, a.k.a. MatPat, released a video explaining why Felix Kjellberg's Fiverr stunt failed as comedy. YouTube, by its nature, blurred the lines between performer and persona. "It's often hard to see where PewDiePie ends and Felix begins," Patrick observed. And that particular joke's supposed target—the abject capitalism of the gig economy—carelessly mixed in shock antisemitism without explanation. Also, the joke punched down: a rich white celebrity made two unsuspecting Indian men its butt. "Risky humor needs to be humor done right," Patrick concluded. "As much as it sucks, words matter, Felix. Especially when you're reaching an audience of fifty million."

YouTube, the company, stayed quiet as all this unfolded. Its executives didn't discuss the episode publicly. Privately, Susanne Daniels, a former MTV executive whom YouTube hired to run its Originals, expressed frustration with Kjellberg's antics and the delay YouTube's leaders took in acting. "They moved too slowly and ineffectively," she said later. Robert Kyncl published a book on YouTube creators later in 2017 that compared Kjellberg's problems to Ted Danson's cringey blackface routine from 1993. The YouTuber, Kyncl wrote, "underestimated the responsibility he had as the platform's most popular ambassador, even if he himself is not a hateful person." Behind the scenes, however, YouTube tried to salvage any damage to its brand. The company arranged a call with Kjellberg; YouTube's policy chief, Juniper Downs; and the Anti-Defamation League (ADL), a prominent Jewish group. During the call ADL staff explained that extremists they tracked used anti-Semitic humor online to justify real violence, and simply casting the material as memes disavowed any responsibility. The group suggested Kjellberg make a public donation or apology to Jewish groups, perhaps a video about tolerance.

One person on the call remembered Kjellberg staying mostly silent, like a bored schoolboy at the principal's office. Nothing came of the meeting. YouTube's hands-off approach to its biggest star and its platform were starting to look irretrievably broken. And its brand problems were just beginning.

CHAPTER 25

Adpocalypse

A month after the *Journal* article on PewDiePie, Jamie Byrne sat in YouTube's Los Angeles office doing his best mea culpa. Byrne was YouTube's old man. Not technically, because he was in his midforties, though he looked younger, with a sprig of blond anime hair and the sun-kissed tan of a Venice surfer, which he was. But he had been at the company forever, since before Google, and knew its customs and tribulations. He had seen it all. He now served as a liaison to YouTube's top talent, which often meant apologizing for things outside their control, or his.

In this particular meeting Byrne was apologizing to a set of LGBTQ creators for YouTube's machines, which made a habit of punishing them. YouTube had long won praise as a bastion for LGBTQ progress. "It Gets Better," an online movement of uplifting testimonials aimed at queer youth, took off on the site. Coming-out videos, like the one from Ingrid Nilsen, had become a familiar genre and these creators exposed swaths of viewers to lives they rarely saw on TV and movies. YouTube wanted to spotlight this kind of material.

But other parts of YouTube kept kicking that effort squarely in the teeth. As YouTube pushed into schools, the company set automated filters to limit what might be regarded as "mature," and its shift toward "ad-friendly" fare further tightened filter dials. The result was a clunky automated system that placed many videos that addressed sexuality, or even mentioned words like "gay" and "lesbian," into Restricted Mode, a feature schools and libraries set to limit YouTube's catalog to cleaner fare. Videos in Restricted Mode were buried on the site.

Byrne had invited a group of prominent queer creators to YouTube's office, where he explained that the company was righting its system and that, honestly, it wanted these creators to make money. By then the PewDiePie fiasco had mostly quieted down, and Byrne ended the meeting feeling pretty good. His job, like others at YouTube, didn't require bouncing from one controversy to the next. Yet.

As he was packing up to leave the office, a colleague grabbed him and asked, "Hey, could you pop into this conference room for a second?"

Inside, Byrne heard news stunning even to him. Almost all of YouTube's largest advertisers were boycotting the site.

• • •

To understand YouTube's situation in March of 2017, you must back up a bit and take a good look at Marc Pritchard.

The marketers who bought the most ads on YouTube typically looked like Pritchard. Solid posture, firm handshakes, nice suits, fluent in tech jargon and TV ratings. Great teeth. Pritchard had worked for Procter & Gamble, the consumer goods giant, since 1982. He climbed its ladder to be chief brand officer and as such had two primary tasks: (1) market P&G's bountiful household products—Crest, Tide, Tampax, and so on—as efficiently as possible; and (2) make sure those brands never looked bad. Initially, Google and the internet made the first part far easier. An ability to target particular strata of consumers and reach the cool kids was a godsend for P&G. Pritchard leaned in. His firm's 2010 commercial for Old Spice (the impossibly sexy man in the shower) won awards and went viral on

YouTube. Wojcicki publicly praised "Like a Girl," an ad from P&G's Always. Pritchard controlled serious money, too: P&G spent $7.2 billion in marketing in 2016, the most of any business on earth. He had begun to do the thing YouTube desperately wanted: moving some of this money from TV to the web.

But the internet also took the handshake simplicity of Madison Avenue—*this* commercial runs on *this* show or *this* billboard—and mucked it all up. Back in 2007, Google bought DoubleClick and the iPhone was born and Facebook exploded. And an entire cottage industry quickly surfaced orbiting these things, existing only to facilitate the automated buying and selling of online ads. It resembled stock trading, and like Wall Street this industry grew needlessly complex. Companies offered auxiliary services to sell ads seconds faster, measure them slightly better, or blast consumers with messages at the exact moments they stood in front of an exact product in the shopping aisle. Fraudsters appeared, devising ways to generate fake website views and ad-clicking bots. This, in turn, inflated ad prices and bothered businessfolk like Pritchard who paid them. Even if actual humans saw his ads, there was the concern over "viewability." Facebook deemed an ad *viewed* (and charged for it) one way; YouTube did another. That was surmountable, but then Google and Facebook, as they went to war, began placing more restrictions on data they shared with marketers. Companies like P&G could not easily see how its web ads translated to sales. Google's algorithmic system for spraying ads around, already murky, became even less transparent. Madison Avenue called this a "walled garden" and hated it. "That became a very big bone of contention with our clients," explained Martin Sorrell, who ran WPP, a massive ad agency that placed commercials for P&G and other big brands. "Because they thought, when the web came along, that they would have direct access to consumers." Turns out, Google had that access. Also, compared with TV, Google offered paltry and inconsistent audience ratings.

So YouTube brass would stand beside ad executives in New York and France and wax on about creativity and tear-jerking commercials, but behind closed doors they would go at it over "viewability" and "measurability" and fraud. "Frenemies" is how Sorrell described the relationship. In January 2017, Marc Pritchard delivered a speech about the frenemies, explaining that he was tired of wasting money on "a crappy media supply chain." His industry now spent $70 billion a year online, more than on TV. "We've been giving a 'pass' to the new media, all in the spirit of learning," he said. "It's time to grow up."

Eleven days later *The Times* of London printed this front-page headline: BIG BRANDS FUND TERROR THROUGH ONLINE ADVERTS. The newspaper had found ads for Mercedes-Benz and a British supermarket on YouTube clips from ISIS supporters and neo-Nazis.

This did not stir immediate alarm at YouTube. The internet had always produced ad placement glitches. YouTube once had to genuflect to Toyota after the carmaker's ad showed up on a viral video of a woman crashing her Toyota into a supermarket. This sort of mistake never involved huge sums of money, and most of YouTube's paying clients knew these flukes were just the price of entry to an amazing automated global commercial colossus—or so the company thought.

And yet, for some time, Google's European salespeople had warned colleagues in private that its flippant, very American approach to speech, privacy, and media did not play well overseas. European staff were "banging on the windows" about ads running on troublesome videos, recalled one former salesperson. Still, Google was unprepared for the onslaught.

In March the company was summoned to the U.K. Cabinet to explain why British advertisers were endorsing terrorism and hate speech. Then a major French ad agency, Havas, told *The Guardian* it was pulling all its spending on Google properties over the issue. Sorrell, the WPP agency boss, fired a broad warning shot. Google and Facebook were "media companies," he told the newspaper, adding, "They cannot masquerade as technology companies, particularly when they place advertisements." A *Daily Mail* cover on the topic screamed, GOOGLE BLOOD MONEY.

This debacle quickly crossed the Atlantic. A *Journal* story the following week identified several brand names on vile YouTube videos, including a P&G ad sponsoring a clip titled, "A 6000 Year History of the Jew World Order." Fortune 500 companies avoided practically any association with politics, particularly its extremes. Trump made all this radioactive. At the same time, marketers were grappling for leverage in their fight with Google over standards and data. These stories, one after another, came right after the PewDiePie mess.

A dam broke.

P&G joined Starbucks, AT&T, Walmart, and dozens of YouTube's biggest advertisers in halting ad-buying until YouTube could offer "brand safety," a guarantee that their businesses wouldn't appear in newspaper reports as sponsors of terrorists or neo-Nazis. "YouTube is un-curated," WPP reminded marketers in a note, which said that the agency was working with Google to find a solution for "un-curated content, if there is one." YouTube had dealt with innumerable flare-ups over controversial videos, but this was the first to really hit its bottom line. A share price slide erased some $26 billion in value from Alphabet, Google's holding company, within a week in March. The company never shared the sales damage of the boycotts, which extended for months, but according to one person familiar with the figures, YouTube forfeited close to $2 billion in revenue.

YouTube immediately tried damage control. It apologized and pledged publicly to throw its "latest advancements in AI" at the problem. It offered rebates and rewrote rules to prohibit ads on any "dangerous or derogatory content." The company tried framing the issue as it framed everything: mathematically. Google's sales boss, Philipp Schindler, an energetic German, explained in an interview that "very, very, very small numbers" of ads ran on these troublesome videos, and a fix was hard. "Take the n-word," he offered. Removing ads from every video that deployed it would hit "a pretty significant percentage of all rap videos." Machines couldn't distinguish rap lyrics from hate speech. "Think about the scale of the problem that we're dealing with here," he argued. Wojcicki invited ad agency honchos over for a détente at Eric Schmidt's Manhattan home. Google tried calming one agency concerned that it had placed ads on videos from the North Korean government, a potential regulatory violation. (This was okay, Google explained, because the clips had come from Mexico, which had diplomatic ties to the Hermit Kingdom.) During these conversations YouTube leaders developed a metaphor: their site had exploded from a small village into a big city, without the traffic lights, zoning, or policing that cities need. Wojcicki tried out this metaphor with Marc Pritchard in a private meeting later that year. Pritchard disagreed. "You went to a large galaxy that was beyond what anyone had ever seen," he said. "I don't think you've realized the impact you've had."

"We want to be on the right side of history," Wojcicki insisted.

At other moments, though, YouTube cried foul. The whole ruckus struck the company as gotcha journalism. Google funding terrorists and Nazis was a helluva story. To keep it going, reporters simply needed to find more ads on vile videos. This wasn't hard: just search for the worst possible videos on YouTube and let the internet do its thing. One *Journal* reporter spent an evening typing racial slurs onto YouTube and immediately found videos mocking Michelle Obama that household brands sponsored. "They were basically hunting," YouTube's Jamie Byrne said later about these reporters. Privately, many at Google pointed to the particular newspapers involved; *The Times* of London and the *Journal* were both owned by Rupert Murdoch, a sworn enemy.

But that did not change the fact that YouTube, an advertising business, could not control where its ads ran and its advertisers had consequently lost faith. And finger-pointing would not solve YouTube's next daunting problem: telling creators why their money had disappeared.

• • •

As soon as advertiser boycotts began, YouTube's pricing algorithms reacted. With fewer ads to show, the algorithms dropped ad rates. When the company adjusted filters to remove ads from any remotely controversial videos, scores of creators making money on YouTube lost great amounts of it—some as much as 80 percent almost overnight. The company warned select creators that the dire situation could spell an end of payments to any YouTuber who wasn't tied to a media company or MCN. "Once we get through this, it will turn back around," Jamie Byrne told some marquee YouTubers. "But if we can't, you know—it's over."

Still, most creators were left in the dark. To them, YouTube seemed to assign economic value to videos by some mystical, veiled criteria. So they had to rely on oracles like Hank Green.

Green had barely aged since his YouTube vlogbrothers debut a decade before. He still wore glasses and had disheveled dirty-blond hair on a lanky frame, like a sprightly middle-school teacher. He filmed his own vlogs in front of a wall of books and a colorful, framed map of Narnia. He still lived in Montana. But he was no amateur. Green managed eight YouTube channels; several charity efforts; the VidCon conference, now held in three countries; and a twenty-person media company, Complexly. Green also had access to Oz. By then he was among a handful of creators whom YouTube brass spoke with personally and often. While Green didn't command PewDiePie-level numbers, he was composed and ad-friendly, a moral authority on the site that YouTube staff respected. And feared. The year before, Green had started the Internet Creators Guild, an effort to organize professional YouTubers. Green made YouTube sit up and listen.

For more than a month YouTube officials said nothing publicly to its creators. Green finally did. He sat in his regular vlogging spot in a striped hoodie, more unshaven than usual. (His first child had been born five months earlier.) Green's video essay unpacked the boycott, even cementing a term for it (Adpocalypse), but its real weight came from his careful, passionate attack on the very commercial model of YouTube and the internet.

vlogbrothers: "The Adpocalypse: What it Means." April 21, 2017. 3:54.

"Here's the crazy thing about YouTube advertising," Green begins. "Every ad that plays on a YouTube video—that person is getting paid like ten times less than if that ad played on TV . . . *Why?* Why are my eyeballs worth ten times more when I'm watching TV than when I'm watching YouTube? They're the same eyeballs, I promise!" And YouTube is active: the viewer is surfing, clicking, leaning in, not splayed on a couch. "There has to be a point where your audience is being valued so low by these people that you're just like, *No!*"

Green gets more animated: "Is this worth it? Is this relationship between content and brands—that was designed, by the way, for radio—is this a useless holdover from another century?"

Ten days later, on May 1, Green and about a hundred other creators arrived at the 1 Hotel Brooklyn Bridge, a swanky new eco-luxe New York hotel that offered a sweet view of the Manhattan skyline. YouTubers gathered in its ornate ballroom for the third annual Creator Summit, where the company had chosen to explain itself to its stars.

The event felt less like a summit than an insurrection. Green arrived in the ballroom thinking most fellow YouTubers understood the company's precarious business position. Not so. Wojcicki stood at the ballroom's front with her top deputies, Robert Kyncl and Neal Mohan, and creators grilled them on why one of the world's richest companies couldn't stanch the bleeding. A staffer there later recalled it as a "high-pitched, worrisome moment." One rising young YouTuber told the executives that her costs now outstripped her video earnings. "How am I supposed to survive?" she asked.

Casey Neistat, a breakout YouTube star, watched from the ballroom seats. A tireless filmmaker and showman, he ran a regular vlog featuring his family, posting well-crafted mini films as clean as daytime Nickelodeon. One hit showed him snowboarding through post-blizzard Manhattan streets in a bright red snowsuit with the YouTube logo. His fellow vlogger Philip DeFranco dubbed Neistat the "golden boy of YouTube." At thirty-six the golden boy was an elder statesman, and younger creators were coming to him in despair. At the summit Neistat asked why the company didn't do more to insulate its creators from fickle advertisers. The executives seemed taken aback by these questions, but argued that YouTube, too, was bent by financial forces beyond its control. "They responded to the tenor of the room in the best way that they could," Neistat recalled.

Internally, YouTube often described its main constituents—viewers, creators, and advertisers—as three legs of a stool, equal and sturdy. The stool was wobbling. Five years after YouTube had let every creator cash in, this system started to look like the tenuous, unsustainable experiment it truly was. During this entire crisis YouTube had righted the *other* issue Jamie Byrne worked on: releasing some twelve million LGBTQ videos from Restricted Mode. But as the boycott unraveled YouTube's creator business, this felt like a Pyrrhic victory. "You're focused on one area and you kind of miss something that's percolating over here," Byrne said later. "Maybe it's a fire. It just needs a lot of focus. So, you take your eye off something and *that* ends up turning into a much bigger issue."

The next fire didn't take long to appear.

CHAPTER 26

Reinforce

On a Friday morning in June of 2017, a month after the Creator Summit, a dozen YouTube employees piled into a drab, windowless basement room at a Los Angeles hotel. They were attending Stream, YouTube's all-staff off-site, where attendees enjoyed team-building exercises, a private Snoop Dogg show, and a Harry Potter theme park nearby. Wojcicki had pulled an unlucky few from festivities that Friday alongside her deputies, publicists, and a top engineer to discuss the sobering news: YouTube had a Code Yellow.

Tala Bardan* felt a little out of place in the room. She was a young junior staffer at Google, unaccustomed to meetings with execs. Raised in an Arab-American household, Bardan planned to pursue a doctorate until a friend suggested she apply to Google, which was desperate for fluent Arabic speakers. Bardan landed overseas on YouTube's "VE" team: company shorthand for those charged with moderating violent extremism. VE was part of Trust and Safety, the division dealing with controversial content online. One colleagued called the unit "a dumpster fire," rife with mismanagement and organizational chaos. Bardan's co-workers dubbed it "the burnout factory"; people exposed to such intense material, day after day, didn't last long. When she joined, she wasn't particularly familiar with Google or the web's carnal corners. During training she was shown a fetish clip of a man at a gym

* A pseudonym.

filming his toes, which curled to masturbation sounds. She learned about "Ivory Coast videos"—sexually compromising footage people uploaded to extort money. "I was raised in a conservative Muslim home," she recalled. "I was not prepared." She learned the YouTube ethos on speech: leave videos up, so long as they didn't show or incite extreme violence. "This is a platform," she was told. "We're not responsible for what's on it."

This approach took a turn that June Friday. A week earlier, three Islamist extremists had killed eight people in an attack on London Bridge. Reports soon showed that one of the murderers was inspired by an American cleric he watched on YouTube. Company management, stunned by the news, decided Trust and Safety needed an overhaul. Bardan was invited to the Code Yellow meeting because she was one of three YouTube staffers who handled violent extremism and spoke Arabic. She sat listening to the plan: YouTube would kick off radical clerics and assign more engineers to build stricter AI systems for rooting out as much Islamist extremism as it could. Bardan's team was immediately put to work screening videos that weekend, while colleagues continued their revelry. One teammate was awoken at 2:00 a.m. to deal with a particularly vexing Arabic video; the moderator watched the clip and fielded calls in a hotel bathroom to avoid waking a roommate. Bardan's manager rewarded her with cupcakes for the weekend work. At headquarters that week, YouTube's policy team whittled down a list of inflammatory names to fourteen, all Muslim men, who would be banned no matter what they posted. Before that month, YouTube had used an absurdist thought exercise to demonstrate its hands-off approach: *if Osama bin Laden had a cooking channel, that would be fine*. Not anymore. Anything with bin Laden would come down. Some videos of sermons, like those from Ahmad Musa Jibril, the cleric who reportedly inspired the London Bridge attacker, were put in YouTube's penalty box. It was a blunt decision. The new machine censors were blunt, too; a staffer who knew Arabic recalled sifting through the system's code to discover that its filtering terms initially included the Arabic word for Allah.

Later, YouTube executives would refer back to the Code Yellow meeting as a critical turning point, when YouTube decided to apply its superhuman intelligence for its moderation, not just for its recommendations. Bardan would remember that weekend differently. For her and a few colleagues, it was when they began to wonder if their system spent all its resources policing Muslims online, leaving extremists from other faiths or stripes untouched.

• • •

Weeks after the Code Yellow YouTube meeting, hordes of teens and tweens piled into security lines to get into VidCon. This VidCon looked very different from the MacGyvered first one held seven years before. The corporations had taken over. Its venue, the Anaheim Convention Center, a modernist hulk a few blocks from Disneyland, displayed three large banners advertising a YouTube Originals show with a noticeably diverse cast (Wojcicki had kept her promise to put a Black creator on a billboard). NBC and Nickelodeon, networks once anemic to YouTube, were event sponsors. Eight months later Viacom, trying to bear-hug the web, would purchase VidCon.

Security at VidCon 2017 seemed way tighter than before. A YouTuber musician had been gunned down during a fan meet and greet in Orlando the previous summer. And other forces charged the atmosphere. That June, Charlottesville renamed its Lee Park, prompting a gruesome rally of right-wing thugs.

Inside the convention center, Akilah Hughes, the creator who inspired the #YouTubeBlack event in 2016, watched four fellow YouTubers sit for a panel billed as a talk on "creating and engaging online as a woman." Many were there to see one panelist in particular: Anita Sarkeesian, a feminist author behind a YouTube series that unpacked sexist tropes in video games. This, coupled with her readiness to condemn online trolls, made her a regular target during the #Gamergate saga.

The VidCon discussion veered into this topic. As the crowd grew raucous, Sarkeesian pointed to a bearded man in the audience. "I think one of my harassers is sitting in the front row," she said. "I hate to give you attention because you're a garbage human."

Hughes knew of this character. Carl Benjamin, a prolific British YouTuber, alias Sargon of Akkad, had even received mainstream press for his treatment of a prominent woman; after a British politician shared how often she heard rape threats online, Benjamin mocked her thusly: "I wouldn't even rape you." Benjamin had made several videos lambasting Sarkeesian, often with her YouTube handle (Feminist Frequency) in his titles, an old search trick. In his videos he was careful not to break YouTube's rules by calling explicitly for harm, but his loathing was no secret. Hughes knew him from Trump's election. She had posted an emotional video days after the vote with footage shot from Clinton's election night event. It was not an impartial vlog: Hughes, devastated about Clinton's loss, wore a black sweatshirt with "Awful" etched in white. She spoke about her personal fears over changes to health policy and broader societal concerns that Black women and many others clearly felt. Sargon of Akkad re-uploaded her clip on his own channel with the derisive title "SJW Levels of Awareness." Hughes learned of this after an influx of missives came her way online—racial slurs and doctored images of her bleeding. She had inked a book deal, and her agent suddenly received confusing notes from strangers calling Hughes "the real racist" and demanding she be dropped.

At the VidCon panel, Benjamin and like-minded YouTubers had arrived early to fill the front rows. While the session ended without serious altercation, tension poured onto Twitter and YouTube, where creator spats were always mined for views. Hank Green was drawn in, and it dawned on the vlogger that his duties as VidCon host now required adjudicating conduct in cultural discourse. Eventually, he released a statement. He didn't want speakers calling audience members "garbage humans." But he also didn't appreciate YouTube acts that just stoked outrage and existed as "base-camps for years of harassment," prompting followers to launch vitriolic attacks "focused not on ideas, but on people (usually women)," Green wrote. "We are all watching as those techniques wear at the fabric of not just internet culture, but our whole world." He told Benjamin that he was not welcome at future VidCons. Benjamin took to YouTube to discuss his treatment, appearing on Joe Rogan's channel and posting a video on his own with a thumbnail of a cartoon depicting Sarkeesian as a sweaty, sharp-toothed Medusa. It would be one of his most popular videos.

For Hughes, the whole incident felt as if YouTube's comments and algorithms had sprung to life. Hank Green, at least, had done something about it. YouTube, she thought, just refused to.

• • •

YouTube *was* trying to improve its algorithms. But that year, as usual, the wild sprawl of YouTube's platform outpaced the company's efforts.

Clearly, YouTube knew its ads were broken. As boycotts dragged on, Wojcicki had given a public apology, in May, at the company's annual advertising showcase, where she promised marketers more fixes. But she also pitched her site's anarchy as an asset, declaring, "YouTube is not TV, and we never will be." (YouTube then announced a new series produced by Ryan Seacrest, the *American Idol* host, followed by live performances from Kevin Hart and James Corden, stars from TV.) Back in San Bruno, anarchy wasn't welcome. Wojcicki had assigned a team of engineers to dig a way out of the boycott. This was a Code Orange—not quite emergency Red, but close. Programmers made tools to generate reports for YouTube advertisers showing when their commercials ran and how much they spent. "It was a little bit chaotic," recalled one person involved, who said YouTube's moderation unit was "grossly undersized for the task." YouTube threw more resources at the effort to prevent sponsorship of troubling videos and gave it a cheeky name: Project MASA (Make Ads Safe Again).

Other engineers were assigned to adjust algorithms under the new mandate after the London Bridge attacks. Any videos with "inflammatory religious or supremacist content" that butted up against its rules, without breaking them, were put in the penalty box. YouTube would develop even stronger tools to bury videos in search results, a process called "whistling" internally, according to a former executive, who described it as a method to "deep-six" videos without deleting them. The company had become more sensitive to removals after criticism from human rights groups that YouTube's blunt moderation was scrubbing valuable archives of war crimes as collateral damage.

Meanwhile, YouTube's top engineers considered what to do with unsavory videos that weren't "inflammatory or supremacist content." Advertisers might shun these clips, but it was harder to write clear rules and code explaining how to detect them. YouTube's solution was to measure *satisfaction.* Plenty of reels were made slovenly, on the cheap, and, engineers believed, viewers could tell. So they used a feature to let viewers rate videos afterward with one to five stars. Andre Rohe, a YouTube engineering executive, later explained this logic. Imagine a video promising to show you the "Ten Deadliest Animals." You clicked, probably even watched for a while. But what if the video failed to deliver on the goods? "After you watch it, you're like, 'Oh boy, that was a wasted seven minutes,'" Rohe said. *One star.* That data point fed into an algorithm, which, in late 2016, began using these surveys and thumbs next to videos to gauge *satisfaction.* To fix its quality crisis years before, YouTube had switched its gears from views to watch time, but that didn't cut it anymore. (YouTube never specified the precise equation for its ranking system to outsiders.) When videos suggested the earth was flat or that vaccines cause autism or that feminism ruined society or democracy needed an antiseptic, YouTube engineers thought its viewers would surely share their distaste and translate this antipathy into the satisfaction metric. Initially, YouTube thought it could root out troubling videos this way, Rohe said years later, after it was clear YouTube couldn't do that.

At the time these algorithm caretakers did wisely surmise that creators distrusted YouTube's formulas chiefly because they were cloaked in mystery. YouTube decided there should be videos demystifying the algorithm. Staff invited in Derek Muller, a flourishing EduTuber whose channel, Veritasium, was known for unpacking dense topics. Muller narrated his videos on-screen; he had solid biceps and a neat black beard and looked like the comic Nick Kroll, only handsomer. Muller had noticed YouTube's Oz routine from his earliest days on the site. *They're doing weird stuff inside that place*, he thought, *and we're sort of riding on this wave.*

During a meeting at YouTube's office, engineers unpacked the machinery for Muller. He was uneasy when they explained how this system was less likely to recommend a creator's videos to someone if a viewer hadn't watched their last video or any over the prior month. *Wouldn't that punish infrequent creators?* He didn't like their explanation about weights the algorithm placed on "session time"—how it valued videos that led immediately to further watching, not those that led elsewhere. If one of his videos on quantum physics or black holes inspired people to search for those topics, shouldn't he be credited for that? This system certainly explained why he saw *TMZ*-style YouTube drama videos going gangbusters. "Guys," he said. "These sound like terrible ideas." He didn't make a video explaining the algorithm.

In August, YouTube released its own. The company's video described the algorithm as a "real-time feedback loop" and offered creators an aphorism that was already gospel inside YouTube: "The algorithm follows the audience." *The audience is king.*

This was a half-truth. As the previous five years had shown, YouTube's algorithmic tweaks determined not only what was popular but what kind of videos were *made*. Also, when the company wanted to, it went in and turned the dials. Consider *Minecraft*. After the watch-time transition, YouTube's audience clearly loved *Minecraft*, heaving the niche game into the mainstream.

At one point, in May 2015, fourteen slots on YouTube's logged-out home page—the version of its site for people not signed in with Google accounts—were devoted to *Minecraft* game play, according to a meticulous video from the YouTuber MatPat. By June, MatPat counted seven slots for *Minecraft*. By September, *Minecraft* disappeared. This sparked a theory in YouTube-land that some company suit, outraged by the abundance of *Minecraft*, ordered up the change. Cristos Goodrow, YouTube's senior engineer, denied this, explaining that YouTube had instead determined that it needed a more broadly appealing welcome mat, so it tilted the algorithm to show videos that "everybody from my daughter to my mom might like."

Either way, traffic on *Minecraft* channels crashed. Had the audience tired of these? Maybe. Maybe not. "Humans watch what's presented to them," MatPat observed in his video. "What's right in front of their faces."

On another occasion YouTube decided it placed too many outrageous and gross-out videos on the home page. So, the company silently tweaked that. Within YouTube the computer model built to do this was called the "trashy video classifier." (Engineers working on this were part of the "trashy clickbait team.")

Still, for the most part, YouTube was satisfied with its system. By late 2017, YouTube's recommendations were running on a new version of software from Google Brain known as Reinforce, named after a branch of AI (reinforcement learning). During a conference that year, a Google Brain researcher described Reinforce as YouTube's most successful new service in two years. It sent overall views up nearly 1 percent, a huge sum given YouTube's size. *The New York Times* later described this recommendation system as "a kind of long-term addiction machine." But YouTube didn't think of it that way. That August the company allowed several staffers to be interviewed by *The Verge* for an article on how YouTube "perfected the feed." Some three hundred tweaks were made that year to tee up personalized videos just so. And, the company boasted, viewership on home-page videos had increased twentyfold over three years. Eventually, YouTube would share that more than 70 percent of its views came from recommendations. YouTube used to take days to inject viewing habits into its algorithm, but now it took just hours or minutes, a YouTube manager told the outlet. It was good at serving people what they wanted. "There's stuff that's closely related to what you already liked, and stuff that's trending and popular," the manager said. "But in between, that's the magic zone."

The algorithm was unbothered by what slipped into the magic zone. The algorithm gauged watch time and satisfaction.

When more than five hundred citizens descended upon suburban Raleigh that November for the inaugural Flat Earth International Conference, an attendee told a curious BBC reporter that she had been convinced of the planet's flatness after viewing more than fifty hours of video. "When you're watching videos at home, it's just you and the screen," remarked a man identified as Happy from Virginia. One conference speaker recounted a YouTube journey he and his girlfriend had taken: "We got into the Bilderberg, Rothschilds, Illuminati. All these general things that one ends up looking into when you go on here, because you look at one video and then another suggestion pops up along the same lines." None of these people seemed dissatisfied with footage they had seen.

Certainly some viewers also didn't register dissatisfaction the following January when they watched a livestream debate between Sargon of Akkad, who played a "classical liberal," and Richard Spencer, an avowed white nationalist. YouTube had placed a "Trending" tab at the top of its home page, an algorithmic collection of red-hot clips, and for a brief moment that month this debate was the No. 1 Trending video.

YouTube would eventually bar flat-earth videos and debates like that from its promotional system as "harmful." But not yet.

Early in 2017 the company invited in a reporter from the business magazine *Fast Company*. Given the rapid social upheaval to come, the resulting article quickly became a fascinating relic. A reporter watched Wojcicki at her all-staff meeting in March welcome ten new recruits (NewTubers). Everyone listened to a presentation from Google Brain on recommendations. *Fast Company* described the scene thusly:

> With charges of sexual harassment at Uber in the news, [Wojcicki] urges employees to report any untoward incidents at YouTube to whoever they feel most at ease speaking with, up to and including her. But she also wryly shares her account of attending the Oscars for the first time the previous weekend (as a guest of producer Harvey Weinstein), which culminated in her guiltily eating a cheeseburger at the *Vanity Fair* afterparty—even though she's a vegetarian.

Once ad boycotts hit, the magazine had a front seat to YouTube's reeling response. Wojcicki lamented that creators were suffering and promised new customer support features, but she also leaned into her service's blemishes. "There's something very human about YouTube," she told the magazine. She sat for an interview in her office at Google HQ, where she worked one day a week, and proudly displayed a small sculpture her nine-year-old daughter made for her of Tinkertoys and cardboard, with little inspirational slogans: *Fairness is for everyone. Don't go backward, go forward. I see the future in your eyes.*

• • •

Down the corporate ladder, Claire Stapleton felt like her company was tipping sideways. She sent off her "Down the 'Tube" newsletter that May weeks before giving birth to her first child. She had left her role curating YouTube's Spotlight, where her managers hoped she could "shape the conversation" on the site. But, to Stapleton, YouTube's enormity made this like "putting a thimble in a gushing geyser." She transferred to the team managing YouTube's social media accounts and, when she couldn't avoid it, dipped onto the site for her guilty pleasure: teen mom vlogging. There was always a new video diary to discover that the algorithm served up, each mom seeming a little younger, a little more performative, a little more extreme.

She opened that May newsletter with a GIF from *The Handmaid's Tale*, a new symbol of women's resistance under Trump. Stapleton and her peers knew Silicon Valley's sexism flowed through Google and workplace dalliances were common. But many at Google believed tech-bro culture was a creature of younger, reckless companies like Uber and that bitter partisan squabbles happened out there in flyover country, far from its solar-paneled campus.

This shared illusion shattered that summer. James Damore, a mid-level Google programmer, sent around a ten-page memorandum titled "Google's Ideological Echo Chamber." Conservatives at Google felt "alienated," Damore wrote, though his central argument was that diversity hiring goals were bunk because they did not comport with his read of the science on gender. He submitted his essay first via "skeptics," a company listserv known for touching third-rail topics. By August his memo had spread company-wide and leaked out. During the summer doldrums it was treated as very big news. Sundar Pichai, Google's CEO, who was on vacation once this scandal boiled over, dismissed Damore, further fanning the flames.

A lurid culture war, seeded on talk radio and cable TV and then ripened on YouTube, had now landed inside Google.

Newspapers and TV stations clamored for an interview with the canned programmer. Damore granted his first two to his favorite YouTubers: Jordan Peterson, a psychology professor who courted controversy and had a huge YouTube following, and Stef.

Stefan Molyneux: "Google Memo Fired Employee Speaks Out!" August 2017.

Damore joins via video call from his apartment, white earbud wires dangle down his long, boyish face. Molyneux's familiar pate is on the right of the split screen. "Give us a sense of your intellectual growth," he says cheerily. "I really like understanding things," Damore replies, grasping for the right words. "In these environments where everyone is just in the same echo chamber, just talking to themselves, they're totally blind to so many things." Molyneux likes this, particularly the part about Google being an echo chamber. They discuss coding and libertarianism. When the YouTuber discovers his interlocutor is not terribly talkative, he fills the air. Molyneux compares Damore's memo and its discontents to forefathers of modern science. "It's the Galileo thing. '*But it moves*,'" Molyneux riffs. "I can't believe they call your stuff pseudoscience. No, no, no. Diversity is a pseudoscience." Damore laughs nervously.

Damore's memo was filled with references to evolutionary psychology, an academic minefield that Molyneux loved. Once examined, though, Damore's analysis fell apart—"at best politically naive and at worst dangerous," *Wired* wrote. A researcher Damore cited called his claims on sex differences "a huge stretch." Google was also facing a fresh federal investigation into "systemic compensation disparities against women" at the company. Damore's memo certainly didn't help.

To distance the company from it, Google deployed Susan Wojcicki. She wrote a note to YouTube staff, which the company shared publicly. It began with a question her daughter posed: "Mom, is it true that there are biological reasons why there are fewer women in tech and leadership?" This question had "weighed heavily" on Wojcicki for her entire career, she continued, noting how reading the memo resurfaced pain. Yes, Google supported free speech, "but while people may have a right to express their beliefs in public," Wojcicki wrote, "that does not mean companies cannot take action when women are subjected to comments that perpetuate negative stereotypes about them based on their gender." To female creators who had been subjected to negative comments over and over again on Wojcicki's platform, this line from the CEO probably sounded tone deaf. In a subsequent interview Wojcicki was asked about Damore's appearances on YouTube. "That's fine," she said. "We enable a broad, broad range of topics to be discussed, from all different points of view."

Stapleton was out on maternity leave during the Damore episode, but several outraged co-workers weren't. They felt attacked and spurred to do something. These Googlers, mostly women, began building a network, convening over encrypted apps and in person, chalking up gender imbalances they saw at Google. They chalked up other grievances against their company, piling them up one after the next.

CHAPTER 27

Elsagate

Greg Chism loved YouTube. When he first discovered it, he was in a tough spot. The single father of two young girls in southern Illinois, he survived on his lawn-care work, his profession for years, same as his father. Chism had a long face, with little splotches of hair on his head and chin. For most of his life he had severely crooked teeth, which made him insecure. Until, just shy of forty, he discovered YouTube and inspiring videos about life improvement. He got braces and started working out. He began a YouTube channel about lawn care, shooting on his janky Motorola phone, and found fellow obsessives. "Freaks," he called them. "I got a sense of community," he told a fellow YouTuber. "A sense of 'I'm not alone in this world.'"

He started posting family videos—clips of his daughters around their house unboxing toys. He tinkered with video titles, the tags, the material. He called his channel Toy Freaks. "I started seeing a pattern—these certain videos were getting more views than the others," he said in 2015. "It's creative. It's rewarding. It can also be financially rewarding, too. YouTube is absolutely amazing." By 2017, Toy Freaks was extremely rewarding. It hit YouTube's top viewership charts (No. 68) and entered its premium ad tier, raking in serious money. Chism was able to move out of a trailer and into a house. He kept following YouTube's patterns. He wore pacifiers with his daughters in videos, playing out scenarios called "bad baby" that got bonkers traffic. He ate gigantic sweets with his kids. He played pranks on them.

YouTube sent him one of its gold Play button plaques, its reward for crossing a million subscribers. The company flew him out to California for an event and assigned a partner manager and gave him the red-carpet treatment. No one complained about his videos—until suddenly everyone did.

• • •

In the years since YouTube shuttered its effort to categorize videos as Delicious and Nutritious, kids' content had mutated into something deeply strange, something beyond categorization. Like every trend on YouTube, creators spotted this one first.

Ethan and Hila Klein were YouTube mainstays behind h3h3, a channel that adapted the absurdist stylings of *Mystery Science Theater 3000* to offer a profane tour through YouTube's backwaters. The company never showcased h3h3 on its stages, but many YouTubers watched religiously. "Today, we're exploring the weird part of YouTube, guys," Ethan Klein told viewers back in the spring of 2016. The Kleins had pinpointed a phenomenon quickly moving to YouTube's foreground: adults dressed as superheroes performing for child audiences. One channel in particular, known as Webs & Tiaras, mastered this form with keyword-mash videos such as "Spider-man & Frozen Elsa vs Joker! w/ Pink Spidergirl Anna & Batman! Superhero Fun in Real Life:)." By June 2016 that channel ranked third in all YouTube-land traffic, trailing only little Ryan and T-Series, an emerging Bollywood hitmaker. The Kleins marveled at this unbelievable viewership. A few months later they highlighted the trend again.

h3h3: "Toy Channels are Ruining Society." January 25, 2017. 13:08.

Ethan Klein sits at his computer screen telling a story. A goonish YouTuber pair, a favorite subject of his, used to post pranks. "Until they discovered the new gold rush," Klein narrates. "Spider-Man and Elsa." We

> see the video in question: ragtime music plays as four adults bop around a room with a pool table. Two are dressed as Spider-Man, one as Hulk, and one as Elsa. A child is there, also costumed. One Spider-Man starts groping adult Elsa's fake breasts. "Wow. This is cool," Spider-Man says. Klein shows more of the channel's work, each thumbnail a candy-bright yellow backdrop, with Elsa and Spider-Man in various lewd acts and stages of undress. "This is your brain on YouTube, kids," Klein concludes. He then begins to fake masturbate to the footage.

Klein's targets of scorn were chasing the masters of this new format. Webs & Tiaras operated out of Quebec City and featured actors performing vaudeville antics dressed in cheap Halloween store getup on a drab row-house street. They staged plots without speaking, usually a romantic narrative between Spider-Man and Elsa, the damsel in distress. The channel's owner identified himself only as Eric, a pseudonym. YouTubers like the Kleins suspected bot traffic. Just as likely, Webs & Tiaras exploited a perfect algorithmic storm: a huge surge in kids' programs plus a continued vacuum of mainstream fare. ("Baby Shark Dance," the earworm YouTube sensation, came out in 2016.) Because *Frozen* and superhero franchises didn't live on YouTube, any parents or kids typing "Elsa" or "Spiderman" into the site were shown popular entries from Webs & Tiaras. Again and again. "Some of these are probably seen by the same child fifty times," Phil Ranta, an executive with Studio71, an MCN that signed the channel, told a reporter in 2017. "It really helps to juice those numbers."

Webs & Tiaras began trying to ape the success of little Ryan, who, by then, was appearing on-screen with his infant sisters playing with boatloads of toys and gigantic foam foodstuff, earning hundreds of millions of views. At some point Webs & Tiaras hit on the popularity of costumes and a strange pairing of two popular kids' search terms. Like any good YouTuber, "you just keep repeating the thing that went viral," Ranta later explained. Ranta didn't mind the anonymity when he signed Webs & Tiaras.

Many kids' YouTubers preferred that, best to avoid any hiccups with children's privacy laws or simply because their chosen trade was exceptionally weird. Webs & Tiaras was deeply weird: videos depicted costumed Elsas with chicken feet and with a "brain belly." But Ranta, a former stand-up comic, insisted that the channel was "pretty harmless," operating like old silent films or cosplay theater for kids. And its plots, like placing characters behind bars, were catchy. "You're a little kid and you're like, 'Wow, I love Elsa. I love Spider-Man. What? They're in jail?'" offered Ranta. "'That's a story I've never heard before.'" *Click.*

With its success came a wave of imitators. Some borrowed tropes from YouTube pranksters, another hot trend, who competed in brinkmanship games for carrying out the most absurd stunts. When pranksters joined, the superhero genre got even weirder. Elsa flushed Spider-Man down toilets, "evil Santa" kidnapped Elsa, Spider-Man injected Elsa with strange liquids. Elsa often gave birth. "You half expect the scenarios to be porn setups," a blogger wrote about the trend in February 2017. A few months earlier, at Maker Studios, David Sievers issued a report for lawyers at Disney, the studio behind Elsa. All of Disney's promotional clips on YouTube, not including Maker hits like PewDiePie, generated about a billion views a month, a solid sum. Amateur videos featuring Elsa, Sievers concluded, had *thirteen billion* monthly views.

That year Harry and Sona Jho, the Mother Goose Club creators, noticed that one of YouTube's top-trending terms was "bad baby." That category included benign animated clips of defiant toddlers and gross-out live-action fare showing kids overeating and puking. Toy Freaks specialized in the latter, with Chism pranking his two school-age daughters dressed as infants. One video showed one of the girls wiggling loose a tooth, screaming, and spitting blood. (In the clip Chism calmly reassured his screaming daughter, but this wasn't evident to people who only saw the video's bloodiest stills.) Toy Freaks kept climbing YouTube's charts.

In March the BBC ran a damning story. Shocked parents had found their toddlers watching violent, nightmarish clips on the YouTube Kids app, marketed as a safe space. An off-brand Peppa Pig tortured at the dentist. Mickey Mouse playing pranks with feces. Minnie Mouse dismembered and bloodied. YouTube's machines saw these only as children's cartoons.

Neural networks, the superhuman AI plugged into YouTube's recommendations, were often described as "black box" systems since they operated in ways humans couldn't fathom. To many at YouTube, the flood of disturbing kids' material reminded them that they didn't have the box's combination. "It took on a life of its own, and no one was really minding the store," one person at the company recalled.

It soon seemed as if everyone watching knew about this mounting disaster except YouTube itself.

• • •

"Bad actors." That's the term Google used for spammers, hackers, and election meddlers—those who made the internet unsafe. On YouTube, people who made bloody Minnie Mouse cartoons, people who took advantage of the company's loopholes or generous parameters, people who posted "problematic content"—those were bad actors.

During the summer of 2017, a team at YouTube started looking closely at videos aimed at kids that felt *problematic*. The company was still wobbling from its ads boycott, but it had updated commercial guidelines and built models to better identify terrorist videos and, the gods of AI willing, to bury that problem for good. One person hired for the team that spring was told that her job soon "might not even be needed anymore."

Once they looked under the hood, this optimism dissolved.

Toy Freaks was not alone. Chism's success had inspired dozens of mimics ("replica content," the company called it). Some used "keyword stuffing" to ride algorithmic waves like "bad baby"—an old spammer's tactic of filling a video with unrelated tags, for machines' eyes only. When YouTube staff watched "bad baby" videos featuring minors, some felt a sickening discomfort. Several of the videos followed a trend of shaving young children's faces on-screen as punishment. (Real or fake shaving? It wasn't clear.) Others had kids gorging to show distended bellies, a trope from porn. The company had long had rules against child exploitation and sexual fetishes; these videos didn't break them but came close. For years YouTube had relied on parents to steer children to YouTube Kids, yet the app's relatively meager traffic showed this wasn't working. Millions of unsupervised kids were not staying off YouTube.com.

Staff invented a new category ("borderline fetish") and wrote policies for moderators and machines to detect videos that fell into it. YouTube made another label for footage that mixed children's characters with "adult themes"—the screwy Peppa Pig fare and legions of Spider-Man–Elsa mashups.

They moved cautiously. While YouTube staff envied Facebook, they liked to say they didn't "move fast and break things"—Facebook's old motto. In the wake of Trump's election, that motto was synonymous with the social network's careless wreckage of democratic norms. When YouTube decided to ban videos giving firearm instructions or selling guns, the company took six months to write rules, stress test them, and develop enforcement protocols. They worried about collateral damage. AI scientists called this "precision and recall." If machines were trained to look for bomb-making videos, they needed to find enough of them (*recall*) without accidentally axing news footage or World War II documentaries or Wile E. Coyote cartoons (*precision*). Google preferred using AI models with high precision and recall. When YouTube staff looked into the strange kids' material, they struggled to draw guidelines for their machines. What about *Adult Swim*–style cartoons? What about parodies? They had a heated debate about one vulgar spoof of *Frozen*'s "Let It Go" before deciding the clip didn't make it clear it wasn't intended for kids. What distinguished people in Spider-Man–Elsa mash-ups from cosplaying Comic-Con attendees? Humans could hardly tell them apart. *How could machines?* And what about Toy Freaks? Couldn't families post fun home videos? Greg Chism said he was the children's dad, though his videos would sure look different if he wasn't. YouTube couldn't fact-check parenthood; it certainly couldn't *at scale,* across thousands of channels.

YouTube was also worried about upsetting its creator class even further. YouTubers had grown angrier as the Adpocalypse dragged on and began to see more ways the platform was stacked against them. After a gunman killed nearly sixty people in Las Vegas that fall, Casey Neistat made a clip supporting a charity for victims, telling viewers ad money made on his video would be directed there. YouTube's systems deemed his video unsuitable for most advertisers, citing the topic, but a clip on the shooting from ABC's Jimmy Kimmel ran ads without a hitch. "It sort of reeks of hypocrisy," Neistat said in a follow-up video titled "DEMONETIZED DEMONETIZED DEMONETIZED." "A lot of the community is unhappy with YouTube," he explained, holding up a drawing of a frowny face for good measure.

In September, Wojcicki scheduled a meeting at headquarters with engineers, publicists, and Trust and Safety personnel, who joined in person or over videoconference. "We're always in crisis," Wojcicki told her staff. They were ordered to come up with plans to handle "problematic content" faster and communicate better with creators. Tech companies often named crisis response operations "war rooms." This new group, executives decided, was to be a "constant war room." One of its first battle plans involved purging the site of "borderline fetish" material like Toy Freaks. Enough people had watched this footage and determined that children were being either ordered to appear on-screen or placed in uncomfortable situations.

By then, Neal Mohan, YouTube's product chief, had started taking a bigger role, appearing at events to field fire for Wojcicki and overseeing more of her company's policing efforts. In October, Mohan signed off on this war-room battle plan. Still, YouTube staff wanted to roll out this new plan carefully. So they waited a few weeks.

• • •

By that fall household brands and ad agencies were gingerly bringing their money back to YouTube. Holidays neared, cutthroat marketing season. And YouTube promised safer ground on Google Preferred, its higher-priced programming slate. Still, some agencies, fearing any further missteps, had started to audit their YouTube ads just to be sure. During one audit that October, a Manhattan agency executive found something disturbing.

The agency had run ads on scores of kids' videos—not surprisingly, because kids were good for holiday sales. But some videos were weird, *off.* Fetishy. There were videos about "learning colors" with a child tied up in colored tape, and another with kids and adults wearing pacifiers and Speedos. Several had kids in swimsuits. Some comments below the footage were obviously sexual—innuendo about milkshakes and whipped cream. To sponsor these videos, YouTube charged rates close to those for TV. The executive alerted contacts at Google, pressing them on why they didn't screen premium footage.

Google reps, usually chatty partners, came back with a form reply, reading it "like hostages," recalled the agency executive, a veteran ad buyer. In a subsequent email Google representatives explained that "user interest" determined footage in Google Preferred. The audience was king. "It's about high-popularity engagement," the email read. "However, we understand that there may always be content that some advertisers might not be comfortable with." Google offered a refund.

This all played out in backroom talks until November 4, when *The New York Times* ran a story on "startling" videos in the YouTube Kids app, like a *PAW Patrol* knockoff showing cartoon pups demon possessed. "My poor little innocent boy," a mother of a three-year-old viewer lamented. The *Times* printed a still from a channel called Freak Family, a "bad baby" clip of a small girl, looking distressed, receiving a forehead shave from a razor, splotches of red on her face. On YouTube's main site this clip had tens of millions of views. A YouTube director told the newspaper that less than ".005 percent" of videos in the Kids app in the last thirty days were flagged as inappropriate, calling them "the extreme needle in the haystack." But the earth had shaken. And two days later, an avalanche hit.

James Bridle, a British author who wrote about drones and warfare, had turned their attention to kids. Bridle published a very long entry on the blogging site Medium with a catchy title, "Something Is Wrong on the Internet." Bridle's writing was crisp and detailed, but their visuals told enough of a story. Bridle's post first displayed stills from surprise eggs unboxing, nursery rhymes, Peppa Pig fakes, categories with tens of billions of views. Next, the Finger Family sensation ("at least seventeen million versions" on YouTube). Many looked robot generated, but it was impossible to tell. "This," Bridle wrote, "is content production in the age of algorithmic discovery—even if you're a human, you have to end up impersonating the machine." Scroll down the article, and everything looked worse. Still after candy-colored still, disturbingly identical, each tailored for YouTube's algorithm: "bad baby" offshoots, demented cartoons, and even more surreal material ("wrong heads"—disembodied Disney figures floating on-screen). There was Toy Freaks, which, Bridle noted, showed children vomiting and in pain, and a vast field of Toy Freaks replicas mixing pranks with Spider-Man–Elsa–superhero strangeness. "Industrialised nightmare production," Bridle called it, before adding the kicker: "To expose children to this content is abuse. . . . And right now, right here, YouTube and Google are complicit in that system."

Longtime YouTubers like the Jhos and the Kleins had seen these trends rise, but most people, parents of toddlers, even Google employees, had no idea this sort of material existed. Within YouTube, which tracked everything online, staff saw a huge, disconcerting spike in Twitter activity about Bridle's post. With Madison Avenue looking skittish again, this felt like another brewing fiasco. Newspaper reporters seized on Bridle's piece, chasing its one recognizable face: Greg Chism, the Toy Freak dad. *The Times* of London wrote a story about advertisers who appeared on his videos angrily pulling their money. Article title: CHILD ABUSE ON YOUTUBE. Subtitle: GOOGLE MAKES MILLIONS FROM DISTURBING VIDEOS.

• • •

That CHILD ABUSE article was the breaking point for Sridhar Ramaswamy. He ran technical operations for Google ads, holding the senior role since his spat with Wojcicki four years earlier. Ramaswamy, a blunt engineer, oversaw the complex systems of auctions and exchanges for Google and YouTube ads, but had to make many decisions about YouTube jointly with Wojcicki. Ramaswamy was certainly involved in the boycott cleanup to Make Ads Safe Again, yet he was not aware of Toy Freaks, its popularity, or its armies of replicas. That London *Times* article, he later confessed, made him decide to exit the advertising business for good. He would leave Google the following year.

In the wake of the *Times* article YouTube considered something unprecedented. On a weekend that November, a small group of executives, including Wojcicki, Mohan, and Ramaswamy, convened over a video call from their homes to strategize on their debacle. YouTube had begun to consult outside experts on these videos, and experts told the company that certain material—colored tape tied on kids and comments left on gymnastics clips—was clear code for pedophiles online. Ramaswamy asked on the call if this was something Google wanted to be associated with at all. YouTube sold two types of ads: "direct response" (coupons and offers) and "brand" (traditional commercials for companies like Tide or Chevy); brand ads brought the most money. At one point in this meeting, Ramaswamy proposed ending all brand ads on YouTube, a potential multibillion-dollar loss, until the site righted its kid problems.

Ultimately, the group voted on a different drastic move. YouTube, which once fought vigorously to keep even some controversial videos up, nuked hundreds of thousands all at once. A few days before Thanksgiving the company removed ads from more than 2 million videos, took down more than 150,000 videos, and deleted more than 270 accounts, including Greg Chism's two channels, which had thirteen million combined subscribers. Some 50 Toy Freaks replicas also came down.

Phil Ranta woke up to frantic calls from his superhero actors in Quebec. Webs & Tiaras was gone. All his efforts to contact staff at YouTube failed. Panicked YouTubers phoned and emailed Melissa Hunter, the doll-review mom who ran a kids MCN. In YouTube's urgency it had deployed its AI filtering system for all problematic children's videos without well-tuned precision and recall settings. "They just went at it with a machete because they had to," recalled Hunter. "That was a really hard time."

April and Davey Orgill, prodigious family vloggers, had added a superhero parody channel to their YouTube repertoire earlier that year. Their dress-up acts with their kids earned more than two million subscribers, but they quit in August. "The videos just started getting weirder and weirder," April told viewers about the genre. "*Ew.*" Still, they kept the channel up, raking in eyeballs and ad money. On November 23, YouTube approved two of its videos for advertising. A day later the channel was gone. "YouTube blames it on these people that were doing it, but for a year their algorithm pushed this content," Davey Orgill told a reporter. "They created a monster." Someone gave this monstrosity and YouTube's machete reaction a fitting name: Elsagate.

Greg Chism's life turned upside down. He thought of Toy Freaks as Bugs Bunny cartoons brought to life, an innocent way to give his daughters a college fund and boost their self-esteem. Suddenly everyone else thought the worst of him. Strangers berated him online. One woman posted accusations that his younger daughter was actually a missing child. When his channel was deleted, he released a statement noting how troubled he was "that anyone would find inappropriate pleasure in our video skits." Law enforcement in Illinois began investigating Chism for child endangerment. "Everyone is disturbed by this," Rich Miller, an Illinois police chief, told *BuzzFeed News*, "but finding the proper criminal aspect to being a bad parent at times is challenging." Ultimately, Chism was cleared of any charges.

Years later he was still rattled from the experience. "It was really bad mental-health-wise. I almost freaking died," he recalled. "I got wronged by the media. YouTube—they were fine. I didn't have a whole lot of contact with them. It just happened."

• • •

YouTube had a terrible Thanksgiving in 2017. Several employees spent the holiday hunched over laptops, writing status reports for Wojcicki and making sure that the sweeping changes that had been set in motion actually worked. As always YouTube had to monitor for people re-uploading the very videos it had just removed. This was a Code Red. Engineers were dragged in to wrangle machine filters as quickly as possible. "It was a frenzy," recalled Jack Poulson, a Google researcher placed on response duty. "To be blunt: people realized you would get a promotion if you did well." Around the holidays YouTube brought puppies into its San Bruno office to comfort rattled staff. Of all the troubles YouTube faced over the years, nothing changed its approach to moderation as swiftly. "It was like I started working for a different company," said one former staffer. "It was really visceral for me," recalled Neal Mohan, a father of three who oversaw the effort. "A lot of the impetus, the passion, and the stress, frankly, came from the fact that this was fundamentally about protecting children."

YouTube made a number of public promises. It would remove ads from inappropriate family entertainment, block indecent comments about kids, consult more experts, release a creator guide on "family-friendly content," and provide "faster enforcement through technology." Wojcicki, in a December blog post, noted how YouTube's video bounty had "helped enlighten" her children. "But I've also seen up-close that there can be another, more troubling, side of YouTube's openness," she wrote. "I've seen how some bad actors are exploiting our openness to mislead, manipulate, harass or even harm." She pledged that Google's moderation workforce that screened its videos for *bad actors* would cross ten thousand by the following year, an eye-catching total.

Wojcicki did not mention most of them would not be working directly for Google.

CHAPTER 28

Bad Actors

Jakob Høgh Sjøberg arrived at the Dublin complex to see two signs at its entrance: an entry for Facebook contract workers, and another for Google. A plastic card he was handed at its front desk permitted him entrance through Google's door. Sjøberg was trim, with red hair and a well-traveled résumé—law degrees from his native Denmark and Ireland, a summer studying at the London School of Economics. But it seemed that the people hiring for this job were primarily interested in his ability to understand Danish. During his hiring, the company conducting the interview, Accenture, which he had never heard of, asked how he felt about working with "sensitive content." Sjøberg considered this briefly. "I'm just as strong as other people," he replied. He was given little further information about the job.

Five others joined him in his orientation in September 2017: a Russian, three Spaniards, and an Irish woman who spoke French. They sat in a sterile classroom going over basics on freedom of expression and protected speech, subjects Sjøberg enjoyed. Then the instructor moved to a big screen fifteen feet from Sjøberg's seat. YouTube videos began flashing.

"This is graphic," an instructor warned. *Next video.* "*This* is ultra-graphic."

Sjøberg felt his heart beating faster. "Ultra-graphic," he learned, meant footage involving people set on fire or mangled body parts. Shootings. Murders. Grim stuff. When a video played of a man screaming while being stabbed repeatedly, Sjøberg had to rush to a bathroom and splash water on his face to stop himself from fainting. *This was horrible.* But he didn't consider himself a quitter. He returned to the room.

His actual work took place in a larger office with rows of computers, each showing a ticketing system—a digital interface that looked like Gmail, only with more labels and folders. A steady queue of videos streamed in. Sjøberg was assigned to moderate violent extremism and focus on Danish material. They had a quota, 120 videos a day. Usually, though, there weren't enough troubling Danish videos to review, so Sjøberg sometimes assessed footage in other languages. He could not handle beheadings. YouTube's machines were by then able to remove most of those automatically, but some slipped past—grotesquely innovative ones involving dull knives. Once, a video in his queue played a familiar scene—a man bound, a message being delivered. Sjøberg felt dizzy. Instinctively, he clicked on a button to remove the footage only to later learn it was a prank: the video terrorist whipped out a toothbrush and held it to a man's throat. Sjøberg's manager reprimanded him for the error. Eventually, co-workers agreed to swap, taking the violent material and letting him screen grim stuff with animals and kids. "For some sick reason," he recalled, "that didn't affect me as much."

Some colleagues made jokes to cope. Sjøberg listened to cheerful Top 40 jams on his headphones and took periodic walks around a pond on campus. He worked at this job for nine months. Only once did he receive an email instructing him to make his office space extra clean. Someone from YouTube was visiting.

• • •

Google had started outsourcing certain jobs, mostly those considered nontechnical, after the 2008 financial crisis. As the company grew, this labor force did, too. Google called them TVCs ("temporary, vendor, and contract" employees). By 2018 it reported more than 100,000 direct staff; it had nearly as many TVCs but did not disclose the number. Some worked on short-lived projects and were paid handsomely. Others cleaned offices or tested not-quite-yet-self-driving cars. At YouTube, TVCs moderated content, working for companies called Accenture, Vaco, and Cognizant, opaque back-office firms—ones that make those recruiting commercials that you watch and still have no idea what they do. Moderators in their employ rarely met senior managers. Even rarer was an encounter with a YouTube staffer. These screeners were the descendants of the old YouTube SQUAD, some of the internet's first frontline workers. But SQUAD members had YouTube salaries and equity and sat at desks right at the front of the office. Now most screeners worked in a sprawling shadow economy of anonymous office parks in Dublin, Hyderabad, and Kuala Lumpur, absent from Google's balance sheet.

Beginning in 2017, with public pressure mounting, Google and Facebook scrambled to expand their screening workforces to avoid further excoriation from the press, advertisers, and regulators. Many contract moderators experienced acute anxiety, depression, and frequent night terrors from their jobs, as the reporter Casey Newton documented in damning investigations for *The Verge*. Some YouTube moderators Newton spoke to who were employed by Accenture in Austin earned $18.50 an hour—around $37,000 annually—with no medical leave. A YouTube staffer was once sent to brief new moderators working near headquarters, where one asked if they would have access to a therapist. The staffer didn't know the answer.

Sjøberg and his Dublin colleagues did not receive health benefits, but they did have a "nap room" where they could take breaks. Eventually, a psychologist was brought in and immediately lined up a long list of appointments.

Not all screeners dealt with gruesome footage. Fleets of contractors handled YouTube's copyright disputes, bizarre feuds that happened layers below big media companies. One contractor in California had to become versed in Cambodian fishing videos, a genre that, for some reason, was rife with copyright complaints. Lawyers for the guru depicted in Netflix's *Wild Wild Country* sent over stacks of copyright takedown requests. Sometimes screeners mistakenly sabotaged YouTube's own business. That year the company's salespeople in the Middle East booked sponsors in Saudi Arabia during Ramadan, a marketing season one employee described as "the Super Bowl times thirty." Commercials were lined up to run on popular Saudi food and entertainment channels, but once the holiday arrived, nearly all these ads didn't. It seemed that YouTube's moderators and machines, trained to prevent any commercial ties with Islamist extremism, had clumsily removed ads from videos in Arabic or with Islamic imagery. (However, according to one executive in the region, YouTube was careful in Saudi Arabia to remove cartoons of Peppa Pig from its recommendations system to avoid offending dogma on pigs in the country.)

Google justified using contractors for their speed; indeed, contract firms could hire much faster than Google, which recruited at a painstaking pace. If Google suddenly needed to police something—say, weird kid videos—then it required an influx of moderators "to help train the system," its machine filtering classifiers, an executive told a reporter. Contractors were well aware of their role as machine trainers. Sjøberg's colleagues called YouTube's algorithm "the robot." Once the robot got good enough, the moderators knew they were out of a job.

Even Google's in-house moderators, who had job security and benefits, felt threadbare during the company's hypergrowth years. One YouTube staffer who joined in 2011 recalled that there were only forty reviewers, who were tasked with screening up to a thousand videos a day, she said later. Daisy Soderberg-Rivkin, a moderator hired in 2015 to review Google image results, watched the one Arabic speaker on her team depart and management make no moves to fill the position. This felt particularly strange at a company worth half a trillion dollars, with kombucha on tap and seemingly endless computing power. *It's just not in the budget,* she was told. "You mean," she snapped at one point, "it's not in *our* budget?" One graduate of YouTube's Trust and Safety team sought PTSD treatment after departing, primarily to deal with their work screening troubling kids' content. "I'd have to take a Lithium or a Xanax every single day," they recalled.

Elsagate had prompted the company to invest far more heavily in its policing operation, adding more resources for efforts the company called "child safety." Mohan, the product chief, would take over Trust and Safety, moving the unit up the company hierarchy. Moderators were given mandates to look more closely for child abuse in videos (although an exact definition remained fuzzy). And YouTube made sure to assign moderators to screen its most frequented sections, such as its home-page Trending tab, which someone somewhere across the globe manned at all times.

Which is why it was so jarring when even Trending went horribly wrong.

• • •

Amid YouTube's child safety disasters, its stars kept rising to greater heights of fame. None went higher faster than Logan Paul.

Paul and his younger brother, Jake, a smaller, brasher replica, had started a YouTube prank channel as tweens in suburban Cleveland. But when Logan entered college they really found their groove on Vine, the six-second video app that captured some of the spontaneous creative magic YouTube had lost. A former high school linebacker and wrestler, Logan Paul looked like a young Matthew McConaughey playing a beefy Disney prince. On Vine his *Jackass*-style antics earned an enormous following. In his clips he entered strangers' cars and wrestled in supermarkets. He often removed his shirt. In October 2016, Twitter, Vine's owner, unable to handle the service, unceremoniously shut it down. Paul led a mass migration of Viners to YouTube, where they made their videos longer and wilder. These creators grew up on

Web 2.0—Paul was born in 1995—barely cognizant of a time before YouTube or internet fame. "Social media influencer" had become a natural career. "I want to be the biggest entertainer in the world," Paul, then twenty, told an advertising magazine in 2016. This seemed plausible. By the following year he had more than fifteen million YouTube subscribers, a TV show with Disney, a film contract, a deal with the MCN Studio71, and a massive following. PewDiePie had his Bro Army; Paul had his "Logang."

In person Paul had midwestern manners and an attention deficit, often bouncing in his chair when network suits got him to sit down. He also had a preternatural grasp for what made people click online. John Carle, a former Studio71 director, described Paul as "relatable," if you had a frat-boy drinking buddy, or "aspirational," if you aspired to that. Paul had a regular posse of fellow internet celebs and hangers-on. Over the holidays, he brought them on a trip to Japan to shoot some vlogs.

Dan Weinstein, a Studio71 executive, was lounging poolside with his family in Turks and Caicos to bring in the New Year when he glanced at his phone and saw ten missed calls from his office. *Dear God.* He rang back.

"Logan just filmed a dead body," a colleague reported.

"I'm at a pool with my child," Weinstein replied. "What do you want me to do?"

Paul was shooting a three-part "Tokyo Adventure" series in the metropolis. In one entry, he visited Aokigahara, an area known as the Japanese "suicide forest," as Paul's New Year's Eve video informed viewers. After a brief teaser of the coming attraction, a YouTube editing staple, Paul put up a warning screen. He appeared in frame to tell viewers he had disabled ads on the footage, given its subject, which, by the way, probably "marks a moment in YouTube history." He added a final disclaimer: "With that said: buckle the fuck up, because you're never gonna see a video like this again."

His posse had packed camping gear and binoculars to stay overnight in the forest, and Paul wore a silly green cap depicting those cute aliens from

Toy Story. A few minutes in, Paul stopped. "Bro, did we just find a dead person?" His camera turned to show a man hanging from a tree. "I'm so sorry about this, Logang," Paul said to the camera. "This was supposed to be a fun vlog." He then gave some unscripted comments about depression's toll, how all his viewers were loved and should seek help if needed. But then the remainder of the video lingered on him and his posse in the forest processing the "way too scarily real" interaction, giddy with adrenaline.

Paul blurred the dead body's face in his final edit. Then he posted. Viewers clicked. YouTube's algorithm, a bug drawn to light, promoted his video to the tenth spot on the site's Trending page.

Executives at Studio71, which took a cut of Paul's video earnings, knew his video violated norms, if not YouTube's rules, and would certainly turn off advertisers, who were still skittish about YouTube. Within a day Paul removed it (after it had accrued millions of views). Yet, as with anything viral, re-uploads flooded YouTube, and press covered the affair. Weinstein dealt with its fallout while on vacation. YouTube stayed quiet. "They wanted to see how big of a shitstorm it was," Weinstein recalled.

Graham Bennett was on holiday in his native United Kingdom when he saw Paul's video on the Trending page and the storm commence. An easygoing, bearded Brit, Bennett served as YouTube's chief liaison to Paul and other megastars. Bennett had joined YouTube in 2007 from the BBC and was among the first there to deal with mercurial YouTubers like Danny Diamond. Paul's ordeal was the "scariest time" in his YouTube career, Bennett later recalled.

Even after the PewDiePie fiasco, YouTube relied on the good graces of its stars to stay within unspoken bounds of good taste and acceptability. The week after Paul's video, Bennett met with YouTube brass, who concluded, *Okay, we can't do that anymore.* "It seems kind of naive now," Bennett said later, "but it was the first time we realized that YouTube creators were legit global stars. And that meant that if they did something out of line or crazy and newsworthy, it will be news everywhere in the world." Perhaps YouTube also knew the sight of an all-American, entitled internet celebrity romping chaotically through foreign lands and cultures did not help the company's desired image as a global business.

Paul, unlike PewDiePie, apologized immediately. On January 2 he posted a short clip ("So Sorry.") looking straight at the lens, a little red-eyed. "I should have put the cameras down," he said, before concluding, "I promise to be better. I will be better."

He was. For a month. But then Paul piled onto a bizarre trend that had caught YouTube and social networks flat-footed—a viral internet challenge to consume tiny, colorful Tide detergent pods. On Twitter, Paul joked about swallowing them. That same day, in a YouTube video shot on a balcony of his Los Angeles mansion, he zapped a dead rat with a Taser gun.

Finally, a line was drawn. Bennett, Wojcicki, and other leaders at YouTube held a "Code of Conduct" review for their troublesome star. Paul dialed in for a videoconference call, lasting well over an hour, detailing the company's decisions. YouTube would now factor in what its creators did off its site, including things like Paul's tweet, and it would tighten rules for what appeared on-screen. Pranks would be no-goes based on how easily a teen could repeat them at home, Bennett explained. Videos featuring setting household fires or popping Tide pods would be removed, though hard-to-replicate stunts like skydiving were still okay. (Paul would later do this on YouTube, naked.) Also, YouTube was, in a first for the company, temporarily removing ads from Paul's entire channel as punishment. By then YouTube had begun using a phrase internally: *making money off YouTube was a privilege, not a right.*

During the call Paul was gracious and understanding. He did not leave YouTube. Without its ads, though, he did lean on another commercial lever: hawking his apparel line, an increasingly common YouTuber trend after the Adpocalypse shriveled earnings. Chris Stokel-Walker, a British journalist, analyzed fifty videos that Paul and his brother posted that February and March. On average the Pauls mentioned their own "merch" once every 142 seconds.

• • •

YouTube's litany of scandals had, so far, stemmed from videos that outsiders (advertisers, parents, journalists, politicos) found unseemly. That winter something shifted. An insider started talking.

Guillaume Chaslot, the French former YouTube engineer, first tried cajoling his old colleagues after he had discovered YouTube's disturbing appetite for conspiracies. Chaslot spoke to a friend at YouTube in early 2017 about this. His friend shared his concerns but suggested these findings were a reflection of human interest or idiocy beyond Google's control. "What can I do?" his friend asked Chaslot. "I wish I could change people." So Chaslot went public. FICTION IS OUTPERFORMING REALITY, read the *Guardian* headline on February 2, 2018. Chaslot presented his case that YouTube was poisoning the well. "The recommendation algorithm is not optimizing for what is truthful, or balanced, or healthy for democracy," he said. Five days later Chaslot's research appeared in a *Wall Street Journal* story on YouTube's habit of elevating blatant untruths. Republicans had just released a document accusing officials of bias in investigating Trump's Russia ties; a search on YouTube for "FBI memo" produced videos from Alex Jones and an account called Styxhexenhammer666. A search for "flu vaccine" produced screeds against medicine and a rabbit hole of similar videos. The article compared YouTube's results with those that appeared on Google, which placed legitimate news outlets and health organizations at the top of results for such searches.

Like all CEOs, Google's Sundar Pichai read the *Journal* regularly and voiced his irritation with this story's findings to YouTube management. Pichai's indictment stung YouTube particularly hard, because the company had assumed that issue had been fixed.

The prior fall YouTube had adjusted its algorithm to surface "more authoritative" news channels. Yet Chaslot's research pinpointed a glitch in its mechanics. "Typically when news breaks, people write stories about it,"

Johanna Wright, a YouTube executive, told the *Journal*. "They don't make videos about it." Cable outlets waited hours or days to post on YouTube, if at all. Accounts like Styxhexenhammer666 did not wait. Ugly stuff on YouTube's Long Tail rushed to the Head. Google knew this phenomenon because it had already been burned by it. Engineers would cite a classic example: Obama birtherism, the racist conspiracy that Trump rode into a political career. People who believed in Obama's legitimate citizenry did not write stories about it. People who disbelieved it (or found grift claiming so) certainly did, pushing their links to the top of Google. Staff called these rare, exploitable holes "data voids" or "evil unicorns" and rushed to patch them after the 2016 election, when a top result under Google searches for "who won the popular vote" momentarily showed an anonymous blog that falsely claimed Trump did.

More than a year later YouTube was either unwilling to tackle this beast or still unprepared for it. But the beast kept rearing its head.

After the mass shooting in Las Vegas in October 2017, some YouTubers filled the data void with crackpot theories about "false flags" signaling that the massacre was staged. That happened again after a November shooting in Texas. Then again in February, when a gunman killed seventeen at a high school in Parkland, Florida. On the internet fringes, theories arose that student survivors from that tragedy, outspoken in their gun reform support, were paid "crisis actors." YouTube thought its systems were prepared to deal with such falsehoods until an account named "mike m" uploaded an old local TV clip of a Parkland activist with the title "DAVID HOGG THE ACTOR . . ." Propelled by conspiracy peddlers, curious onlookers, or some combination thereof, this clip went viral. At first glance the repurposed news footage didn't seem to violate YouTube's rules, so a moderator somewhere cleared it for YouTube's Trending page, where it sat for a time on the No. 2 spot, making it appear to the outside world like a tacit endorsement. "Dear God," a YouTube public relations staffer texted a colleague. "What do we do?"

Some YouTube contractors were all too familiar with this experience. Those overseeing copyright disputes had an entire section of their rule book devoted to Leonard Pozner, a father from the 2012 Sandy Hook school shooting who had become a frequent target of internet hoaxers. Pozner fought back against mocking and conspiratorial videos with copyright claims, an approach that didn't always work. YouTube screeners felt terrible for Sandy Hook parents mobbed online, one moderator recalled, but when claims came in, screeners were allowed to email only rote, legalese replies. Sometimes, they could bold certain portions, like a line advising people to "go somewhere else" for help.

• • •

That March, a month after the Parkland video mishap, Susan Wojcicki sat onstage in Texas for a long interview that featured plenty of awkward moments. Wojcicki's public appearances no longer focused on all the ways YouTube upended TV. Nick Thompson, *Wired*'s editor, instead grilled her on YouTube's strategy for handling conspiracies and election meddling. Facebook and Twitter had recently disclosed that Russian agents had flooded their sites with bots and ads in 2016, setting off a public frenzy about social networks as consequential geopolitical forces. The wave of school shootings and malicious conspiracies made those tracking social media think that YouTube could be such a force, too. Suddenly everyone realized how little they had heard from YouTube's boss.

When Wojcicki considered a hard question, she had a habit of pausing, darting her eyes up, and keeping her hands, which she used to talk, poised in her lap. Onstage in Texas she explained YouTube's penalty box—how her company kept certain videos up but didn't recommend or run ads on them. She listed some categories that warranted this purgatory.

"What about falseness?" Thompson interjected.

Wojcicki held her considering pose. "I mean," she began, "falseness is a very hard value to put as an absolute. Because that would require *us* to determine whether something is true or not. And, in general, we don't think we should be the one determining it, that's why—"

Thompson interrupted. "You determine whether something is hateful? Or whether somebody is nude?"

"Well, nudity is usually clear," Wojcicki replied. Judging hate was harder but feasible, she added. Judging truth wasn't. "Look, I studied history as an undergraduate," she offered. "That's what history is about. It's different interpretations. Who was a hero? Different people will say different things happened." She cited flat-earthers, a conspiracy the company frequently referenced, perhaps because it felt relatively harmless. At the interview's outset Wojcicki had shared another personal anecdote: her grandmother had run the Slavic department for the Library of Congress. Libraries celebrated banned books and free speech, kind of like YouTube.

"We're really more like a library," Wojcicki said.

This was a strained analogy for the chief of a division that would net $11.2 billion in sales that year by placing ads in front of its library material. But it was very Googley. Wojcicki then laid out her Googley fix for YouTube's conspiracy problem: the company would introduce "information cues," little boxes of text beneath videos about flat-earth and other "well-known internet conspiracies." It would scrape this material from Wikipedia, as Google did on search, relying on this user-generated, nonprofit site that managed to hew consistently to the truth. (Wikipedia, after Wojcicki spoke, said it had not been informed of this plan.)

Wojcicki also introduced a term she had begun using frequently at YouTube. Its algorithms favored watch time, daily viewers, and satisfaction, but they had added a fourth metric. "We're starting to build in that concept of *responsibility*," she told Thompson. "We're still in the process of figuring out exactly what that means." When the interview ended, one YouTube staffer in attendance privately felt relieved that the *Wired* editor didn't pull up searches on YouTube like "flu vaccine," which were rife with conspiracies.

Another person on YouTube's policy team later said that Wojcicki argued against limiting such videos, citing friends of hers who subscribed to alternative health beliefs that shunned vaccines—a stance she changed once a global pandemic struck.

Still, Wojcicki ended the interview on high notes. YouTube had implemented a series of updates she said would purge *bad actors* abusing its systems. Videos like the Parkland Trending clip were removed for breaking rules against harassment. With new ad filters and controls in place, Wojcicki had persuaded nearly all boycotting brands to spend again.

And most critically that winter she made the difficult decision to undo the great equalizing experiment YouTube had begun five years earlier. There would be no more money for everyone. To earn ad sales, YouTubers now had to have a minimum number of subscribers (one thousand) and hours of footage watched (four thousand). These thresholds were chosen at levels where ad payments were "sort of meaningful" to someone's income, YouTube's Mohan later explained. With this restriction, YouTube hoped it would stop rewarding so many unsavory characters and bring an end to all the publicity headaches of the prior year. Almost overnight YouTube went from paying around six million channels to about twenty thousand.

Advertisers welcomed all these changes, and YouTube believed most big creators understood them, too, although the company knew some would not. No one was prepared for how far one would go in response.

CHAPTER 29

901 Cherry Avenue

YouTube headquarters did its best to fit into its strip-mall suburban home. Two vine-covered outdoor staircases jutted out from the office's front, curved outward like wings. A company logo was etched into the south staircase, colored to match the building's eggshell white. The office sat on a corner facing Cherry Avenue, a four-lane road that passed under a freeway near a Carl's Jr. and a parking garage. Sandwiched between this garage and YouTube's entryway corridor was a small courtyard, filled with chairs and bright red sun umbrellas, where YouTube held its Friday meetings.

On Tuesday, April 3, 2018, employees streamed off shuttle buses in the morning, as usual, taking the mossy stairs up to the second floor, which housed their desks and the big red slide. Around noon they spilled out onto the courtyard for lunch. Kurt Wilms sat inside at his second-floor desk. Wilms had been at YouTube for seven years, decades in tech time, cycling through its various projects. He now worked on its "living room" division, crafting YouTube's experience on video game consoles, smart TVs, and other devices. His was the YouTube of normals: of cooking instructions, sports highlights, *SNL* skits, and chess game commentary (his favorite), far from the beheadings and suicide forests and Spider-Man-Elsas. Wilms was a happy-go-lucky dude. He said "learnings," a gerund the tech industry deployed to give gravitas and gusto to "lesson," which Wilms used in sentences like "A good learning for me: I'm going to try to stay chill."

During his career at YouTube, his co-workers treated their space like an open, inviting college campus, as Google and other companies did. Staff brought in friends and family to see the micro-kitchens and rooms named after viral videos. This didn't change after Wojcicki's arrival, when YouTube swelled in numbers to more than a thousand people in its San Bruno office. Wilms, who once knew everybody, suddenly asked himself, *Who are all these people?* This was a good sign, though; it meant growth.

On that sunny April Tuesday, some construction clanged outside. A little before 1:00 p.m., a noise interrupted Wilms while he typed an email.

Pop.

"What was that?" a co-worker next to him asked.

"Oh," Wilms replied, turning back to his screen. "It's construction."

Pop. Pop. Pop.

Wilms swiveled to his co-worker. No, that was the distinctive sound of gunfire, a sound very close by. Wilms stood and shouted the first thing to come to him: "Run!"

• • •

Nasim Najafi Aghdam was a thirty-eight-year-old YouTuber living near San Diego. Young Iranians might have recognized her as Nasime Sabz or Green Nasim. She was a minor social media star in Iran and its diaspora, a strange internet personality who had made videos for Persian satellite TV and many, many for YouTube.

Born in Iran near the Turkish border, Aghdam created her videos in Turkish, Persian, and English, often discussing her family's persecution for its minority Baha'i faith. In 1996 they moved to California, and Aghdam soon became deeply passionate about animal rights. At twenty-nine she went to Camp Pendleton, a Marine Corps facility in Southern California, where PETA activists were protesting the use of pigs in training exercises. Protesters stood outside the entrance with signs that read STOP TORTURING

ANIMALS. Aghdam, thin with jet-black hair and sharp features, showed up with a plastic sword, wearing black gloves and jeans painted with dark red droplets resembling blood. She had painted two red droplets on her chin. "For me, animal rights equal human rights," she told a reporter there. Jena Hunt, a PETA staffer who organized the protest, later said she had to ask Aghdam to leave. "The poor thing seemed very mentally ill," Hunt recalled.

Aghdam turned online to advocate and find solace. She lifted weights and identified herself as "the first Persian female vegan bodybuilder." She started a website and a nonprofit called Peace Thunder. Animal rights groups had trouble delivering their message, she told a wellness publication in 2014, "because many media or even internet websites only care about their own financial profit." The publication printed a photo of Aghdam flexing her biceps in a bedazzled neon-green tank top bearing a butterfly.

She took to YouTube to spread her message. She made videos in her parents' house in Menifee, a town at Los Angeles's outer edge, shooting before a dark blue wall opposite a small bed and a mannequin dressed in sequins. She ran at least four YouTube channels, uploading frequent, strange clips that melted into the site's vast tapestry—workout routines, bizarre musical parodies, and graphic videos documenting cruelty against animals. One did well in Iranian internet circles: Aghdam danced in it, wearing a flamboyant purple dress and a breastplate with a plunging Elvira neckline. As she swayed, she removed the breastplate, and a caption appeared on-screen, "Don't trust your eyes." She was occasionally subjected to ridicule for her odd antics and deadpan delivery. Viewers asked about her mental fitness. In videos Aghdam started to complain about life in America, that people taking on the system and big companies were "censored." That it felt as bad as Iran. "There they kill you by axe," she told the camera. "Here they kill you with cotton," an Iranian expression for a death at the hands of something seen as innocent.

Starting around 2017, she began to complain about YouTube. "I'm filtered on YouTube," she said in one video. "And I'm not the only one." On her website she documented this corporate crackdown, which she viewed as retaliation for her outspoken challenge to the meat industry. She posted three screenshots of her YouTube dashboard, showing watch time, views, and subscribers on her videos and how they kept falling. One post listed 307,658 minutes of watch time and 366,591 views. "Your estimated revenue," the YouTube dashboard read, "$0.10." This she circled in red pixels. "There is no equal growth opportunity on YouTube," her website blared in bright, frantic text. "Your channel will grow if they want to!!!!!"

Aghdam had meanwhile moved to live with her grandmother near San Diego. On January 2, 2018, she purchased a 9 mm Smith & Wesson from the Gun Range, a San Diego retailer. She picked it up two weeks later, the same day YouTube announced its big ad policy change. She drove north.

On Monday, April 2, Aghdam entered 901 Cherry Avenue during the lunch hour. She approached YouTube's front desk and asked a receptionist about job opportunities. She left within ten minutes.

That night, police found her in a white sedan parked on the street thirty miles south in Mountain View, where Google had its headquarters. Aghdam opened her driver's door wearing a light hooded sweatshirt, hood pulled over her head. A roll of toilet paper sat in the passenger seat.

"Are you taking any medication at all?" a female officer asked her.

"No," Aghdam answered.

"You don't want to hurt yourself, do you?" the officer tried. Aghdam thumbed her phone. She looked up and shook her head no. "And," the officer continued, "you don't want to hurt anybody else?" Aghdam looked down at her phone and gave a fainter nod. The officers left and notified Aghdam's family.

Her brother later told reporters that once police called and he googled his sister's location, alarm bells went off. "She was always complaining that YouTube ruined her life," he recalled. He said he called police back to warn them that "she might do something." Mountain View police would deny this.

That morning Aghdam went to a local shooting range. She returned to YouTube's office shortly after noon, parking in the adjacent garage. An employee stopped her at an entrance and asked for ID. Aghdam pulled a pistol from her purse, sending the employee fleeing to call 911. Aghdam continued to the courtyard.

Dianna Arnspiger, a YouTube project manager, saw this dark-haired stranger spraying bullets with a gun. On instinct Arnspiger shouted, "Shooter!" A pedestrian nearby would tell TV cameras, "Oh, man. It didn't stop. It was no mercy. No mercy." Inside the office, a YouTube manager inside peered down a staircase to see drops of blood on the floor.

Kurt Wilms had bolted to a door by his desk and raced down the stairs, his eyes folding in tunnel vision. Suddenly he stopped: he looked down to see the entryway lobby, usually buzzing with activity during lunch, totally empty. He turned around to discover several of his colleagues had followed him in panic. They all ran back up the stairs, through the door, and into a conference room, where they flipped a table against the entry as a barricade and waited. Wilms took deep breaths, fully expecting the shooter or shooters to charge in.

Police found Wilms and his huddled colleagues before any shooter did. Employees were escorted out with their hands up. Others had taken refuge in a nearby shopping center or sprinted over a fence toward the freeway. Three YouTube employees were wounded in the gunfire and sent to a San Francisco hospital in fair, serious, and critical conditions. (They would all survive.) Aghdam had fired twenty shots, including one killing herself.

Susan Wojcicki, who was in a meeting on the second floor as the shooting began, walked out of the office in a borrowed black overcoat, trailed by her staff and reporters who had raced to the scene. "We will come together and heal as a family," she wrote on Twitter soon after. Police had arrived within three minutes of the first 911 call, followed by TV cameras, helicopters, and FBI officials. "It was very chaotic, as you can imagine," Ed

Barberini, San Bruno's police chief, said at a hastily assembled press conference in a nearby parking lot.

All sorts of speculation flew around in the hours before Aghdam's identity and story became public. Once it did, critics used her tragedy as an abject lesson for YouTube about its capricious, unreliable machines. But Aghdam's tale was fundamentally an American one, a saga of inadequate mental health care and easy access to firearms. Three months before Aghdam bought her pistol in San Diego, the seller had advertised a "12 Guns of Christmas" sale. The store clerk who sold her the weapon reported that this transaction did not stand out as unusual.

CHAPTER 30

Boil the Ocean

On Wednesday, a day after the shooting, Susan Wojcicki held an emotional company town hall where she shared plans to tighten office security immediately. A colleague told Kurt Wilms afterward that YouTube should drop any pretense of being a freewheeling tech campus and behave more like TV networks and newspapers, which were prepared for these kinds of lone-wolf attacks. "We should be treating ourselves as a major media company," the colleague told Wilms. "We essentially are."

Google had long been concerned about the safety of its executives; in its Brazil office, the company had installed a secret back door for quick departures, should a video or corporate decision inflame citizens or officials there. Days after the shooting, handymen came to install a bulletproof wall encasing Wojcicki's office, which staff could enter only after swiping corporate badges. Her security detail grew, adding members from a Texas firm that recruited former marines, who also stood guard outside her home. A tech conference had invited Wojcicki to speak; she wouldn't attend unless the conference added armed security. If Wojcicki's security detail wanted to be alert to threats to her life, they needed only to look on her website. Irate YouTubers and edgelords learned they could pin their frustrations on the company's chief, a Jewish woman, no less; many relied on tropes about both identities. A year later a fourteen-year-old YouTuber, known for vulgar, anti-Islam videos, would directly threaten Wojcicki's life on camera.

The shooting reminded everyone at YouTube of the gravity of their responsibilities—how they controlled a system that had paid millions of people, giving them a stage with few rules and limitations, and then had swiftly taken much of that away. Jennie O'Connor, a YouTube director, had moved a month before the shooting to manage a new division dealing with the site's problems and threats. She had called in sick that Tuesday, so watched the horror unfold from her house. "It just sharpened the importance of the decisions we make," she recalled. "We have to really tread that balance of 'Get YouTube to a safe space' without over-enforcing. There are actual, real consequences to that."

Wojcicki was particularly concerned about placing more restrictions on creators. In the days after the shooting, Google's CEO, Pichai, and co-founder Sergey Brin came to visit YouTube's office and met with a small team of leaders. As CEO, Pichai was careful not to meddle much in YouTube affairs, partially, according to one associate, out of respect for Wojcicki's close ties to Google's founders. Of all company leaders, YouTube's CEO seemed to be the only one on equal footing with Google's chief. Inside the office, Pichai proposed adding further measures for advertisers to keep them clear of potentially troublesome YouTubers. Wojcicki replied bluntly, "We've done more than enough." That settled the matter.

But the shooting did have a chilling effect on one of Wojcicki's plans. For years, the company had prepared to "boil the ocean" and rework creator payments with its Moneyball project. This concept was partially driven by a desire to fund admirable videos that scared advertisers, like those about sex education or suicide awareness. Under the plan, YouTube would stop paying creators for each commercial that ran on their videos and use an overall pool of ad money instead, doling out checks based on engagement—the likes, comments, and watch time videos got. This felt fairer and more sustainable.

YouTube briefed a few creators on its ambitious plan. In March, Wojcicki presented it to her staff, telling them, "Please don't leak this."

The project didn't leak. But it also didn't hatch. Politicians and critics had been hammering social networks for their allegiance to "engagement" at the expense of accuracy, civility, and all else. They hammered YouTube, too. "What we are witnessing is the computational exploitation of a natural human desire: to look 'behind the curtain,' to dig deeper into something that engages us," the sociologist Zeynep Tufekci wrote in a March *New York Times* op-ed titled "YouTube, the Great Radicalizer." After signing off on YouTube's Moneyball plan, Pichai reversed his position for fear of exacerbating this concern. Some staff had calculated that the new engagement model would have paid more money to brazen stars like Logan Paul than most news outlets on YouTube.

And then, the office shooting—a violent creator aggrieved by sweeping service changes—ended discussions about "boil the ocean" for good.

• • •

That year another force shocked YouTube and its corporate parent. Newspaper columnists and politicians weren't the only ones tearing into the corporation. Its employees began doing so too.

After her maternity leave Claire Stapleton felt as if she had returned to a different company. Everyone was suddenly walking on eggshells, a little on edge. She couldn't pinpoint an exact moment of the turn, but it might have been Valentine's Day, before the shooting, during Black History Month, when her team sent this tweet from YouTube's account:

> Roses are red
> Violets are blue
> Subscribe to black creators.

It was cute, Twitter-savvy, about *values*. Constantly her team had heard an edict: *We need to be out there for our values.* When Patagonia, the apparel line, launched its first big TV ad campaign the previous summer, a plea for public lands (and an unsubtle stab at Trump), YouTube's marketing chief sent an article about it around, adding, "*Love* this." Stapleton was told this edict came from the top: "Susan wants to be *out there.*" Stapleton's team had to script plans for Wojcicki to lean into gender equity as a core value. "As a global tech company with a female CEO, we're ready to be a leader on this issue," read a document the brand team prepared in 2018. It also called for the company to take stands on mental health, immigration, LBGTQ affairs, and racial justice. Stapleton's team often discussed ways to counter the algorithm's habit of surfacing creators of color less often. YouTube invited speakers on topics like systemic racism to offsites. The Twitter poem seemed to fit nicely into the Black History Month plan.

And yet, when it yielded a predictable reaction—"I'll support all creators," a white dude avatar wrote back; one popular reply simply read, "Nah"—the company objected. Ariel Bardin, an executive frequently dispatched to deal with creators scorched by YouTube's changes, worried aloud about the backlash. "Why are we wading into this?" he asked. From then on Stapleton's boss had to approve every single official tweet. A marketing colleague later recalled a different YouTube tweet, about transgender creators, that required thirty employees to workshop and approve.

For International Women's Day, YouTube hired an ad agency to make a promotion that appealed to America's heartland. Once it arrived, more complaints came. "Too polarizing," Stapleton's bosses concluded. To Stapleton this felt less like the company's being out there than sticking its head in the sand. "We were disillusioned and confused about what YouTube was supposed to stand for," she recalled.

Two years into Trump's America, conversations about gender, race, values, speech, and power had changed rapidly and intensified. Reports on Harvey Weinstein's grotesque abuses rippled across Hollywood, spurring a reexamination of the men in charge of cultural production. Other media moguls fell swiftly. Stapleton and some colleagues followed these developments at the internet's normal speed, a mile a minute. For YouTube, stormer of gates, the #MeToo movement looked redemptive: it turned out that some of media's biggest gatekeepers were wretched men who held women's careers hostage. But Google had less interest in steering these conversations anymore, much less weighing in. As critics punched from the left, conservatives laid into Google with charges that its lefty workforce censored videos and views from the right.

So far, Google had managed to avoid its deepest concern: the menace of politicians. While Google continued to have regulatory problems in Europe, at home the company was mostly left untouched, despite clearly dominating the markets for online ads, mapping, email, web browsing, video, and information.

How had Google dodged political fire? For one, the company spent $17 million in 2017 lobbying Washington, D.C., more than any other corporation. A Googler offered another explanation by way of an old joke about two hikers encountering a bear: One of them starts sprinting, while the other bends down to tighten his laces. *What gives?* asks the sprinter. *I don't need to outrun the bear,* the shoe tier explains. *I only have to outrun you.* Mercifully for Google, its hiking companion, Facebook, was spectacularly clumsy.

By late 2017, Congress, the press, and a special counsel were all furiously excavating social media for signs of Russia's election meddling. Facebook's hands looked reddest. After dragging its feet, Facebook had finally disclosed more than $100,000 in Russian-bought political ads, reaching some 126 million users on a network that so easily spread conspiracies. Google acknowledged only $58,000 in Russian ad spending. But YouTube had a Russia problem. Russia Today, the state-backed TV network that had rebranded as RT, was huge on the platform: it had more than two million subscribers, just shy of CNN's totals, numbers that everyone could see. Only Google salespeople could see that RT was also a major YouTube advertiser, spending loads to promote its videos in several channels and markets. European YouTube officials met privately with RT leaders to nurture the relationship. As Russia tightened its grip on internet censorship, Google worried that the nation might follow China and oust it. "We couldn't afford to lose Russia," a former sales director recalled. YouTube's Kyncl flew to Russia in 2013 on a goodwill tour to court broadcasters. He appeared on an RT segment to celebrate the network's milestone of one billion YouTube views, praising RT for being "authentic" and not pushing "agendas or propaganda." U.S. politicians disagreed. After Trump's election, federal officials forced RT to register as a "foreign agent." Senator Mark Warner, a Virginia Democrat, called YouTube RT's "go-to platform" and a "target-rich environment for any disinformation campaign."

YouTube, feeling the heat, removed RT from its premium slate and added labels for all state-backed media outlets. Yet just as pressure on YouTube mounted, Facebook tripped again. Four days after Wojcicki was grilled in Texas over conspiracies, the Cambridge Analytica scandal broke. A consulting firm had scraped Facebook data to make psychological profiles for Trump's campaign. Outrage and attention turned back on the social network. Google found it best to lie low.

Perhaps this reflected the temperament of its leader. Pichai, Google's soft-spoken, ruminative CEO, vastly preferred consensus over confrontation. For some at Google, this read like indecision; one former executive ascribed Pichai's decision to can YouTube's Moneyball project to his timidity. But Pichai managed to keep Google from Facebook-sized scandals.

Well, until the summer of 2018.

When Pichai took over Google, he determined that its future lay primarily in two fields: business software sales, via cloud computing, and emerging market internet consumers, which he called the "next billion users." Google had signed a contract with the Pentagon to provide drones with computer vision, paving the way for lucrative government cloud deals. In June, after weeks of raucous internal protests over the Pentagon deal, Google caved and pledged not to renew its contract. Then, that summer, employees discovered a shocking part of the "next billion users" plan: Google was building a search engine for mainland China with censored results. Both projects angered vocal swaths of Google staff, which saw them as betrayals of company values.

They also angered D.C. lawmakers and military officials, who saw Google's moves—bailing on a Pentagon deal and then getting in bed with China—as sacrilege and a convenient political football. Trump latched onto Google's Pentagon mess and to unsubstantiated claims that the company manipulated information to censor conservatives. "They better be careful," the president huffed in August, "because they can't do that to people." Trump allies attacked Silicon Valley platforms for abusing their protected status under Section 230, the law that shielded user-generated websites from liability. Ted Cruz, a Texas Republican, berated Mark Zuckerberg in a hearing for failing to operate Facebook as a "neutral public forum," as Section 230 dictated. In fact, the law did not dictate this, but this blustering threat still worked. Within Google, members of its policy team were instructed to be extra cautious about anything Section 230 related and anything that might make Google appear as if it were taking on the role of a publisher.

At YouTube the severe caution from the Viacom lawsuit era returned. One engineer tasked with cleaning videos for advertisers went hunting on his own for troubling clips, until YouTube lawyers stepped in to stop him. A YouTube sales rep who asked to handpick clips for its premium slate (because algorithms, he recalled, picked only "pranks and pick-up artist videos") was told this would undermine the company's legal standing. A senior executive started collecting a spreadsheet of worrisome videos, including scores he described as "deeply racist," before being told YouTube shouldn't proactively search for such things. After the campus shooting, another employee tried to measure how often shooting threats appeared in YouTube comments, until this too was thwarted. Publicly, YouTube leaders downplayed its role in politics.

It did help that Trump, a Twitter fanatic, wasn't terribly active on YouTube, but his most extreme supporters were. Becca Lewis, an academic, began looking into YouTube's right flank after suspect videos about Hillary Clinton's health started playing on cable news before the election. Tracking fifteen months of footage starting in 2017, Lewis found that more mainstream YouTubers, like Joe Rogan, gave airtime to fringier figures. (Rogan had Stefan Molyneux, the Canadian guru, as a guest for multiple three-hour talks. Molyneux, in turn, invited a hipster-clad Austrian YouTuber known for refashioning white nationalism as "Identitarianism.") Right-wing YouTubers, Lewis concluded, flourished because they functioned much like mainstream ones. "YouTube is built to incentivize the behavior of these political influencers," Lewis wrote. Many fringe YouTubers used reliable search engine tricks. As early as 2015, David Sherratt, the men's rights YouTuber, saw maybe one out of every ten videos on immigration or "Western civilization" use tags like "white genocide" and "Great Replacement." After clicking one of those videos, he was mostly fed more of the same.

YouTube leaders struggled to distinguish fringe or dangerous material from normal contentious political fare. ISIS was a clear bad actor; right-wing shock jocks weren't. Besides, numerically, neither had a large viewership. "It's important to remember," Wojcicki told *The Guardian*, "that news or news commentary [is] a very small percentage of the number of views we have."

• • •

While news or commentary might have had a relatively small YouTube audience, some commenters flourished so well online that they took their acts on the road.

During the summer of 2018, Stefan Molyneux went on tour. The guru paired with Lauren Southern, a younger Canadian YouTuber and alt-right staple known for her disdain of multiculturalism and her self-described "gonzo" confrontations with feminists. Trump's White House had given her a press pass. A reporter who visited her Toronto home that July described her walls as bare, save for a plaque from YouTube congratulating her on 100,000 subscribers. In Sydney that month Molyneux and Southern spoke to a fully booked auditorium. Australian regional governments had recently proposed treaties with Indigenous populations, sparking a national debate. Molyneux, who then had around 800,000 YouTube subscribers, reportedly told the audience such treaties were unnecessary because Aboriginal peoples sat at "the lowest rungs of civilization."

In August the YouTubers arrived in Auckland, New Zealand, where their venue, a well-known music hall, canceled their appearance. This played well to their posture as defiant free-speech radicals, and the pair spoke to the TV station Newshub about the saga.

Newshub: "Full interview: Lauren Southern and Stefan Molyneux." August 3, 2018. 13:46.

> "This country is known as a melting pot," begins Patrick Gower, the TV anchor. He asks his guests how his nation should receive their message that diversity is a "weakness." How great is that melting pot, Southern asks, if it runs "against everything that has created the most beautiful culture in the world: the West"? Gower, stunned, pauses for a few moments. He turns to ask Molyneux about his position that some races are genetically weaker than others. "Never said that," Molyneux replies. He is in his element, an argument. "The most established metric in social science is IQ," the YouTuber says. Mid-speech, Gower cuts him off, justifying his interruption by calling Molyneux's claims a "rant." "I was thinking of the audience," Gower says. "Oh, trust me," Molyneux replies. "The audience is very interested in what we have to say."

After the TV station posted the interview on YouTube, a YouTuber who marketed his channel as a place to "learn the advanced social skills you need to get what you want out of life," uploaded his commentary on the exchange. "Brutal!" the video title read. "Stefan Molyneux & Lauren Southern DESTROY Patrick Gower (Body Language Breakdown)." Soon, that video nearly doubled the original's view count.

• • •

As the year dragged on, Google was forced to address politics. In December 2018, Sundar Pichai finally appeared on Capitol Hill. Congress had invited him in September for a hearing on election meddling, privacy, and other perceived ills of Silicon Valley. Both Pichai and Larry Page had refused the invitation, prompting lawmakers to stick an empty GOOGLE placard on the table between testifying executives from Facebook and Twitter, probably causing more political damage than attending would have. So now, after Google was panned for its absence, Pichai was in Congress alone for a three-hour grilling.

He sat ramrod straight in a dark blue suit, his hands kept steepled on the congressional table, never correcting legislators when they butchered the pronunciation of his name. An activist dressed as the board-game Monopoly Man sat a few rows behind Pichai. "In general," the CEO replied to a question about Russian interference, "we are not a social networking company." He even tried a joke about its failed efforts with Google Plus. Some Google officials sat dutifully behind their CEO, well aware that for the matter at hand YouTube *was* a social network (user generated, lightly governed, massive). YouTube had the same problems politicians were charging Facebook and Twitter with: ISIS recruitment, Russian propaganda, conspiracies, and a moderation approach inviting cries of political bias. Some Google political staff privately complained about the disproportionate amount of headaches YouTube was causing the company. Google's search ads, its primary business, were mostly based on people's queries and locations and didn't rely on browsing histories or the murky, privacy-invading tracking that angered critics. YouTube's business model did, though. On one internal conference call, a Google policy employee raised the question of spinning YouTube off from Google. Some familiar with internal debates said this was seriously considered, although others dismissed it as "overblown" idle chatter.

No one asked Pichai about this on Capitol Hill. Lawmakers, like many older Americans, probably visited YouTube rarely and certainly didn't rely on it for news. (One Google operator in D.C. said the company struggled to gin up interest for events when Wojcicki came to visit because few capital powerbrokers knew who she was.)

But then Jamie Raskin, a Maryland Democrat, threw Pichai a curveball: "Do you know what Frazzledrip is?"

Yes, Pichai did, sort of. His staff had briefed him on it that day, he explained. Raskin opened a copy of the prior day's *Washington Post* and read, "YouTube recently suggested videos that politicians, celebrities and other elite figures were sexually abusing or consuming the remains of children, often in Satanic rituals." He looked up. Certain videos claimed that Hillary Clinton and her top aide assaulted a young girl and drank her blood. This was Frazzledrip, a bizarre cousin of Pizzagate, a theory that had, by then, morphed into QAnon, the cultlike conspiracy theory and movement. "What is your company policy on that?" Raskin asked.

At that time YouTube was working on a major overhaul of its recommendation engine to bury conspiracy clips and other footage deemed "harmful" in its penalty box. But this change wasn't ready for public consumption, so Pichai didn't mention it. "We are looking to do more," he replied.

"Is your basic position," the congressman pressed, "that there's just an avalanche of material and there's nothing that could be done?"

"We do grapple with difficult issues," Pichai understated, before answering no, in a roundabout way. "It's our responsibility, I think," he said, "to make sure that YouTube is a platform for freedom of expression but that it's responsible and contributes positively to society."

• • •

Claire Stapleton's brand-marketing team managed to steer clear of Frazzledrip and YouTube's wacky fringe, and most politics, actually. In 2017, Trump's White House had solicited the company to send YouTubers to promote a coding project in Detroit; YouTube couldn't recruit any. A co-worker once told Stapleton that conservative shock jocks were big YouTube moneymakers, and more spirited Googlers started calling YouTube "CNN for Nazis." To Stapleton the shock jocks swirled around in the rest of YouTube's vast ocean. "There was so much going on," she recalled. "It just felt like a toxic cesspool."

She did have a front-row seat to another crisis that spring. In June, just before Gay Pride festivities, which counted YouTube as a corporate sponsor, queer creators complained again about having videos restricted and stripped of ads. Worse, some ads that *did* run were for gay conversion therapy groups. Hank Green called them "despicable and gross and disgusting." YouTube apologized and explained that while these commercials didn't violate policies, YouTubers could keep certain types of ads off their channel. (Although many creators didn't fully understand how to do so.) In the aftermath Stapleton opened a compelling document from YouTube's policy team unpacking the episode. With an automated system sorting a messy Long Tail of videos and ads, such crises would keep arising, the document concluded. To untie its Gordian knot, YouTube had to pick a side—either by rewriting its rules definitively to appease certain stakeholders, like queer creators, or by going fully hands off. YouTube couldn't be all things to all people. Or in Stapleton's words, it couldn't be "liberal icons, *yas queen*, #Pride and libertarian Peter Thiel–ville."

But so often her company wanted to be both. Google and Silicon Valley, after all, were all about abundance. This contradiction grated on Stapleton in the months after the Pride debacle.

Then she learned of a time when her company *had* picked a side. And that pushed her off the ledge.

On Thursday, October 25, she left YouTube's Chelsea office for her Brooklyn apartment, put her toddler to bed, poured a glass of red wine, and opened her laptop. She clicked to a company listserv for moms that she frequented and found a link to a morning headline from the *Times* about her company. *Click.*

> Google gave Andy Rubin, the creator of Android mobile software, a hero's farewell when he left the company in October 2014. . . . What Google did not make public was that an employee had accused Mr. Rubin of sexual misconduct.

The New York Times reported that a woman had accused Rubin, a longtime Google chieftain, of coercing her into oral sex in a hotel room. Security staff had found bondage sex videos on his work computer, and the newspaper unearthed an email he sent a woman that read, "Being owned is kinda like you are my property, and I can loan you to other people." All that and then the *Times* noted, "Google could have fired Mr. Rubin and paid him little to nothing on the way out. Instead, the company handed him a $90 million exit package." (Rubin denied coercing the woman to have sex and dismissed the allegations as "part of a smear campaign" from his ex-wife amid divorce proceedings.)

Stapleton finished the article and immediately waded into Google's internal chatter. Pichai had apologized over email, writing that the article was "difficult to read" and informing his staff of the shocking fact that Google had fired forty-eight people for sexual harassment over the past two years. The *Times* reported that Larry Page and Google's board had signed off on

Rubin's $90 million. Some at the company justified this as a standard exit package designed to keep c-suiters from heading to rivals. Laszlo Bock, then Google's HR chief, later said he advised Page to send Rubin off with nothing.

At 7:58 p.m., Stapleton typed a note to the moms' listserv. "Google women (and allies)," she wrote, "are REALLY rage-fueled right now, and I wonder how we can harness that to force some real change." An open letter or a strike? she asked. A walkout? Staff were still raging the next day at work. That afternoon Stapleton did what her company had trained her to do in the face of controversy: she started a listserv, "womens-walk," and invited others to join. When she awoke on Saturday, more than two hundred people had signed up.

Everything moved fast. One Googler proposed a list of demands for management and another started collating those pouring in. Men joined the listserv, and Stapleton broadened its scope. Seasoned protesters behind the revolts against the Pentagon and China projects came aboard. Several blamed Google for being too chummy or spineless with Trump; some were upset that Google contractors, like YouTube's outsourced moderators, worked with crummier pay and benefits. All the employees had watched the #MeToo reckoning bounce from one industry to the next, accelerated by social media, and arrive now for tech.

Five colleagues joined Stapleton to form an ad hoc organizing committee. They picked that Thursday, November 1, for the walkout. They chatted over an encrypted messaging app, but also used Google docs and calendars, their own tools. (Stapleton proposed calling the protest #MeGoo, but this was vetoed.) HR and publicity staff joined the listserv, which seemed fine; Google had always encouraged employees to voice concerns.

Stapleton dressed for work on Monday in her "The Future Is Female" T-shirt, "like Joan of Arc or something," she later said. Her group had by now cleared one thousand. She sent another email, asking, "Why are you walking out?" Hundreds of stories of sexism, racism, and harassment poured back. This, Stapleton would later write, "was a monument to disillusionment."

On Tuesday, Pichai sent a company-wide email that his prior apology "wasn't enough" and, in what was effectively an endorsement, assured staff that Google would support Thursday's walkout. *If you can't beat 'em,* Stapleton thought, *join 'em.*

Early Thursday morning Erica Anderson, a Google manager and walkout organizer, left for work with a good-luck bag of apple cider doughnuts from her girlfriend, the YouTube beauty guru Ingrid Nilsen. Nilsen had followed Google's activity and thought it mirrored dynamics on YouTube. She saw LGBTQ and sex-positive creators constantly struggle for funds and viewership, while bullies and troublemakers soared in its charts. "People who were actually causing harm were not only rising to the top but getting paid millions," she recalled. "It just seemed so backward."

Shortly after 11 a.m. that morning, Stapleton, draped in a green army jacket, led her colleagues, more than three thousand strong, out of their office to a small park near the Hudson where they gathered with megaphones. Others had marched out in London, Singapore, and Zurich. Protestors cried at Google HQ listening to a female engineer describe being drugged by a colleague at a corporate event. Googlers held signs that read: TIME'S UP, TECH and DON'T BE EVIL. In all, more than twenty thousand employees protested in fifty cities, a groundbreaking moment of white-collar activism, of Trump-era catharsis. A movement that could boil an ocean. The techie rebellion grabbed national headlines and, it seemed, made the company proud. Ruth Porat, Google's CFO, would describe the event as "Googlers doing what Googlers do best."

This halo of approval, Stapleton soon discovered, would last about two months. Google's c-suite and its insurrectionists were, like the cultural warriors and loudest voices on YouTube, talking past each other.

CHAPTER 31

The Master's Tools

The Google walkout had shown cracks in the diehard faith many Googlers placed in their corporation. Something similar was happening on YouTube's platform.

The people who made and populated the site in its earliest days *believed* in YouTube, as a shared project and shared community. As YouTube grew and splintered, the faith of its loyal creators and fans gradually eroded. By now, plenty had given up on the institution entirely, a widespread contagion in the Trump era. They were prepared to channel their fury.

It started with Will Smith.

YouTube: "YouTube Rewind 2018: Everyone Controls Rewind." December 6, 2018. 8:13.

The former Fresh Prince kicks off YouTube's annual year-in-review clip, the one YouTubers and fans pick apart, mining for faces, trends, and phenoms. Images whiz by. A cast of dozens of YouTubers celebrate K-pop, *Fortnite*, ASMR, "Baby Shark," charity, drag performers, "all women in 2018 for finding their voices."

To a YouTubeland outsider, this video might seem harmless.

Not so. After two years of economic turbulence and rapid upheavals,

YouTubeland picked this moment to let loose. The video showed old-media dudes (Will Smith, John Oliver, Trevor Noah), an affront to many native YouTubers. It featured many creators who didn't produce videos in English—mega-YouTubers from Korea, Brazil, and India—who were unfamiliar to boisterous American fans. And it glossed over big, indecorous moments from the year: beauty guru feuds and Logan Paul's newfound, hyped-up boxing career. The 2018 Rewind felt alien, ad-friendly. *Corporate.*

The mob spoke. Within a week more than ten million people clicked a thumbs-down icon on the footage, making it the most disliked video ever. Naturally, YouTubers made videos about the Rewind video. In one of his, PewDiePie told viewers he found YouTube's marketing reel "so disconnected from its community and creators." Although, he added, there were now too many YouTube stars to manage; some two thousand channels had more than one million subscribers: "To really please everyone, it's going to be impossible."

Since the DEATH TO ALL JEWS disaster, Felix Kjellberg had gone wilder on-screen. He started a new format, "Pew News," riffing on media critics and fellow YouTubers, raging like *Network*'s Howard Beale. He grew his beard out to Tolkien-dwarf length. He dropped the n-word in a gleeful moment on a video game livestream off YouTube, prompting an apology on YouTube ("I'm an idiot") and another critical news cycle. In one "Pew News" clip, he dissected Logan Paul's apology tour: after his Japanese forest stunt, the star went on daytime TV and made a doe-eyed video on suicidality. People had advised Kjellberg to do the same, but that felt "very disingenuous," he told viewers. "I would rather just show people that I've changed through my videos and time."

Not taking an apology tour may have hit his earnings, but not his audience size. By the fall of 2018 PewDiePie had more than sixty million subscribers whose loyalties had cemented during his publicity woes. Still, his audience wasn't growing fast enough for him to keep his crown.

That August, Social Blade, a sort of sabermetrics for YouTube, published a chart showing PewDiePie on track to lose his title as most-subscribed YouTuber. His challenger: T-Series, an enormous Indian record label, Bollywood studio, and entertainment juggernaut. T-Series started posting frequently on YouTube as cheap smartphones spread across India, introducing the internet to tens of millions of Bollywood devotees. Much of this audience had never owned computers or even TVs. T-Series was the music hitmaker, box-office mogul, and pop-culture factory that YouTube's Head had forever courted, all wrapped into one. Being an Indian hitmaker also dovetailed nicely with Google's fervor for its "next billion users."

And yet, for much of YouTubeland, T-Series was an invader: big, corporate media that pumped out dozens of polished YouTube videos a month. To be honest, few Americans had heard of the channel or cared about it until it began to encroach on PewDiePie, the embodiment of YouTube's freewheeling culture. Somehow PewDiePie, a Swede whose entire career depended on advertisements sold by one of the largest global corporations, became an antiestablishment figurehead. Kjellberg rose to the occasion. In October he made a dis track video, "bitch lasagna," addressed to T-Series, rapping with an Eminem inflection and edgelord lyrics. ("I'm a blue-eyes white dragon while you're just a dark magician.") It was classic PewDiePie—ludicrous, laden with internet in-jokes ("bitch lasagna" referenced a meme of an Indian man failing to court a suitor), a caricature of *something*, though it was hard for outsiders to tell what.

The rallying cry formed: "Subscribe to PewDiePie!" Its force shocked YouTube, which had by now grown all too accustomed to shock. An impish YouTuber, part of a Paul brother posse, purchased a Times Square ad for the cause, and Logan Paul himself commanded his Logang to support PewDiePie. Jimmy Donaldson, a.k.a. MrBeast, a rising YouTube phenom known for on-screen stunts of charity and excess, purchased billboards in his hometown of Greenville, North Carolina, that read, CALLING ALL BROS! YOU CAN SAVE YOUTUBE. SUBSCRIBE TO PEWDIEPIE. Others conveyed that message by hacking printers, streaming devices, and (poetically) smart cameras from Nest, which Google owned. A self-propelling internet meme took flight. PewDiePie added millions more subscribers.

In December, after the botched YouTube Rewind, the company decided it needed to appear hip to the criticism coming its way. Wojcicki told staff even her kids found that Rewind video cringeworthy. As an act of self-awareness, YouTube's marketing team prepared a playlist of top videos reacting to Rewind, a task that fell to Claire Stapleton. There was much consternation about including PewDiePie's piece, which obviously was popular. Kjellberg had a YouTube senior partner manager, Ina Fuchs, a German executive, who kept in touch with him, but since the 2017 fiasco the company had severed public ties with its biggest star. (YouTube continued to run ads on his suitable videos, of course.) To YouTube brass this felt increasingly untenable, particularly given the groundswell of support for him, which often landed as condemnation of the company. Fuchs and other colleagues argued that Kjellberg had been misunderstood and deserved more corporate support. Such decisions were not made lightly.

Stapleton was interrupted during a massage on Google's campus with missive after missive from her boss about including PewDiePie in the Rewind playlist and getting the messaging just right. Email debates ensued over whether YouTube's Twitter account should click the ♥ below one of his tweets. Stapleton believed it shouldn't. The YouTuber, she argued, was "irresponsible with his influence." She refused to put his video on the playlist.

But it appeared anyway. Her manager had asked another marketer to add it, going around her entirely.

• • •

Stapleton should have seen it coming. After the walkout, when she appeared in the *Times* and on TV broadcasts, a colleague warned her that such visible actions, all planned and unspooled inside the company, invited a response. The colleague quoted the civil rights activist and writer Audre Lorde: "For the master's tools will never dismantle the master's house."

Stapleton later recalled the lesson this way: "If you become inconvenient, then your days are numbered."

She had become the face of the walkout, along with Meredith Whittaker, a Google researcher and ringleader of protests against its Pentagon contract. Whittaker argued fervently that her company was taking dangerous ethical missteps with its artificial intelligence. She had been at Google since 2006, a veteran, like Stapleton, which made them useful advocates. They were also both white. A YouTube colleague who wasn't once told Stapleton they agreed with her on issues like PewDiePie but didn't have the privilege to go nuclear on the bosses.

Google's c-suite support for the walkout did not last. Its organizers hadn't stopped at a march but presented five demands, which included an end to pay inequity and an employee seat on the corporate board. Shortly after the walkout Stapleton and a few other women from YouTube held a private meeting with their CEO. Wojcicki had confided to some staff that she knew nothing of charges against Andy Rubin and felt disgusted by them. In this meeting her employees raised concerns about YouTube's gender pay gaps and its scarcity of Black leadership. Wojcicki indicated ignorance of these gaps and told them YouTube would right these disparities. After the meeting disbanded, a colleague turned to Stapleton and said, "She's completely lying." Wojcicki knew the data, they concluded, but was deflecting responsibility. "It was just lip service," Stapleton recalled.

That would be her last sit-down with the CEO. In January, Stapleton's manager informed her that her role was being "restructured." Officially, this was an employee "reorg," a regular occurrence at Google, but Stapleton lost several responsibilities and half the staff she managed, so she suspected other motives. She took her frustrations up the corporate chain, where she was advised to "rebuild trust" with her manager, maybe take some days off. The writing appeared on the wall.

March came, and she was invited to fly to California for a Well-Being Code Red retreat. "Oh this will be fun," she wrote to a colleague, in email deadpan.

• • •

On Thursday, March 14, Jennie O'Connor arrived at YouTube's "intelligence desk." The company had formed this division to mitigate risk in early 2018, after the Elsagate crisis. As its leader, O'Connor was responsible for looking past YouTube's Long Tail to spot troubling threats on horizons and anticipate messes so YouTube's moderators and machines could adequately address them. O'Connor recruited former intelligence officials and creator managers to keep a finger more firmly on the site's pulse. A twelve-year company veteran, she could "speak Google," said an old colleague. Most critically, she had worked on *product* as a deputy for Neal Mohan. "Unless you're in product or you're writing code," explained Hong Qu, a former YouTube designer, "you can't influence anything."

In her new job O'Connor got up to speed on ISIS. A former high school math teacher, she also got acquainted with the wild things kids were now up to. Like the "condom challenge" (dropping a prophylactic filled with water over a person's head to form a fishbowl helmet, good internet fodder). Sometimes O'Connor's unit was caught off guard, as in February, when a YouTuber exposed how pedophiles used coded links and phrases in comments below clips of children, sparking another wave of advertiser boycotts. O'Connor's team moved fast. Within two weeks it stripped comments from millions of videos, released a more efficient AI comment sorter, and set harsher penalties. O'Connor had set up a rotating global team of "incident commanders" to be on call for such immediate disasters. War rooms, intelligence, incident commanders—the militant language made everyone feel as if YouTube were battling adversaries.

That Thursday, U.S. senators rejected the president's emergency measure to build his border wall, prompting Trump to tweet, "VETO!" A Googler in Japan broke the Guinness World Record by counting pi to thirty-one trillion digits. O'Connor's workday ended without incident. She left YouTube's office and began to settle in at home when the emails about New Zealand arrived.

• • •

The terrorist was Australian and twenty-eight years old. He grew up in a city north of Sydney, playing video games and perusing message boards like the backwater forum 8chan. His parents divorced, and his mother entered an abusive relationship. Before the terrorist reached twenty, his father had died of cancer following an asbestos exposure, which resulted in a settlement and left his son significant money. The terrorist traveled often, usually alone, and, according to a later government report, "he did not form enduring relationships with others." He was white and considered himself European, believing both to be signs of superiority and identities under dire threat from rising immigration levels, particularly of Muslim migrants, his version of the Great Replacement theory that had spread online.

He watched YouTube and subscribed to channels. He posted in the Lads Society, a far-right internet clubhouse relegated to a private Facebook forum after the social network swept its public groups clean. This was likely the reference of his final line before committing mass murder: "Remember, lads, subscribe to PewDiePie." There was no evidence he watched PewDiePie's material or felt inspired by it in any way. This line was spoken to get attention.

In early 2017 the young man did make donations to a U.S. white nationalist think tank and to Freedomain Radio, the network from Stefan

Molyneux. (In a statement, Molyneux said he "immediately condemned the New Zealand terrorist.") On a trip through France that spring, the terrorist saw migrants walking in a shopping mall. "Invaders," he called them, later writing online that this epiphany led him to violence. But even before, there had been signs. His family would tell authorities that in late 2016 he returned from a trip "a changed person"—hardened, extreme, often discussing how Muslim migration heralded the demise of the West and world. His mother worried about his mental health. "Patriots and nationalists triumphant," he posted online after Trump's election. Later, in his Lads group, he wrote, "Our greatest threat is the non-violent, high fertility, high social cohesion immigrants." He read and absorbed material on the "Great Replacement." Books, forums, 4chan, Facebook groups.

Above all else, one service had undue influence, according to a report from the New Zealand government, which interviewed the terrorist after his act. This report concluded, "The individual claimed that he was not a frequent commenter on extreme right-wing sites and that YouTube was, for him, a far more significant source of information and inspiration."

Two days before his attack, he posted on his Facebook page several dozen links, including fertility rate figures and a British tabloid on Asian gang violence. He linked to many YouTube videos: speeches from a 1930s British fascist; news footage of Europe in disarray; Russian bombers over Syria; a biped robot marching to a German military soundtrack. Next to a Latvian folk song video, he wrote, "This is what they want to destroy." He later told investigators he used YouTube tutorials to assemble firearms for his attack.

He had moved to Dunedin, in southern New Zealand, in 2017 and lived there without incident. After his crimes New Zealanders scrambled to make sense of the mass murderer. "He was a total nothing," recalled Kirsty Johnston, a reporter who investigated his life. "A garden-variety racist. He had money and time."

Haji-Daoud Nabi had an adoring family and an admiring community in Christchurch. Nabi was seventy-one, a grandfather who had moved from Afghanistan in the 1970s and still wore traditional *pakol* caps. He fixed old cars and liked driving guests visiting the city to his mosque. He mentored fellow migrants in New Zealand but had also embraced his adopted land. His funeral featured a caravan of Harley-Davidson motorcycles, which he loved. "He was as Kiwi as he was Afghan," a friend remembered. Nabi called everyone brother.

On that fateful afternoon of Friday, March 15, Nabi stood at the door of his mosque, Al Noor, to welcome fellow faithful. Shortly after 1:40 p.m., he encountered a man carrying an AR-15 and a body camera set on record who would take Nabi's life and those of fifty other souls at Masjid Al Noor and another Islamic site. Nabi greeted him warmly, "Hello, brother. Welcome."

• • •

Jennie O'Connor opened her work laptop on her kitchen counter soon after, still Thursday evening in California. Colleagues had notified her of a mass shooting in Christchurch, where the attacker had streamed his murders live on Facebook. And that the stream had arrived on YouTube.

A protocol was in place. YouTube would quickly categorize the footage on its violence gradient and write corresponding rules for moderators and machines. O'Connor decided the video should come down, set those wheels in motion, and, as the night wore on, eventually tried to get some rest.

Tala Bardan, the YouTube violent extremism specialist, awoke early to see news of the shooting on her Instagram. She immediately wept. She wiped her tears, opened her computer, and began watching the terrorist's footage. It was even harder to watch the bystanders—clips of worshippers and neighbors crying in disbelief. Bardan helped write the guidance for moderators: scrub any re-uploads or clips praising the violence, but be careful not to delete news coverage. She took a taxi to the office to work uninterrupted; she would spend all weekend at home reviewing violent footage, her husband bringing plates of food to her desk. YouTube was bombarded with tributes to the Christchurch death reel; hatemongers or devious trolls had spliced the footage in ways to outsmart machine detectors. Bardan and her co-workers in Europe and Asia frantically tried to put out fires while California slept.

O'Connor awoke Friday morning to learn the protocols weren't working. At first she had thought YouTube needed more screeners. "We don't have enough reviewers," she pleaded overnight. But by morning even that solution wasn't adequate. At one point YouTube saw one new video replicating the shooting appearing every second. An executive later said that the alarming speed of reuploads led some inside the company to suspect that a state actor was involved. Mohan, O'Connor's boss, would describe it as "a tragedy that was almost designed for the purpose of going viral."

Virality had been a gift to YouTube so many times throughout its history. YouTube was designed as the internet's bottomless repository—videos first broadcast elsewhere, like the Christchurch livestream, could easily jump to YouTube and take off. To remain relevant, YouTube had rewired its algorithms to promote more breaking news footage so people who once flipped on the TV after a mass shooting would open YouTube instead, which they did. Even oddities of YouTubeland, like the "Subscribe to PewDiePie" cry, were now major news events. Unlike social networks, which had clunky search features, YouTube offered remarkably easy ways to find anything. All those mechanisms that let YouTube flourish as a business, tools created with little regard for unintended disasters they might bring, had combined into a nightmare fuel that the company couldn't turn off.

O'Connor was on a call en route to work when the decision was made. YouTube would remove any footage showing the Christchurch shooting, not just exact re-uploads. And it would cut the ability for viewers to search for the tragedy at all, removing an entire category from its searches for the first time. The company called off human reviewers, who couldn't move fast enough. YouTube turned its filter dials up and handed over control to its machines.

Part IV

CHAPTER 32

Roomba

In May of 2019 a billboard went up in San Francisco, near the freeway many Googlers used to commute to work. The billboard declaimed BREAK UP BIG TECH.

Presidential contender Elizabeth Warren had purchased the ad, but the sentiment crossed political lines. That summer Trump's Justice Department would open a monumental case accusing Google of being a monopolist. Senators devoted a panel to the dangers of unchecked artificial intelligence in social media. "Companies are letting algorithms run wild," declared one lawmaker. Republicans and Democrats, otherwise acrimonious about everything, agreed that internet gatekeepers like YouTube were too large and influential.

By that summer, YouTube had mitigated many of its major crises—brands didn't accidentally sponsor fringe videos anymore (or at least didn't make headlines for it); freaky kids' content disappeared; the company had, by and large, bottled up its irascible stars and brought its business back from the brink after advertising boycotts. But *this* crisis didn't have a clear resolution. Governments had decided Silicon Valley needed to be governed. In Europe officials passed Article 13, a sweeping law holding website owners more liable for copyright infringements, undermining YouTube's process for managing rights. Many countries, caught flat-footed by social media's rise, started drawing plans to regulate the industry. Even the Davos set began calling Facebook the new tobacco. While YouTube usually evaded that label, the company knew lawmakers wouldn't write just Facebook laws—they would come for YouTube, too.

The stakes had grown larger, particularly as the world moved closer to 2020. A *New York Times* investigation in 2019 detailed how medically dubious videos handicapped efforts to fight the Zika virus in Brazil and how YouTube's ubiquity there fueled the election of its right-wing president, Jair Bolsonaro, who would raise dubious claims about a virus to come. Micah Schaffer, an early YouTube employee, blasted his old company for propping up dangerous theories, like vaccine conspiracies, in ways he said wouldn't have happened during his tenure, when there wasn't such hunger for profits. "We may have been hemorrhaging money," he told a reporter. "But at least dogs riding skateboards never killed anyone." In June of 2019 a queer creator, Carlos Maza, publicly shamed YouTube for failing to halt what he considered racist and antigay slurs from another YouTuber, the right-wing pundit Steven Crowder. YouTube waffled on the dispute, which outraged some employees. At the San Francisco Pride Parade, where Google was an official sponsor, a few dozen Googlers marched in protest, carrying signs that read, YOUTUBE HARASSMENT KILLS US.

During a meeting that year, an employee asked Wojcicki to name her biggest fear. She replied quickly, "Regulation."

YouTube, which had so often moved slowly to respond to criticism from its creators and staff, began to move fast, embarking on a frantic effort to regulate itself, before governments did. In June the company rewrote its hate-speech rules to prohibit videos "alleging that a group is superior in order to justify discrimination, segregation or exclusion." Gone was anything glorifying Nazis or denying "well-documented violent events" like the Holocaust and school shootings. After Christchurch, the company banned footage of a "deadly or major violent event" filmed by the perpetrator. YouTube updated harassment policies to outlaw threats against other creators; at first, this included an exemption for "public figures," though the company struggled with that, having created a service designed to make everyday people famous. (The policy was later extended to cover public figures.) YouTube recruited childhood development experts and started an internal program (code name: Crosswalk) to promote more educational and wholesome videos for kids. One employee remarked that the world would soon look back on YouTube before all these overhauls the way it viewed cars before seat belts, a public hazard.

But the new safety measures were not universally admired. Weeks after announcing its hate-speech update, YouTube wrote on Twitter that it applied such policies "without political bias," to which Donald Trump Jr. replied, "No one believes you." After YouTube removed ads from the channel run by Crowder, the conservative comic, Senator Ted Cruz demanded the company "stop playing God." To the Trumpian right, "hate speech" was just a cover for Silicon Valley to force its liberalism down everyone's throats. Faced with this bare-knuckle politics, YouTube never punched back. Instead, it doubled down on a belief in its machines; the company wrote rules but left enforcement to its automated systems, which it believed could move faster, more efficiently, and without the taint of human biases. Machines could *scale*. Machines had no predisposed feelings about Donald Trump Jr. YouTube stuck with this logic in ways that sometimes strained credulity. Back in 2018, the *InfoWars* shock jock, Alex Jones, was wiped from the mainstream web after Spotify, Apple, Twitter, Facebook, and YouTube all banned him within a month. It looked as if major platforms, after broadcasting Jones's conspiratorial rants for years, decided he had crossed a collective line. Not quite. YouTube actually froze Jones's account for showing a minor being bullied, a breach of child endangerment rules. (The video was unrelated to his school shooting conspiracies.) Jones then tried to maneuver his way around the freeze and upload on another account, which got him expelled outright. YouTube took him down on a technicality, like Al Capone busted on tax evasion. (Jones had more than two million YouTube subscribers at the time.)

Despite its rapid overhauls, YouTube got busted by Uncle Sam that autumn. The site had amassed a voracious audience of children as big as television's (or maybe even bigger) without being bound by any of the same rules. YouTube's standard company line—kids watched only with parental supervision—no longer held water. In September 2019 the FTC charged YouTube with violating COPPA, the children's privacy law, which outlawed targeted ads on "child-directed" media, and fined the company $170 million, the largest sum ever for such a case. Afterward, YouTube had to split its site in two: every video was either marked "Made for Kids" or not; videos marked that way couldn't run higher-priced commercials based on viewers' browsing and personal details, kneecapping income for thousands of creators.

Even YouTube cheerleaders began to express doubts. After leaving his role running the company's European operations, Patrick Walker still praised YouTube for inventing "a whole new language of storytelling." Yet he decided never to leave his young daughter alone with the site. "We didn't really anticipate the dark side," he recalled. "These platforms became so compelling that people start to lose their sense of agency." Walker went on to form Uptime, an educational company designed to get people to stop mindless scrolling, the inverse of the goal his old employer kneaded into its algorithms.

• • •

Susan Wojcicki developed a ready response for the mounting attacks, a go-to company mantra: "the Four *R*s of Responsibility." This was a vow to *Remove* material that broke rules, *Raise* "authoritative" sources in rankings, *Reward* "trusted" creators, and *Reduce* borderline footage.

Wojcicki recited these maxims in meetings and during a goodwill tour of interviews she gave to popular YouTubers. She recorded her first that April, a month after the Christchurch tragedy, in a conversation that revealed the company's logic and its political strategy.

Wojcicki was in India, where she had announced that YouTube had over

265 million monthly visitors, its fastest-growing market—and where a new competitor, TikTok, had suddenly exploded in popularity. She spoke with Prajakta Koli, a multilingual comedian behind the channel MostlySane. Koli asked the executive what she watched on YouTube. "I think I'm a typical user," Wojcicki replied. "I love to do yoga videos." She mentioned cooking and crafting. (YouTube had initiated an internal effort to recruit more women to the site, which skewed heavily to male viewers; that may have explained Wojcicki's calculated reveals.)

Wojcicki cycled through all the ways YouTube strove to support creators. And then she tried shedding light on why the company had grown much stricter on what creators could broadcast and how they were paid. "If I were to say I had one focus," she said, "it would be responsibility." But here, as in future interviews, Wojcicki was very careful to explain that she and her company weren't calling the shots. Viewers were. When YouTube ranked and recommended videos, it relied on clicks, survey responses, and eyeballs. "So, it's not us making those decisions," Wojcicki told Koli. "It's what our users are telling us." This, she added, was a way of "really highlighting the content that is useful to our users and good for society. And again, we don't want that to be *us*."

YouTube wanted to be seen as *responsible*, as a platform propping up creators that respected shifting norms of speech, tolerance, and decency (and avoided firestorms). YouTube *did not* want to be seen as responsible for setting those norms—or even for doing the propping. That might risk losing liability protections, pissing off conservatives crying bias, and violating the company's sacrosanct belief that its audience was king. This stance could be confusing to outsiders but made perfect sense to YouTube.

Yet YouTube did call plenty of shots.

For one, it determined how *responsible* each video was, ranking them accordingly in its promotional algorithm. This worked a bit like Uber rankings: YouTube counted when viewers gave videos four or five stars (out of five) in viewer feedback buttons, and fed those clips—along with the surveys, likes, and smorgasbord of metrics—into an undisclosed formula to determine what they called "valued watch time." More responsible videos brought more valued watch time. But this was an inexact science. One engineer who worked on the features recalled that viewer surveys had "alarmingly low response rates," around 2 or 3 percent, and chiefly came from male twentysomethings. According to this engineer, executives okayed algorithmic tweaks that improved *valued* watch time so long as they didn't dent *regular* watch time. A usual rule of thumb traded a 1 percent increase of valued watch time for anything less than 0.2 percent of overall viewing. It was always a judgment call. (A spokesperson said the company didn't have any "hard and fast rules" for this process.)

To monitor YouTube's revamped policing system, Wojcicki had created a standing Friday meeting with senior staff, which she called Roomba, after the little robotic vacuum that sweeps up floors on its own, purring in the background. (Wojcicki gifted the actual devices to team members.) Employees who attended these meetings remembered frequent debates over specific videos with political or cultural weight, such as whether those from Steven Crowder, the popular pundit fond of yellow-face routines and the insult "anchor baby," were satire or harmful racism. Deputies praised Wojcicki for bringing in advisers beyond tech's typical set of product and engineering. Enforcement squads within YouTube were no longer as understaffed or sacrificed on the altar of growth. Still, Roomba, a governing board for more than two billion global viewers, reflected both Google and its industry as a whole: mostly white and Indian American, many advanced degrees, wealthy. "The people in that room do not look like America," recalled one former executive.

During Roomba debates Wojcicki rarely argued forcefully, preferring to arrive at decisions by consensus. She voiced displeasure with applying different judgments to similar videos, such as placing one video deemed satire in its penalty box but not another. YouTube considered such evenhandedness a point of pride. Managers argued that its rules were designed to judge a video's context, not the speaker behind it, a clear dig at Facebook, which tended to carve out ad hoc exemptions for a certain bloviating president.

"Everybody is treated equally," explained Neal Mohan, YouTube's product chief. "Why should a head of state get a pass when you and I don't get a pass?" Indeed, in the coming years YouTube would pull videos from Trump, Bolsonaro, and several elected officials for violating its rules.

But to some this insistence on applying even standards could lead to dithering. "Whataboutism," a former executive called it. *What about this video? What about that one?* Another equated the decision process to "death by a thousand cuts." Several former staffers lamented that Wojcicki and her leadership were indecisive, responding only when there was bad press or financial threats, and favoring consensus over action. "It's ridiculous. Everyone needs to agree," recalled Susanne Daniels, a veteran executive. "It slows down the process." Daniels, who left YouTube in 2022, said the company had made progress since the PewDiePie drama, but concluded: "It's a company that was not and still isn't wholly prepared to react to the potential negative consequences of hosting an open platform."

Jennie O'Connor, the YouTube director, disagreed. She dealt often with Wojcicki on thorny decisions, like in early 2019 after a YouTuber exposed how comments beneath videos of minors were a haven for pedophiles. In crisis mode O'Connor's team voted to purge all kids' videos of comments. She fretted before taking their decision to Wojcicki, nervous about proposing an end to what had been a core feature of YouTube for its entire life. "That's just what we need to do," Wojcicki agreed. "She's the calmest person ever," O'Connor later said of her boss. "She's pretty decisive."

Certainly, every decision Wojcicki made had massive ripple effects. Advertisers, for instance, welcomed YouTube's move to strip comments from kids' videos. ("I've never seen Google respond or react so quickly," recalled one ad agency director.) But affected YouTubers complained about the loss, because comments were the main avenue for audience feedback and an engagement metric YouTube had rewarded for years. Major decisions never pleased everyone, and YouTube began to accept this. "There are no right or wrong answers," O'Connor said. "There are just trade-offs."

• • •

YouTube did believe, however, in right and wrong regulation. The severity of Europe's copyright law particularly shocked the company. During 2019 YouTube reshuffled its policy team and spent considerable energy and resources combating Europe's plan, starting a campaign hashtag (#SaveYourInternet), and asking YouTubers to inveigh against the measure. Many did, including Felix Kjellberg, who by the year's end had returned to YouTube's good graces.

On April 28, six weeks after Christchurch, Kjellberg uploaded a video addressing viewers as himself, without slipping into his persona's screech. He asked fans to end the "Subscribe to PewDiePie" meme. "To have my name associated with something unspeakably vile," he said, "has affected me in more ways than I've let show." YouTube had assigned a team to handle the delicate race to 100 million subscribers between PewDiePie and T-Series, drawing up scenarios for possible reactions from YouTubers, media, and rabid fans. T-Series crossed the milestone in May without incident. That month a document circulated around YouTube about Kjellberg with notes from Ina Fuchs, his partner manager. The star, it read, would like the company "to recognize him more again since he feels that he has been publicly ignored." Fuchs praised PewDiePie's newfound success with meme reviews, including collaborations "with other top creators (such as jacksepticeye and Elon Musk)." The document favorably mentioned Kjellberg's comments against Europe's copyright measure. (After Brexit, the U.K. decided not to implement the law.)

On July 25, YouTube invited Kjellberg and eleven other European creators to London's Victoria and Albert Museum, where they were given a special tour of a Christian Dior exhibit. The company planned a roundtable discussion followed by a private reception and dinner. Susan Wojcicki flew in, although the advance schedule specified she would not be attending the dinner. Yet it did list the following meeting from 5:00 to 5:30 p.m.: "Susan and PewDiePie." Ahead of the roundtable event, YouTube prepared a list of talking points to reinforce:

1. Responsibility is our top priority as a company.
2. Creators are at the core of everything we do.
3. Regulation will continue to progress.

In the months that followed, Kjellberg would stay out of headlines. He started bleeping out profanities in his videos and even posted footage playing *Minecraft*, a return to his earliest form. He participated in another budding YouTube genre: filming his reaction while he watched his old videos, a nostalgia reel that let his aging viewers relive their own youths. By the next spring, he would sign a contract with YouTube for gaming livestreams with little fanfare, his first official business tie with Google in more than three years. Back in the fold.

• • •

As Kjellberg drew closer to Google, Claire Stapleton's career there unraveled. She had hired a lawyer after her job responsibilities were cut, prompting some reversals, but she continued to feel iced out, left off email chains, unwelcome. She was outvoted on PewDiePie. After Christchurch, Stapleton had assumed that her position on Kjellberg—that the star's influence was corrosive—would win out. But YouTube disagreed. Google, meanwhile, had no plans to reengage with its walkout instigators. Stapleton, pregnant with her second child, worried about the toll work stress was taking on her body. She anguished. ("My life-force has dwindled to the approximate size and consistency of the slimy nib at the bottom of a kombucha bottle," she typed in one email newsletter.) "Why don't you just quit?" her husband asked during one tormented conversation.

"Google is more than just a job," she replied. "It's my home."

Stapleton had never rebelled before the walkout, but she felt whiplashed seeing Google's values and view of her change so quickly. Though perhaps its values hadn't actually changed, but she had. Stapleton, a former colleague said, would not have so vocally protested her company "if she hadn't gone to YouTube." If she hadn't been working for the site so close to the churn of humanity's vile parts and so adept at amplifying that vileness right back. Half of the Google walkout organizers worked for YouTube. That April, a month after the wine country retreat, Stapleton and Meredith Whittaker, another sidelined walkout organizer, wrote to co-workers detailing their ordeals and calling for more activism. They used a loaded word for Google's actions: "retaliation." Stapleton now heard Google's message clearly: *You don't belong here anymore.*

She finally left in June, rather unceremoniously. Some colleagues planned a farewell in the Chelsea office, where a security officer waited in the wings to confiscate her corporate devices. Most of her marketing team was out on another off-site retreat, in Southern California, bonding over sessions of goat yoga. In her farewell email to colleagues, Stapleton wrote that she was branded with a "scarlet letter." She kept up her newsletter and, a couple weeks later, let loose there: "If youtube looks from the outside like a rudderless ship without a clear point of view on its absolutely consequential role in the general geo-socio-political landscape—well, yes, it is that."

And yet, a day before, while surfing the web, she stumbled on a clip she enjoyed. Then another. "I have to admit," she wrote. "THE VIDS REMAIN GOOD." She linked to several before signing off: "Kill the youtube recommendation algorithm!"

• • •

Eight months later YouTube's algorithm caretakers saw the unsettling activity begin. On February 9, 2020, there was a sharp rise in people searching the site for anything on the terrifying new virus. Engineers working on

Google search saw the same thing. Google closed its offices on March 6, days before much of the country shut down.

By May, as the grim reality of the pandemic settled in, Susan Wojcicki arranged a video chat with Hank Green that he would broadcast. The pioneering YouTuber had already addressed COVID-19 on his science channel and his personal vlog ("The Anxious Scroll"). He welcomed Wojcicki, who looked into her webcam in front of a massive white inlaid bookshelf stacked neatly with paperbacks and family photos. "Well, let's get this out of the way," Green began. "How is quarantine?"

"It's tough," she replied.

Wojcicki was facing all the c-suite lockdown challenges: managing a business remotely, caring for stressed-out staff, and bracing for economic nosedives. And she had to deal with the videos that had rushed in about the coronavirus. It was a hoax, a nightmare, a Chinese plague, a product of secret puppet mastery from Bill Gates and Big Pharma. It was one silver cure away from disappearing. Doctors uploaded hours of footage on virology. One physician posted exhaustive instructions for washing groceries, which, it turned out, wasn't effective against the virus at all; the video went viral anyway. Everyone trapped indoors, scouring online for news, was hit with what health officials called an "infodemic." In March, YouTube halted some human moderation until lawyers sorted out how reviewers could screen "egregious content" from homes.

YouTube, Wojcicki explained, leaned on its new self-regulation systems. Engineers built a "shelf" to prominently display select videos about COVID-19 on the site and rejiggered code to promote videos from established news outlets and medical authorities higher in rankings. In April, YouTube wrote a new rule banning videos with "medically unsubstantiated" claims, and the company said viewers were primarily fed "authoritative" footage on the pandemic. Weeks before Wojcicki sat with Green, Brits who had been stirred up by web conspiracies, which claimed 5G networks spread the virus, attacked several cell towers. YouTube scrubbed videos touting those conspiracies with uncharacteristic speed. "We had to take incredibly fast action," Wojcicki told Green. She touted her moderation division and intelligence desk, sketching out these operations in the broadest of strokes.

"I agree that these are the right steps," Green replied, before moving delicately to his next point. A few people atop a private company were making decisions affecting most of humanity behind a thick curtain. "But it also seems," Green went on, "that this is so much power for one organization to have."

Wojcicki immediately swung into defensive mode. YouTube, she insisted, had plenty of powerful competitors.

A month later YouTube held a virtual event for advertisers highlighting how viewership had exploded during quarantines, when people were stuck at home. On TV screens alone, not including computers and phones, humanity watched 450 million hours of YouTube a day, up 80 percent from the year before.

CHAPTER 33

Which YouTube?

The pandemic was very good for Google's business. It suffered some in the first uncertain months as fewer people searched for things to buy and do, but eventually the world moved online for commerce, work, and comfort, and that only helped the internet's front door. From March of 2020 to the fall of 2021, Google's share price nearly tripled.

YouTube's value to Google also grew tremendously. The video site had finally started reporting ad revenue: it would post $19.8 billion in 2020, more than double its 2017 figures and just $6 billion shy of the annual intake of its old foe Viacom. During the pandemic billions frequented YouTube to ease boredom or simply get through life, devouring how-to tutorials on haircuts and meditation. When Hollywood shut down, YouTube became the de facto media. Housebound late-night TV hosts fumbled it at first, shooting from webcams angled awkwardly upward, displaying their nose hairs, and keeping in pauses for audience reactions instead of the jump cuts YouTubers knew online attention spans demanded. "Y'all seem to be failing pretty hard when it comes to doing our job," the YouTuber MatPat teased in a video. As COVID-19 shutdowns spread, the *Office* star John Krasinski launched a YouTube series called *Some Good News*, which was purchased within two months by—wait for it—Viacom.

YouTube disclosed few statistics, but those it did—views of videos about sourdough grew more than 400 percent in the pandemic's first months—revealed an unprecedented explosion in usage. YouTube delighted ever more people every day.

Yet as the world kept changing in 2020 and the company tried to keep pace, it became clearer that YouTube still could not please everyone.

Internally staff often discussed this dilemma with a series of thought exercises posing one main question: *Which YouTube*? Would it be the place agreeable to employees, advertisers, and liberal values, a sort of Disneyfied version of the web? Or was it a wild playground agreeable to all sorts of speech and scale? Staff considered this identity crisis as a tug-of-war between its "brand" and its platform. *Which YouTube should YouTube be?* The answer was never clear. Ideally, YouTube wanted to be both.

But as the year progressed, the company had to reckon with this question again. When it did, those affected by the answer, as usual, found it capricious and unfair. Others felt it came far too late.

• • •

America's tense summer began when a Minneapolis police officer knelt on George Floyd's neck and took his life, igniting the nation's largest mass protests since Vietnam. YouTube attempted to rise to the occasion. It put "Black Lives Matter" on its home page and managers spoke to the historic moment. (Sometimes clumsily: in one company meeting about the protests a white executive recruited to lead Trust and Safety told staff that he loved John Legend and that one of his groomsmen was Black.) The company earmarked $100 million for Black creators. Most took the money.

But not all. On June 2, the YouTube director Malik Ducard reached out to Akilah Hughes about the fund. Hughes, who had spent more than a third of her life on the site, had not posted for over a year. Since then she had taken TV roles and begun podcasting, an increasingly popular format for online creators. She felt little residual warmth for YouTube. After the vlogger Carl Benjamin, a.k.a. Sargon of Akkad, repurposed her 2016 election video, Hughes sued him for copyright infringement. She lost. Hughes referred to Benjamin as a "white supremacist"; Benjamin denied being one. Several YouTubers weighed in on the case, and Hughes felt inundated with invective online. She heard nothing from YouTube staff during the episode and concluded that they didn't care. Only now, when every Fortune 500 was embracing racial justice, did YouTube reach out.

Hughes emailed back, thanked the executive, and then turned blunt. "Until YouTube commits to ridding this site of White Supremacists and their communities we will continue to have desensitized white people killing us," she wrote. "YouTube is *fully* complicit in the moment we are in. Run that up to Susan." Hughes declined the offer. "They want to make a lot of money, where everyone's safe and fun, like the Disney channel," Hughes said later about YouTube. "And they want none of the heat for the fact that they have absolutely allowed white supremacy to spread."

On June 29, a day after Trump called Joe Biden "a Low IQ person" on Twitter, YouTube purged the channels of several inflammatory white men. The purge's full extent was not specified, but it wiped out such prominent figures as the former Klansman David Duke; Richard Spencer, a white nationalist who once delivered a fiery "Hail Trump" speech; and Stefan Molyneux, who had uploaded thousands of videos to the site over the course of fourteen years. YouTube made no public report or comments to specify which videos violated its rules and how. Molyneux said he received no word from the company explaining the deletion. "My account was in perfectly good standing before being deleted," he said later. All the videos simply disappeared.

From the outside it looked as if YouTube had awoken to the moment. Officially, however, the company attributed the purge not to protests but to its hate-speech update from the prior year. Rewriting policies didn't have an immediate effect. YouTube had to figure out enforcement, how to translate those policies into rules for moderators and code for machines, which then had to make their way down the site's Long Tail. YouTube only evaluated videos uploaded after its policy change and gave a channel a strike if it broke a rule. Like baseball, channels had three strikes before going out. "That just generally takes time," said O'Connor, the YouTube director.

This had become a standard company response to criticism, sometimes delivered with a tint of frustration that those outside were unable to see just how massive, nuanced, and unwieldy YouTube could be. "It's not like we pick our heads up and say, 'Oh my gosh, there's hate speech on YouTube. We should really get on top of that,'" offered O'Connor. Definitions and standards were a moving target. The company consulted experts to determine if videos discussing immigration were touting ethnic superiority or simply recounting political debates. After all, the U.S. commander in chief had referred to Mexicans as "rapists" and several nations as "shithole" countries. (YouTube declined to share the experts it consulted.) "It's a very hard policy to get right," O'Connor added tactfully, "particularly because of its adjacency to political speech."

Still, some inside YouTube saw a glaring double standard in these practices.

That June, Tala Bardan, the violent extremism staffer, worked with colleagues to create a presentation showing the discrepancy between YouTube's treatment of Islamist hate and white hate. Slides revealed how every clip featuring radical Islamist clerics was taken down, even scriptural sermons, while footage from neo-Nazis and their ilk (including the "Unite the Right" Charlottesville rally organizer discussing "white genocide" on camera) was left untouched. After the Charlottesville rally, some inside YouTube proposed identifying accounts as domestic terrorists to include in the more strict "violent extremism" category, but the company never did. Bardan's violent extremism team had labored to achieve a near-perfect 98 percent "quality score"; the team moderating hate speech never got close to such results. "Because hate is hard," she recalled. A colleague on that team once confessed to Bardan that they were so swamped with material that they rarely touched videos marked as white supremacist. In one meeting, Bardan argued that videos touting the Great Replacement theory should come down since they had a clear link to real-world violence, but she felt that executives didn't grasp the issue. "Clear policy lines. Poor detection," read the presentation Bardan co-wrote. It listed recent attacks carried out by avowed white nationalists in Christchurch, Wisconsin, South Carolina, and Texas. "List goes on."

That document had included recommendations for improving the enforcement of rules against hate speech. By the end of the year, Bardan had left YouTube and hadn't heard any response from leadership.

When YouTube did address this issue, it used the same arguments as Facebook and other social networks: eradicating Islamist extremism online was easier because national governments agreed on how to define it. There were terrorist registries and sanctions. YouTube relied chiefly on the proscribed terrorist organization lists from the U.K. and U.S. governments. "On a relative basis, it's simpler," O'Connor acknowledged. Nothing similar existed for white nationalism, she added.

At one point, some at YouTube proposed basing policy on the hate group and actor classifications from the nonprofit Southern Poverty Law Center. But the SPLC had become a prominent Trump adversary and YouTube leaders, worried about political risk, declined to do so, according to two people familiar with the decision. SPLC "became a dirty word inside YouTube," recalled one former executive.

This approach was not unique to internet companies. Many governments tended to prioritize a certain form of terror over others. In New Zealand's official analysis of the Christchurch mass murder, the report concluded that the nation's security apparatus had focused "almost exclusively" on Islamist extremism—an approach that "was not based on an informed assessment of the threats of terrorism."

• • •

As 2020 dragged on, YouTube's business achieved a remarkable stability. Wojcicki's effort to Make Ads Safe Again had largely worked, with YouTube delivering enough "brand suitability" guarantees and reliable eyeballs to satisfy advertisers. In the summer of 2020 major brands boycotted Facebook for failing to handle hate speech; they didn't leave YouTube. The company had taken over Google's music app ($9.99 a month) and launched YouTube TV, a streaming service with dozens of cable channels ($64.99 a month). Lumped with ad sales, those products helped YouTube rake in more than $20 billion that year, hitting the goal Wojcicki set back in 2015.

And YouTube began spreading its wealth more widely again, quietly expanding the pool of creators eligible for payments since slicing that number in 2018 after its scandals. During 2020 countless new broadcasters turned to YouTube for money, a side hustle in a profession less than two decades old. By the following year, the company would announce that more than two million creators were in its partner program—not quite the sum before the 2018 reduction but still one of the largest, most complicated payments systems on the planet.

YouTube reignited a marketing program, "Creators for Change," that handpicked stars to talk about issues like bullying and racism. And for other radioactive topics the company ignored, creators did the work themselves.

Natalie Wynn had come to YouTube a decade before, plunging into the skeptics world of raging atheists. Wynn, who is transgender, took a break from YouTube to study philosophy and returned under a new persona. Neighbors in Baltimore wouldn't recognize her as such, but once she applied makeup, set the lighting, and donned an elaborate costume on camera, she was a demi-celebrity.

ContraPoints: "Men." August 23, 2019. 30:34.

"What are we going to do about men? Because, no offense, as a group you guys kind of seem like you're not doing okay?" Wynn wears a black blouse, red lipstick, and a huge black fedora. The video follows a familiar arc: she dissects books, vlogs, and message board posts like a waggish philosophy lecturer. Chapters appear on-screen. "Proposition II: Diary of an Ex-man." She mentions two more mass shootings that happened over the weekend, both carried out by white men. "We tell them they're broken without really telling them how to fix themselves," she concludes. "As long as this male identity crisis goes on, I don't see any end to these problems."

ContraPoints was often grouped with "LeftTube," an amorphous batch of Bernie Bros and pundits who punched against the site's right using reliable search engine tricks and bombast. Wynn preferred theatrical pageantry. Press dubbed her "the Oscar Wilde of YouTube." (A YouTube commenter called her "the leftist PewDiePie.") She criticized purists both on the right and left, and her dense tutorials on the radioactive vastness of edgelords, incels, and angry men online had a ready audience. "Gender Critical," an opulent ContraPoints video on transphobia, claimed nearly half a million views in its first day. She was widely recognized for an ability to "de-radicalize" viewers, chiefly men, drawn to extremes on the internet.

On-screen Wynn borrowed from horror auteurs like Dario Argento. But, really, ContraPoints was pure YouTube: meta and surreal, mixing thick coats of irony with sincerity, dick jokes, and a running, intimate conversation with viewers. Movies and TV had started depicting meaningful trans characters, though they were often played by cisgender actors. YouTube let people play themselves. It's hard to imagine CBS or Netflix broadcasting a trans woman unpacking Hegel in lingerie or the Hammurabi code in cat-eye contacts. Wynn, like all good YouTubers, watched everything: shock jocks, beauty gurus, competitive eaters, ASMR whisperers. She devoured YouTube's full madness, its stuff that "no gatekeeper would ever permit," she said. "I loved that."

In 2020 one of YouTube's most chaotic stars surprised everyone with a political gesture. Early in June, Logan Paul uploaded footage of his podcast *Impaulsive* on YouTube; since his debacle in Japan, the Gen-Z Adonis had added podcasting and boxing to his YouTube repertoire. George Floyd had just been murdered. Paul titled this video "America Is Racist" and went straight to his point: "I'm embarrassed that it has taken me twenty-five years to realize this: it is not enough to be 'not racist.' You have to be anti-racist." Paul read off a script but delivered his lines passionately. "Half of the reason I'm able to get away with some of my hooligan shit on my vlogs is because I'm a white kid," he said.

His speech went viral, propelled by its message and shock that the YouTuber, famed for his recklessness, would wield his influence this way. Afterward, Graham Bennett, his manager at YouTube, started touting Paul's "newfound maturity" internally. Bennett did not consider YouTube's issues with incorrigible creators solved. "In an open, free-speech platform, people can still be idiots and racists," he said. (Paul's redemption was marred a bit after his YouTuber brother, Jake, appeared in the news for promoting coronavirus hoaxes and looting during summer protests.) But Bennett was certainly pleased with Logan's turn.

The Paul brothers were the most famous members of a YouTube cohort that, unlike prior ones, wasn't rebelling against staid forms of media. They didn't watch TV growing up. They watched YouTube. The Pauls had spent much of their sentient lives on-screen, part of a generation with the means and desire to share endlessly for the public record. Someday the Paul progeny might, if they like, leaf through not just old photo albums or even Facebook posts but a living, breathing, moving representation of their existence.

Paul's pivot made Bennett recall something Sergey Brin, one of Google's founders, said early about YouTube. If you think about it in its totality, which Google often did, YouTube was pretty close to a collective human memory. That one day you could go on there and find not just any how-to clip and music video but any event and lived experience you'd like. "We actually have created—not deliberately by any means," Bennett said, with a chuckle. "We created *accidentally* a visual repository of human memories. Which is kind of crazy when you think about it."

• • •

It is crazy. YouTube's repository, inconceivably vast before 2020, expanded during the pandemic at rates even the company couldn't imagine.

But not everyone kept uploading memories. Many YouTubers left the profession before even entering middle age. By 2020, Freddie Wong, the early YouTube enthusiast, had stopped making videos for YouTube. He had determined that the site did one thing well, something he was no longer interested in. "If you want to create content, then go for it," Wong concluded. "But if you're more of an artist type or a little more fickle in terms of what makes you happy with your creative output, it's a stifling place to be."

In Wynn's life as ContraPoints, YouTube took its toll. She was doxed repeatedly. She felt the strain of persistent exposure and output familiar to most YouTubers. "Let's be honest," she said. "This is not good for anyone's mental health." She developed an opioid addiction during the pandemic—and called her YouTube career "a contributing factor." Wynn earned Google ad money, but most of her work was funded from Patreon, a service that let fans pay creators directly. She never heard from anyone at YouTube.

Scores of other creators complained of burnout from the platform's unending demands. The veteran YouTuber Derek Muller once explained the phenomenon using empirical research: To be an expert at something, people needed ample practice, timely feedback, and a "reliable environment." YouTube offered the first two, but its algorithm changed so frequently that its environment was never reliable. "So you never feel like an expert," Muller said. "You never quite know what's going on."

Ingrid Nilsen, the petite "Lifestyle" vlogger, had spent all of her twenties uploading, blurring the boundaries between her private and her public selves. She had flourished in a career that barely existed before she began and recounted much of it with pride. But there were moments, countless ones, she wished she could undo—filming her drive to lunch, loading the dishwasher, doing the laundry, all this normal stuff most people don't broadcast to millions. "I didn't give myself those mundane, regular moments," she recalled. "I felt like I had to—I had to share everything."

She had abandoned her second YouTube channel, the GridMonster, named after the site's endless scroll of shelves, and quit Vlogmas, her daily holiday posting tradition. By June she had decided to participate in another YouTuber ritual: the tearful farewell. She placed a camera set to a nice soft focus on a tripod in her apartment.

Ingrid Nilsen: "This One Is For You (My Last Video)." June 30, 2020. 48:25.

"It is so hard for me to make this video because I do feel like we have grown up together," Nilsen says into the camera, then cries. For this departure, she tells her life story—all the highs, lows, shames, and joys. She won't be leaving the internet, but she is done influencing. "Posting will be more on *my terms* because my mental and emotional and financial well-being will no longer be tied to, essentially, how much people like me online. And that feels like freedom." As the video ends, Nilsen thanks the viewers for the "best decade" of her life. "We did this and we did it together."

She leaned forward and turned the camera off.

• • •

That summer employees on YouTube's Trust and Safety team, settling into their work-from-home routines, got a fresh "p-zero." That's Google-speak for priority zero, a pressing issue facing the company. Three summers earlier, after the London Bridge attacks, p-zero was violent extremism; then it became child safety. Now, with a deadly pandemic spreading and a contentious

U.S. presidential election on the horizon, p-zero was combating misinformation.

There were some flare-ups that summer and fall over controversial coronavirus videos, but for the most part those didn't boil over into publicity disasters. Ahead of the election, Donald Trump's campaign had asked to use a new YouTube feature: video ads that livestreamed at the top of YouTube.com; Trump had wanted to broadcast his live commentary during the Democratic debates, according to a YouTube employee involved in the discussions. YouTube declined, but it did sell his campaign multiple days of home-page ad realty, including slots on election day. The company determined that Trump's ads, characteristically bombastic, did not break its misinformation rules.

As November neared, Trump and his surrogates began heaving more doubt on the integrity of America's electoral process. YouTube bosses displayed a steady calm publicly. They had outlawed videos misleading viewers about voting or inciting them to interfere with the process but decided to permit clips they defined as "discussions of election results." As the election approached, YouTube managers argued that its existing system was prepared for any deluge.

On November 9, when the final election vote count had dragged on for six tense days, a news anchor appeared on-screen. She wore a cream jacket and a lapel mic in front of a green screen of American flags and the Capitol. She looked official. She was from One America News Network, a jingoistic cable outlet that had seen a steady boost in YouTube traffic after Fox News had outraged Trump acolytes by calling votes accurately. "Donald Trump won," the network's anchor began in this YouTube video. "And Democrats have made a nasty play to steal the presidency."

That Monday, Republican senators refused to recognize Joe Biden as the next president. Trump had just fired his defense secretary, who opposed sending troops against unarmed protesters. Foreign news stations declared that Americans were bracing for a "civil war," and it didn't sound implausible. On YouTube dozens of videos echoed One America News Network, making claims of shady election software, hidden counts, Sharpies ruining ballots, the lying media, the truth you had to see to believe.

But these videos were hard to find. YouTube's algorithm didn't serve them up in the sidebar. They didn't appear in searches for the election or even for the phrase "Trump won." After the vote Trump's YouTube channel posted speeches and TV clips promoting an unhinged case that the election had been stolen from him. These got relatively few views. A vast majority of videos appearing in searches and recommendations about the election—all the media YouTube's machines promoted—came from "authoritative" sources, YouTube contended. Fringe outlets like One America News Network, shock-jock bomb throwers, and even the president screaming fraud got no algorithmic boosts. Any views they had arrived from links on social networks or right-wing forums or just audiences finding them all on their own.

Weeks later, after all the legal challenges to the election failed, after insurrectionists stormed the Capitol, YouTube would rewrite its rules and algorithms once again. But that Monday in November, it stuck to them. A company spokesperson sent out a statement about the "Trump won" videos: "Our systems are generally working as intended."

The machines had done what they were told.

Epilogue

From an insurgent underdog in entertainment, a money pit, something of a joke, YouTube had become one of the most dominant, influential, untamed, and successful media businesses on the planet. In less than two decades. At times Steve Chen could hardly believe it.

Back when Chen wrote YouTube's first lines of code, he struggled to get the audio and visual files to sync. When he left the company five years later, an astounding one hundred hours of footage were uploaded every minute; that figure, by 2020, was well over five hundred hours. In the intervening years Chen's health improved, and he had teamed up again with Chad Hurley on a digital media start-up. When that didn't pan out, Chen settled into a role as a wizened industry sage, reminiscing about the era when playing video on desktop computers required great engineering feats. He had moved to Taipei, his birthplace, with his young family, and watched with awe as taxi drivers streamed YouTube on their phones. In his son's elementary school class, all but two kids said they wanted to be YouTubers someday.

YouTube's troubles with bad actors, conspiracies, political speech, and irate politicians, its sheer operational bulk—all that was beyond anything Chen ever imagined. "To be honest," he admitted, "I kind of congratulate myself that I'm no longer with the company, because I wouldn't know how to deal with it." Jawed Karim, YouTube's third co-founder, had become an investor and only commented on his old company when it made changes that irritated him, like removing the number counts on the video "dislike" button.

During the pandemic, Chad Hurley, like many accomplished, restless men, took to Twitter. He posted inane jokes and slung insults at Trumpies and tech bros with the gleeful abandon of someone who no longer had a corporate job. Hurley financed companies and basked in the glow of being a father of the creator economy when creators were all the rage.

As the Trump era began to fade in 2021, suddenly every company wanted in on the industry YouTube had built. TikTok, the red-hot app, began paying select video makers, triggering a deluge of young fame seekers. Rivals Twitter and Snapchat nervously followed suit. Facebook, yet again, made a push to recruit influencers, pledging to spend more than $1 billion on creators and vowing not to take commissions for multiple years. Spotify paid hundreds of millions to recruit podcasters like Joe Rogan, who had used YouTube to build a media powerhouse entirely outside the mainstream. Venture capitalists went wild for "web3," an internet model based on cryptocurrencies that imagined regular people owning and profiting from their online activity; this was YouTube's creator economy taken to its next extreme. Sequoia, YouTube's first investor, minted its 2005 YouTube investment memo as a "non-fungible token," which a crypto enthusiast purchased for $863,000 in digital coin.

There was business logic behind this rush to embrace creators. The pandemic turbocharged online entertainment and commerce. At the same time, the Web 2.0 model of targeted advertising was being dismantled by regulators; companies had a harder time marketing things online. Creators were great marketers *and* salespeople. But perhaps the internet businesses, facing continued political scrutiny, had also decided that paying people who produced the content that made them so rich might look good, optics-wise. Or perhaps the pandemic's upheaval—like the financial crisis, which had helped jump-start YouTube's economy a decade before—had convinced enough people that working for these platforms, even without security, benefits, or guarantees, beat a day job.

And so the world that YouTube ushered in—of abundant content and creativity, of influencers and online hustlers, of information overload and endless cultural wars—became more of our own.

• • •

All this renewed competition only underscored YouTube's unrivaled power. The company's battle scars from copyright fights, ad boycotts, and countless creator turmoil produced a compensation system that worked like nothing else. No other platform distributed video and money as effectively. Creators tinkered on TikTok and Instagram, sometimes cashing in handsomely, but they made reliable money on YouTube. Other companies trying to replicate YouTube's creator economy now had to deal with YouTube's old firestorms. TikTok stars appeared in the tabloids. Spotify saw weeks of outrage over Joe Rogan's comments on COVID-19, while YouTube, Rogan's homebase for years, skated by untouched. U.S. Senators tore into a Facebook official over Instagram's damage to teenagers on the same morning that Kyncl, YouTube's chief business officer, cheerily briefed reporters on a study claiming how great his business was for the economy.

During its history, YouTube tried to push or position its platform as something that it wasn't—a *premium* service, a destination for Hollywood, a manageable and sanitary place with only a few "bad actors," a great equalizing force. That tension between what the company wants and what it has will never end. But YouTube has learned to live with it, or at least run a prosperous business from it. In the summer of 2021, YouTube posted its greatest quarterly ad haul ever, more than $7 billion, on par with Netflix's sales. YouTube announced it had paid broadcasters more than $30 billion in a three-year span. (Although it didn't specify how much of that bounty went to creators rather than media companies and record labels.) For the first time YouTube even began running ads on channels that didn't qualify for its partner program, confident it could open up its Long Tail without brand safety disasters. YouTube kept every dollar from these commercials.

After some rocky years YouTube's c-suite drew closer to its stars. Big-name YouTubers praised Kyncl, who lavished attention on a creator class long neglected by the company; Casey Neistat called him "wildly proactive." Matthew Patrick lauded Ariel Bardin, the executive he once chided for not understanding YouTube, as "an incredible advocate for creators." (Bardin would leave YouTube in late 2020.) YouTube gave creators more ways to make money beyond ads, like fan sponsorship and merchandising. Company managers asked creators for advice on reducing burnout and commissioned a therapist to post videos on the topic. YouTube even managed to improve its comments section. "It's gone from the lowest level of hell on the internet to a fairly pleasant experience," said Natalie Wynn.

Many creators got enough steady viewers and cash that they no longer felt pulled to Hollywood. Lucas Cruikshank acted in three movies as Fred Figglehorn, his squealing YouTube persona, but after growing exhausted from performing in front of a crew, a director, a whole to-do, he went back to posting solo on YouTube, where, he said, "there's no pressure at all." Justine Ezarik, a.k.a. iJustine, marked her seventeenth straight year on YouTube, still scripting, producing, acting, and starring in each video herself. Movies and TV never offered that degree of creative control. "You don't own that," she added.

Wojcicki began referring to creators as "the heart of YouTube." And it seemed her company had begun to appreciate their non-pecuniary value. YouTubers spotted each corrosive trend—the troubling kids' stuff, the bullies, hucksters, con men, and extremists—before the company did. "You *have to* watch your platform," Patrick once admonished YouTube in a private meeting. It seemed like they were.

And yet, YouTube still followed the Google playbook. Graham Bennett, a senior partner manager who worked closely with stars, described his role as "the least Googley part of YouTube"; his job couldn't scale. YouTube's multichannel networks once fulfilled some of these managerial duties, but they had all withered or collapsed. Bennett wished YouTube could do more for creators, although from Wojcicki's perspective, he said, deciding between hiring one more version of him or hiring another engineer "is tough." (YouTube doesn't share how many engineers and senior partner managers it employs.) Efforts to organize YouTubers, like Hank Green's Internet Creators Guild, had died. And to some, as long as YouTube continued to be an ads business that demanded mass audiences and expand like a universe, the company's pledge that it was a stool standing on three equal legs—viewers, advertisers, and creators—just wasn't true. Creators always got the short end. "It's like in *Animal Farm*," said Andy Stack, a YouTube manager who left in 2015. "Some are more equal than others."

By 2022 YouTube had revamped its content strategy again, dropping its program to fund subscriber-only shows starring creators—let Netflix, Disney, and Amazon duke it out for paid streamers. Instead YouTube shifted resources into Shorts, a feature for bite-sized videos. It was an obvious TikTok clone and attempt to fend off the threat it posed. Old-school YouTubers likened TikTok's playful canvas to early YouTube, that long-gone era, where creative types could experiment and flourish. ("It's just come out of nowhere," Wojcicki admitted about TikTok in 2020, even though Google had previously tried to buy Musical.ly, the company that would become TikTok.) YouTube launched Shorts in India, where TikTok was banned, and started a $100 million fund bankrolling creators of these brief clips. It would sort out the business model later. Nearly a decade after tilting its system toward longer videos, YouTube was now paying for shorter ones. Of course, the main algorithmic metric for Shorts, like that for all of YouTube, remained watch time.

Most signs indicated that TikTok did chip away at YouTube's dominance. A 2021 report revealed that for the first time Americans watched more TikTok than YouTube on their phones. But thanks to its smart-TV app and streaming service, YouTube was growing enormously on television screens. YouTube's sales team still focused on eating into TV's market share, not TikTok's, and its product team tinkered with ways for TV viewers to like, comment, and subscribe, making TV even more like YouTube. Besides, TikTok didn't have stockpiles of yoga videos, bread-baking tutorials, "Let's Play" gamers, beauty gurus, and billions of hours of toddler fodder. Only YouTube did.

Other tech platforms (namely, Facebook) panicked about the TikTok generation and disaffected citizens quitting its platform. Viewers complained about seeing YouTube's frequent or annoying ads, but they rarely stopped watching. Throughout all its years of tumult, YouTube never fretted about people fleeing.

As one employee put it, "How do you boycott electricity?"

• • •

Kids certainly remained loyal to YouTube.

Harry and Sona Jho, the veteran nursery rhyme showrunners, braced for a painful shock as 2020 began. YouTube's settlement with the FTC meant they couldn't run higher-priced ads on "child-directed" videos anymore. And, once the pandemic hit, marketers paused commercials everywhere, unsure how consumers would proceed. The Jhos watched ad rates crater. But quarantines, it turned out, were very good for their viewership. Kids stuck at home watched like crazy. By the end of 2020 the top-five most viewed channels on all of YouTube were preschooler fare. A year into the pandemic Harry Jho cautiously admitted the audience surge had helped his business. "It's not rosy, but we're not laying people off," he said.

And he thought YouTube's machines had become far more attuned to quality. After the FTC case, YouTube stopped treating its Kids app like an algorithmic free-for-all and assigned staff to curate the selection, like the coolhunters once did on YouTube's home page. The company started a fund for kids' YouTubers and told creators it would finance videos that "drive outcomes" associated with subjective traits like humility, curiosity, and self-control. YouTube said its system would reward clips that encouraged young viewers to go do things offline.

"This is about as healthy of an algorithm environment that I've ever seen," said Jho. It felt as if YouTube had relinquished some of its blind faith in machines. It felt as if human beings there were actually involved.

During the pandemic, kids' YouTube also became ground zero for new-media moguldom. Moonbug Entertainment, a digital studio, purchased three behemoth YouTube channels, assembling an arsenal (seven billion views a month) rivaling anything on cable. By 2020 little Ryan Kaji was a nine-year-old seasoned pro. He had mostly left toy unboxing, the phenomenon that skyrocketed him to fame, for videos on science experiments, "challenge" gimmicks ("Edible Candy vs. Real!!!"), and exercise tips. He got into video games. Early in the pandemic Ryan and his parents posted footage of their chat with a health official about COVID-19. Ryan, a tireless performer, displayed the exaggerated emotional reactions of someone who had spent most of his life in front of a camera.

Still, Ryan's back catalog continued to put him and YouTube under scrutiny. A 2020 *New York Times* headline asked, ARE "KIDFLUENCERS" MAKING OUR KIDS FAT? and printed a still of an old video of Ryan playing a McDonald's cashier. An advocacy group accused him of breaking laws against deceptive advertising to kids. Chris Williams, a former Maker Studios executive, had started PocketWatch, an entertainment company that worked with Ryan and other YouTube child performers. Williams found such critics of his star misguided, comparing them to the scolds who raised a moral panic about video games and rap in the 1990s. Critics, he believed, failed to see the benefit for young audiences of having a relatable figure on-screen—even one as famous as Ryan. "Really, what they mean is, 'This isn't *Sesame Street*,'" said Williams. "If that were the bar, kids would never watch anything."

When Ryan first exploded on YouTube's charts, his parents set up a production studio to capitalize on his success. They sold Ryan-branded toys, clothes, and bedding in Walmart and Target. They made an animated Ryan character to continue his legacy if, one day, he stops YouTubing. That character appeared in the Macy's Thanksgiving Day Parade. *Forbes* magazine annually listed the richest YouTubers and, starting in 2018, Ryan topped the list. His estimated earnings in 2020 neared $30 million, a sum that shocked practically everyone who heard it. *A nine-year-old made how much?*

The world still couldn't grasp how media worked in the age of YouTube. From Williams's vantage point, Ryan wasn't just a nine-year-old YouTuber but the centerpiece of a business juggernaut. "I'm from Disney," Williams said. "The delta between $30 million and Mickey Mouse is pretty big." There was plenty of room to grow, so long as kids kept watching.

• • •

With such an enormous captive audience and payment system, YouTube could count on keeping a firm lead in its industry for years to come. It could count on another advantage, too: being inside Google, the undisputed leader in artificial intelligence. That capability let YouTube build a machine system that, by 2020, could detect obvious red flags—a Nazi symbol or a sexual comment about kids—as quickly as it could spot a copyrighted song. This, YouTube said, meant most "violative" footage came down without a human involved at all.

But Google's superhuman AI couldn't solve a messier issue—the endless quagmire about truth and misinformation online.

YouTube tried throwing computer science and its rulebook at this problem. Like other tech platforms, YouTube outlawed certain topics when they became politically untenable. A month before the 2020 election, YouTube banned videos promoting QAnon, the extremist pro-Trump movement.

When COVID-19 vaccines arrived, YouTube removed footage that questioned official scientific guidance. (Several Trump videos came down for this reason.) After Russia invaded Ukraine, YouTube zapped Russian state media channels for violating rules against "trivializing well-documented violent events." (Russia blocked Facebook after the invasion but not YouTube, which was massively popular in the country.) YouTube scrubbed more than a million videos for containing "dangerous coronavirus information." Under a system called "Golden Set," staff gave thousands of examples of clear lines in the sand to machine detectors—*this video on COVID-19 contains lies; this one does not.*

Yet YouTube knew this process wasn't perfect. "People might think we have this great AI that can drive cars and everything," said Goodrow, the veteran YouTube engineering leader. "But I don't think right now we could even simply identify what specific claims are being made within a video."

Even if it could, few outside the company agreed on the precise definitions of misinformation or disinformation, so debates on the matter typically circled into political stalemates. And, by and large, YouTube stayed out of the fray. Outrage in the press and halls of power usually focused on the social networks, not the video site. In 2021 President Joe Biden chastised tech platforms for promoting vaccine hesitancy, specifically accusing Facebook of "killing" people with lies. Ire on the right was largely aimed at Twitter, which had booted Trump off the service after the January 6 insurrection. Mark Zuckerberg and Jack Dorsey, the chiefs of Facebook and Twitter, testified multiple times before Congress; Susan Wojcicki never did. YouTube was the sleeping giant of social media.

There were many reasons for this. YouTube was better situated to avoid information warfare. You might see your cranky uncle rant about vaccines on Facebook or Twitter, but probably not YouTube. Political content repeatedly topped Facebook's popularity charts. YouTube was still dominated by music, gaming, and kids' videos. YouTube, like Facebook, had indefinitely banned Trump without clear plans for his reinstatement, but Trump wasn't as big on YouTube, and his absence there drew less attention. And Facebook remained an easier punching bag; in the fall of 2021 a whistleblower released a series of damning documents, including evidence Facebook hadn't acted swiftly enough to combat lies about COVID-19 vaccines. Some believed YouTube was simply a better managed company.

Or it could be that YouTube was more difficult to see into. It's fairly simple to identify when someone touts dubious claims about vaccines in a tweet or Facebook post, in text; it's much harder to do in a long video. YouTube shared relatively little data with outsiders, which helped the company evade scrutiny. After 2020 YouTube started revealing more about its algorithms and disclosing metrics that showed progress on its goals: views of borderline footage and videos that broke rules, before they were deleted, were low and falling. But YouTube was grading its own homework. No external group audited the results. Take Biden's vaccine crusade against social media, which was based on findings from an advocacy group that examined statistics Facebook and Twitter shared; the findings excluded YouTube simply because YouTube did not share comparable data. The Facebook whistleblower revealed that Instagram ignored internal research on the damage its app had on the mental health of teenage girls, spurring waves of criticism of Facebook. Afterward, multiple people who worked at YouTube said their company either didn't share this type of research widely or simply didn't conduct it.

"YouTube is really opaque," said Evelyn Douek, a Stanford Law assistant professor who studies content moderation. "It's much more fun for me to lob stones from the outside. This stuff is hard," she added. "That doesn't mean that they don't have responsibility."

Privately, people at YouTube, like their peers at Facebook, complained they were being scapegoated for the collapse in democratic norms brought about by cable news, inequality, and God knows what else. One longtime YouTube executive put it bluntly: "Don't blame the mirror." It was a common refrain in Silicon Valley, that platforms merely reflected the society that used them.

But YouTube never reflected everything in society. And as regulation dragged on, it started reflecting less and less. In the fall of 2021, YouTube banned false claims about any vaccine and stripped ads from videos denying the reality of climate change. Some praised the moves; some saw them as overreach. Others asked the obvious: What took so long?

Under Biden and the Democrats, who asked that kind of question, YouTube leadership started pushing back. Wojcicki authored an op-ed comparing overly intrusive online moderation to censorship in the U.S.S.R., where her grandparents had lived. Neal Mohan claimed YouTube had seen "disturbing new momentum" in government takedown requests for political reasons. YouTube, he wrote, could aggressively attack COVID-19 falsehoods because health agencies gave official guidance, but it needed to tread carefully on other topics. "One person's misinfo is often another person's deeply held belief," he blogged. He took for granted the assumption, less than two decades old, that any person was entitled to broadcast their deeply held beliefs over mass media.

Still, Mohan was right that YouTube saw disturbing trends. Officials in Russia and India began using the terms "fake news" and "extremism" to demand YouTube remove videos from critics and opposition figures—in effect, forcing the company to choose between its stated values and its desire to be everywhere. Other countries will likely follow suit. In nations rife with conflict and contested elections, YouTube, like other social media, hires few people who understand the language or political terrain.

YouTube leaders often discussed how their world-class AI software, while imperfect, was the only system capable of handling this enormity it had created. This was presented as an inevitability, but it was a choice. YouTube had once placed its coolhunter community managers and partners like Storyful, the Arab Spring newsroom, as editors on the frontlines to vet, verify, and make sense of the information flood. "We kept trying new things, which the culture of YouTube encouraged," recalled Steve Grove, its former head of news and politics. YouTube chose to stop many of those experiments. It chose to pursue scale instead. This trade-off kept humans, as flawed as they are, away from one of humanity's biggest problems—creating a shared set of facts and truth. "Misinformation online remains a threat to democracy," said Grove, who left Google to become a state official in Minnesota. To address the threat, he added, "curation, in various forms, will always have to play a key part."

That's a solution without a clear mathematical formula or ability to scale. Not very Googley.

• • •

A few people at YouTube were consumed with these unending debates. But most were busy running the world's largest online video business.

Claire Stapleton had thought about this more since leaving. She had returned to newsletters, starting an advice column for disgruntled Silicon Valley employees called "Tech Support," where she ripped into her old employer and still linked to marvelous YouTube videos she liked. She had changed her mind about the PewDiePie affair, seeing now how those drawn-out debates about ♥ ing his tweets were a distraction when a rot festered underneath. "It was a futile fight," she recalled. "We got so caught up on the aesthetics of the brand when there was a refusal to discuss the real issues around YouTube." At work they rarely asked the sorts of questions she now did. *How had her company, so devoted to organizing the world's information, built a megaphone and payment system for conspiracists, cranks, and hatemongers? What did it mean that YouTube enticed teen moms to place their entire lives on-screen? Does everyone really need to broadcast themselves? And why couldn't she look away?* For Stapleton, this begged a deeper question: "Is YouTube net negative or net positive for society?"

Everything she put in the positive column—the site's treasured communities and delightful brilliance—didn't come from the company. "YouTube doesn't foster creativity," she concluded. "People do!"

• • •

Susan Wojcicki entered her eighth year as CEO in 2022 stepping back from the stage. She gave few public interviews. Advertising and media partners who met her frequently during YouTube's crisis years said they saw her less often, now that those crises were over. Most YouTube viewers probably couldn't name her. "She is not charismatic," said Kim Scott, a former Google colleague. "But to her credit, that's not a bad thing at all. Especially in her role, I think a charismatic leader of YouTube might be a disaster. Charismatic leaders kill these businesses because it's all about them."

Wojcicki and her husband, another Googler, ran a foundation that gave grants to several Jewish and interfaith groups, including the ADL, and environmental nonprofits like Earthjustice and Environmental Defense Fund. She was careful not to speak of her personal opinions. When she did appear publicly, she deployed a new talking point: YouTube's "responsibility" efforts were good for business. Indeed, YouTube's ad sales nearly doubled in the two years since its major overhauls in 2019. That year Google's founders, Page and Brin, retired in their mid-forties, leaving Sundar Pichai in charge of Google and Alphabet. The Justice Department proceeded with its antitrust case against Google, and Congress introduced bills to regulate tech competition and "malicious algorithms," which included YouTube recommendations. Google's gravest threat, though, seemed increasingly unlikely: the company wouldn't be broken up. In response to political heat Pichai positioned Google as being "helpful," a utility people loved. He pushed to shift its ads business, facing serious backlash, into a powerhouse for e-commerce, an underdog to Amazon. YouTube, a place overflowing with how-tos and commercial influencers, became central to both strategies. When Google insiders speculated about who might take over if Pichai leaves, Wojcicki's name always appeared on the shortlist.

Some observers believed Wojcicki's sober disposition and careful management explained why YouTube hadn't come under the same scrutiny as Facebook. "She cares," said Jim Steyer, the founder of Common Sense Media, an influential advocacy group. Steyer has hit tech platforms hard on children's issues, lobbying for regulation to curb tech's addictive power and business practices. He no longer trusts Facebook. Where YouTube is concerned, "the jury is out," he said. Though he added, "When Susan took over, it changed my attitude."

Few in Silicon Valley or Hollywood would describe Wojcicki as a visionary or today's YouTube as a hotbed for innovation. It's a tanker, an enormous business and institution steered with small, careful turns. Even if she wanted to, Wojcicki probably couldn't steer it entirely in a chosen direction. She is a steward of a platform with a life of its own. Running YouTube means dealing with an entity that is "inherently indefinable and ungovernable," explained one company veteran. "You're holding on to the reins." Even so, Wojcicki has managed to contain certain parts of YouTube that had spun out of control. And she still runs a global mass media and economic giant with little transparency and accountability.

"If I was going to make a list of people to have the amount of power that Susan Wojcicki has, she might be on it," said Hank Green. "But I would just rather not see someone with that much power, especially someone who is unelected."

• • •

When Green sat down with Wojcicki in the early days of the pandemic, the Nerdfighter vlogger managed a rare feat: he got YouTube's leader to reveal something. Wojcicki was explaining how YouTube bucketed broadcasters into three groups—creators, music labels, and traditional media—when she shared that YouTubers made up "about half" of all viewership.

Green tossed his hands up at the revelation. "Wow! That's huge!" He beamed.

He had less luck getting somewhere on another front. During their conversation, Green pressed Wojcicki on YouTube's strategy of financing channels. Why, after failing to get Hollywood on board, had the company given grants to only the biggest YouTube stars to make shows like TV? Why not fund smaller creators to run their own media businesses? "There's a potential for a large middle class of YouTubers," Green argued. "I've always been of the opinion that YouTube should lean into what YouTube is and—"

Wojcicki interrupted, "We agree with that!"

Green laughed. He knew YouTube better than anyone who didn't work there, and probably many who did, and he was speaking with the person who had run YouTube longer than anyone else. And yet, it seemed the pair was talking past each other. "We may not agree on what exactly YouTube is," he replied. "Shocker! Not that anybody does."

By the end, their conversation lasted nearly an hour. Not long ago no one would have published a video that long on the internet, let alone expected an audience or financial reward for doing so. Green simply uploaded his clip, adding another hour to the billions already accumulating on YouTube that day.

Acknowledgments

I began working on this book in late 2019, right before the world shut down. The book wouldn't exist without the hundreds of sources who lived the YouTube story and shared their memories, thoughts, documents, and time with me during an endless pandemic. Many took professional risks doing so, and, while I can't name them here, I am deeply grateful for everyone who did.

Claire Stapleton, still a Bard, was refreshingly candid and tolerant throughout this entire process. Brendan Gahan shared his notebooks and wisdom on early YouTube. No one gets YouTube like YouTubers, and I learned so much about the platform from watching MatPat, Hank Green, Casey Neistat, Veritasium, and ContraPoints—readers should go watch them all. Every exceptional creator who took time to speak with me made this history far richer. Thank you all.

Viking has been the perfect publishing partner. Rick Kot was a patient and thoughtful editor who dealt graciously with a rookie author and a beast of a story. Andrea Schulz, Hal Fessenden, Shelby Meizlik, Julia Rickard, and Camille Leblanc all believed in this book and worked diligently to make it a reality. None of this would have happened without Ethan Bassoff, my agent at Ross Yoon, who convinced me to pitch something, made that pitch work, and has stuck with me through every neurotic bump along the way.

Kelsey Kudak and Sean Lavery were a phenomenal fact-checking duo and total life savers. Sally Weathers provided early research, which required watching ungodly amounts of YouTube. Carrie Frye, a wonderful human, gave exceptional writing counsel that kept me sane during the first drafts.

At YouTube, Jessica Gibby, Andrea Faville, and Chris Dale opened doors at the company, knowing I would scuff up the furniture, and responded to endless fact-checking. True pros, all of them.

There is no finer crew in business journalism today than the one at *Bloomberg Technology*. Brad Stone, our peerless leader and my reporting idol, gave incredibly helpful feedback on early manuscripts. Tom Giles, Jillian Ward, and Sarah Frier were amazing, supportive managers. Olivia Carville introduced me to a wealth of helpful resources in New Zealand. Kurt Wagner, Ashlee Vance, Joshua Brustein, Emily Chang, Felix Gillette, Josh Eidelson, Ian King, and Lizette Chapman offered reporting assistance, guidance, and moral support. Max Chafkin talked me down from more panic attacks than I will admit. Emily Biuso and Ali Barr edited the articles that inspired this book and gave me endless encouragement. Thank you, Ellen Huet, colleague.

My professional better half, Lucas Shaw, has co-authored some of my favorite articles and put up with me for years. Anything in this book about Hollywood that's good is only good because of him. Your turn, Sundance.

So many other great journalists have inspired and helped me along the way. Ken Auletta's *Googled* and Steve Levy's *In the Plex* were my bibles for studying Google history; Keach Hagey's work on Viacom was a tremendous resource; Kevin Roose has done the best reporting on YouTube's culture and impact, and kindly let me pillage so much of it. I am in debt to the work of Becca Lewis (a scholar with great journalistic instincts) and other fine researchers. Kara Swisher and Ken Li took a chance on an untested, unkempt reporter to cover Google and taught me nearly everything I know. James Crabtree mentored me when I was a clueless stringer in India and modeled great, lucid business writing. Alex Kantrowitz and Eliot Brown returned every frantic text and call in the last three years. Peter Kafka,

Jason Del Rey, Johana Bhuiyan, Maureen Morrison, Anna Wiener, Cory Weinberg, and Teddy Schleifer all made me a better journalist.

My beloved friends carried me through these years of writing and social distance. Brendan Klinkenberg, Jackie Arcy, and Nico Grant read early versions of the book and offered sage advice. Jane Leibrock, my messaging guru, distilled the book's themes into digestible pieces in ways I never could have. Will Alden and Dan Sawney, terrific writers, workshopped chapters with me over long afternoons, which I loved dearly. Brad Allen, Danielle Egan, and Dan Gorman indulged me with lengthy reading sessions and gave wonderful feedback. Waheguru Khalsa, Sara Heller, Yaw Asare, Suzanna Scott, Austen Leah Rose, Loreal Monroe, David Vigil, Stewart Campbell, Colin Nusbaum, Harry Moroz, Regal Johnson, Matt Armsby, and Brian Stromquist all lent advice and let me talk incessantly about YouTube for years on end. Michael D'Arcy remains my favorite interlocuter.

My family has showered me with love and encouragement through this whole process. Jack and Fran, dedicated readers, provided counsel and support. Sarah and John gave us light after bright light during the darkest days of the pandemic. My sister, Amy, my very favorite writer, is my inspiration every time I sit down to type. My parents taught me to love a good story and meet the world with empathy. Mom, I know Dad would have been so proud.

And finally, Annie, my quarantine partner, my first and last editor, love of my life and my very best friend: I could not have done this without you. It's over now. How cool is that?

Sourcing Note

Everything in these pages actually happened. The reporting in the book is based on public records, correspondence with sources, documents I obtained, and many hours of interviews I conducted from late 2019 through early 2022. And, of course, YouTube videos.

In my reporting, I interviewed more than 300 people who were part of YouTube's history. That includes nearly 160 current and former employees of YouTube and Google; several business partners, managers, consumer advocates, regulators, and researchers; and dozens of YouTube creators. Many conversations were on the record, but others were conducted anonymously and off the record given the strict confidentiality rules Google asks its employees, alumni, and partners to follow. Emails cited in the book came from documents from the Viacom lawsuit against Google, congressional investigations, or internal company records that I obtained from sources. Some dialogue in the book was re-created using the recollections of people in those conversations or in the room to hear them. Readers should not assume that the person quoted in the pages spoke to me. However, readers should know that I worked to corroborate every detail in the book with multiple sources. Information that came from one source is noted accordingly. The book has been thoroughly fact-checked. And I followed a cardinal rule of journalism I practice at *Bloomberg*: "no surprises." Each person mentioned in the book was made aware of their portrayal and given a chance to comment.

Some key figures in YouTube's history declined to speak with me. In each case, I went to their representatives to confirm facts and request comment or made every effort to do so. YouTube arranged sanctioned interviews with more than ten employees and participated in fact-checking. Comments from YouTube's spokespersons are included in the book or endnotes. After multiple requests for an interview, a representative for Felix Kjellberg wrote, "We respectfully decline your request." Eric Schmidt, Google's former CEO, would only comment through representatives. All three people who have run YouTube—Chad Hurley, Salar Kamangar, and Susan Wojcicki—declined to be interviewed. Larry Page and Sergey Brin, Google's co-founders and majority shareholders of its parent company, Alphabet, have not spoken to a journalist in several years. I was no exception. In response to the bulk of my questions about Page and Brin, YouTube and Google simply didn't reply.

On one occasion I used a pseudonym for a former YouTube employee who asked not to be identified by name for fear of retaliation by the company or others. I corroborated portions of what they shared with documents and other sources and left out certain details to protect their identity. I believe their experience is incredibly important for readers to understand the genuine story of how YouTube works.

Notes

Prologue: March 15, 2019

2 **in *The New York Times***: Farhad Manjoo, "Why the Google Walkout Was a Watershed Moment in Tech," *The New York Times*, November 8, 2018, https://www.nytimes.com/2018/11/07/technology/google-walkout-watershed-tech.html.

3 **had recently returned:** Ellen Huet and Mark Bergen, "Google Talent Advantage Erodes as More Workers Doubt CEO Vision, *Bloomberg*, February 1, 2019, https://www.bloomberg.com/news/articles/2019-02-01/google-talent-advantage-erodes-as-more-workers-doubt-ceo-vision.

4 **unofficially designated:** Interviews with Claire Stapleton and other anonymous YouTube employees. A YouTube spokesperson said the event was a "normal annual team offsite."

4 **"Our Brand Mission":** YouTube shows the length of the video, the time stamp, on each one.

5 **some 1.7 billion people:** This statistic came from a former employee familiar with the numbers. A YouTube representative declined to comment on the figure.

Chapter 1: Everyday People

19 **once told a reporter:** John Cloud, "The YouTube Gurus," *Time*, December 25, 2006, http://content.time.com/time/magazine/article/0,9171,1570795,00.html.

24 **raised the issue publicly:** A spokesperson for Facebook, which later renamed itself Meta, declined to comment.

27 **"CRAZED NUMA FAN !!!!":** Brodack has since deleted this and all her earlier videos, though copies exist elsewhere online.

Chapter 2: Raw and Random

28 **endless rodent problem:** Amici's owner didn't recall mice on the premises, but the building was shared with another restaurant, which has since changed hands.

29 **YouTube could net at least:** Botha also wrote that Reid Hoffman, a Flickr investor and former PayPal leader, had assured YouTube that Flickr would not be trying video anytime soon.

30 **later call YouTube:** Steven Levy, *In the Plex: How Google Thinks, Works, and Shapes Our Lives* (New York: Simon & Schuster, 2011), 249. Additional context about Amazon, Microsoft, and Google from Sequoia Capital.

32 **she cried once it began:** Catherine Buni and Soraya Chemaly, "The Secret Rules of the Internet," *The Verge*, April 13, 2016, https://www.theverge.com/2016/4/13/11387934/internet-moderator-history-youtube-facebook-reddit-censorship-free-speech.

33 **Nike's marketing department:** The clever marketing stunt became the first on YouTube to cross a million views. Botha took the founders up to visit Nike officials in Portland afterward.

34 **YouTube did pull the clip:** Micah Schaffer first ran the website of the group behind "Lazy Sunday," which included his older brother, Akiva. When it became a YouTube hit, Schaffer connected with Maxcy at YouTube and asked if the company was hiring. After the NBC fracas, Schaffer's dad joked that one son would make a video on Saturday and the other would delete it on Monday.

36 **had recently run an article:** Brad Stone, "Video Napster?," *Newsweek*, February 28, 2006, https://www.newsweek.com/video-napster-113493.

37 **a 1994 essay:** John Perry Barlow, "The Economy of Ideas," *Wired*, March 1, 1994, https://www.wired.com/1994/03/economy-ideas/.

40 **go viral and spread:** Wong would write a college thesis on this subject.

41 ***Wired* would write:** Joshua Davis, "The Secret World of Lonelygirl," *Wired*, December 1, 2006, https://www.wired.com/2006/12/lonelygirl/.

Chapter 3: Two Kings

45 **once told a reporter:** Levy, *In the Plex*, 121.

46 **explained an early investor:** John Doerr, *Measure What Matters: How Google, Bono, and the Gates Foundation Rock the World with OKRs* (New York: Portfolio, 2018), 6.

46 **would say her opinion changed:** The Try Guys, "Eugene Interviews the CEO of YouTube," YouTube video, December 16, 2019, 46:15, https://www.youtube.com/watch?v=fKIsuulxJ1I.

50 **Lucas declined:** YouTube made its board offer to George Lucas in 2006, according to Chris Carvalho, then the head of business development for Lucasfilm. Representatives for Lucasfilm, Lucas, and YouTube declined to comment.

50 **outed the aspiring filmmakers:** Virginia Heffernan and Tom Zeller, "'Lonely Girl' (and Friends) Just Wanted Movie Deal," *The New York Times*, September 12, 2006, https://www.nytimes.com/2006/09/12/technology/12cnd-lonely.html.

51 **"the secret president of Google":** Levy, *In the Plex*, 235.

52 **One banker pondered:** The pitch, while thankless, was hyperbolic: YouTube did have some revenue.

52 **Bathrooms lacked paper towels:** The office had only one men's and women's bathroom apiece. At one point, some men, frustrated by the waits and their larger numbers, declared them both dual-gender units, until a female staffer alerted human resources and this ended.

Chapter 4: Stormtroopers

57 **Peter Bjorn and John:** Harper's favorite was "B-Boy Stance," a sketch about a 1970s hip-hop hype man who had surgically wrapped his hands across his chest, from a comic named Donald Glover. Eleven years later, Glover's video "This Is America," a scathing social commentary, would be watched more than twelve million times on its first day.

58 **But the gimmick worked:** Once, in 2008, the editors were joking about Rickrolls—an absurdist prank, born on YouTube, where a person was sent an important-sounding web link, only to open it and find the video for "Never Gonna Give You Up," the earworm hit from the British rocker Rick Astley. It never failed to amuse. "Wouldn't it be cool if we could Rickroll the home page?" Harper suggested, half in jest. Her colleague Michele Flannery called Astley's agent, begging him to participate. He finally gave his blessing on the stunt but declined to join. On April 1, all of YouTube was Rickrolled.

59 **brought Kate Bohner:** Bohner and a spokesperson for Schmidt declined to comment. A source close to Schmidt said he brought multiple people into the YouTube office to receive advice.

60 **His holdings included assets:** Technically, the movie studio behind Gore's *An Inconvenient Truth*, Paramount Pictures.

61 **book on Redstone:** Keach Hagey, *The King of Content: Sumner Redstone's Battle for Viacom, CBS, and Everlasting Control of His Media Empire* (New York: HarperCollins, 2018).

61 **scoped out South Central:** Matthew Belloni, "The Man Who Could Kill YouTube," *Esquire*, August 15, 2007, https://www.esquire.com/news-politics/a3131/youtube0707/.

62 **an admirer swooned:** Keach Hagey, "The Relationship That Helped Sumner Redstone Build Viacom Now Adds to Its Problems," *The Wall Street Journal*, April 11, 2016, https://www.wsj.com/articles/the-relationship-that-helped-sumner-redstone-build-viacom-now-adds-to-its-problems-1460409571.

64 **He was a YouTuber:** Ryan Singel, "YouTuber Warned of Finnish Gunman in June, But No One Listened," *Wired*, November 8, 2007, https://www.wired.com/2007/11/youtuber-warned/.

Chapter 5: Clown Co.

67 **cats on acid:** The fan also declared, "Adding all that shit to TV would take out all the fun, suck the life, the essence out of YouTube."

67 **The YouTubers received $15,000:** The company that made the Zvue then employed Carl Page, the older brother of Larry Page. The company went out of business in 2008.

68 **told Steve Chen:** Richard Nieva, "Inside YouTube, Leaders Look for 'Balance' After Scandals," CNET, July 11, 2019, https://www.cnet.com/tech/services-and-software/features/inside-youtube-leaders-look-for-balance-after-scandals/.

70 **raked in an estimated:** Louise Story, "DoubleClick to Set Up an Exchange for Buying and Selling Digital Ads," *The New York Times*, April 4, 2007, https://www.nytimes.com/2007/04/04/business/media/04adco.html.

72 **had convened a gathering:** Michael Wolff, *Television Is the New Television: The Unexpected Triumph of Old Media in the Digital Age* (New York: Portfolio, 2015), 1. Events described in the book were confirmed with additional sources.

73 **would urge his industry:** Brian Stelter, "Serving Up Television Without the TV Set," *The New York Times*, March 10, 2008, https://www.nytimes.com/2008/03/10/technology/10online.html.

75 **the website Jezebel asked:** Dodai Stewart, "'Abortion Man': The Worst, Supposedly-Funny Video You May Ever See," *Jezebel*, April 23, 2008, https://jezebel.com/abortion-man-the-worst-supposedly-funny-video-you-m-383043.

76 **old media lacked technical chops:** YouTube engineers once had to jury-rig a test for Time Warner, which wanted to run its own video-playing box on YouTube. After the test showed YouTube's player working much faster, Time Warner relented.

77 **elaborate steps to get around copyright:** YouTube worked to remove them, but fans found a work-around: they tagged the clips with the coded term "cheese soufflé," instead of "WWE" or "wrestling," to make them harder to discover and remove.

77 **drag them to Chicago:** Two different public relations staffers said the co-founders were reluctant to go on Winfrey's show. When asked about this recollection, Steve Chen said, "We were excited about the Oprah taping! It was a little awkward that we did all the rehearsal and preparation days before the actual show."

Chapter 6: The Bard of Google

81 **an office eatery:** Levy, *In the Plex*, 133–4.

82 **Al Gore dialed in:** Ken Auletta, *Googled: The End of the World as We Know It* (New York: Penguin Press, 2009), 59.

86 **resources to YouTube's SQUAD:** In the San Mateo office, the moderators had to share one email account on a ticketing software system to answer complaints because one account was all the start-up felt it could afford.

87 **an exposé of its malfeasance:** DeKort uploaded his grainy footage on YouTube after two newspapers turned down his whistleblowing account. After the video blew up, DeKort reached out to YouTube, which invited him to its San Mateo office and, unsure what to do, handed him a shirt and some stickers.

88 **Googlers would join his White House:** Nicole Wong, the Decider, became Obama's deputy chief technology officer.

Chapter 7: Pedal to the Metal

94 **objecting to the promotional stunt:** The event, according to several involved, was also an organizational nightmare. They tried to recruit Dave Chappelle but failed. One staffer recalled seeing the marketing manager behind the event faint from stress.

96 **Walk once greeted:** Asked about this later, Walk said, "I'm sure sometimes I could be a jerk, even in jest." He added, "I do think a product lead has to take some amount of arrows in order to protect the user, who is not in the room for these discussions. But I wish I was better on being 'hard on the problem, not the person,' and I'm sure I pissed off some ad folks at times."

99 **swat it back to the courts:** When a higher court ruled in YouTube's favor in 2013, Hurley wrote a note on Twitter addressed to the Viacom CEO, Philippe Dauman: @Chad_Hurley, "Hey Philippe, wanna grab a beer to celebrate?!," Twitter, April 18, 2013, https://mobile.twitter.com/chad_hurley/status/324986303072575489.

99 **wrote a blog post:** Zahavah Levine, "Broadcast Yourself," YouTube Official Blog, March 18, 2010, https://blog.youtube/news-and-events/broadcast-yourself/.

99 **Another brief post:** Kent Walker, "YouTube Wins Case Against Viacom," YouTube Official Blog, June 23, 2010, https://blog.youtube/news-and-events/youtube-wins-case-against-viacom/.

100 **skin-detection algorithms:** A new coder for this, a less socially adept fellow, once greeted the YouTube designer Jasson Schrock with a shout across the street outside the offices, "Hey! I'm the porn guy!"

Chapter 8: The Diamond Factory

105 **network brass boasted:** Melissa Greggo, "Latenight Laffers," *Variety*, November 16, 2000, https://variety.com/2000/tv/news/latenight-laffers-1117789313/.

106 **"Danny Diamond Gay Bar":** Zappin's original YouTube footage has all since been removed. The account Sleight0fHand uploaded some of Zappin's material as a montage.

113 **the digital era's Haight-Ashbury:** Eriq Gardner, "Maker Studios Lawsuit: Inside the War for YouTube's Top Studio," *The Hollywood Reporter*, October 24, 2013, https://www.hollywoodreporter.com/business/business-news/maker-studios-lawsuit-inside-war-650541/.

Chapter 9: Nerdfighters

115 **Epic Pictures:** They made movies like *Bear*, a campy tale of two couples forced to fend off a clever grizzly, shot with a live beast. Afterward, Wong concluded that "real grizzly bears are not an animal to fuck with."

118 **explained in one video:** vlogbrothers, "How To Be a Nerdfighter: A Vlogbrothers FAQ," YouTube video, December 27, 2009, 3:58, https://www.youtube.com/watch?v=FyQi79aYfxU.

118 **Their brand appealed to the awkward:** On the third page of the VidCon program, Hank inserted references to *Harry Potter, The Matrix, The Princess Bride*, and *The Lord of the Rings*.

120 **said on a podcast:** Kevin Nalty, "Ray William Johnson Is YouTube's First Millionaire Creator," Will Video for Food, April 1, 2011, http://willvideoforfood.com/2011/04/01/ray-william-johnson-is-youtubes-first-millionaire-creator/. Johnson claimed he earned over seven figures in annual ad sales; a YouTube spokesperson told Nalty at the time that the company could not confirm this.

121 **viewership began to slip:** Brian Stelter, "Nielsen Reports a Decline in Television Viewing," *The New York Times*, May 3, 2012, https://mediadecoder.blogs.nytimes.com/2012/05/03/nielsen-reports-a-decline-in-television-viewing/.

Chapter 10: Kitesurfing TV

123 **forgot his password:** Wayne Drehs, "How PewDiePie Gamed the World," ESPN, June 11, 2015, https://www.espn.com/espn/story/_/id/13013936/pewdiepie-how-became-king-youtube.

125 **a cut of earnings:** Most gamers were able to elide concerns over depicting a copyrighted game on video with a "fair use" qualifier, but this was shaky legal ground, particularly for twentysomething or teenage YouTubers.

126 **he once dated Ivanka Trump:** Untrue, though they were friends, according to sources close to Kamangar.

127 **said in an interview:** Danielle Sacks, "How YouTube's Global Platform Is Redefining the Entertainment Business," *Fast Company*, January 31, 2011, https://www.fastcompany.com/1715183/how-youtubes-global-platform-redefining-entertainment-business.

127 **told an audience:** Shishir Mehrotra, "What Will Software Look Like Once Anyone Can Create It?," *Harvard Business Review*, January 30, 2019, https://hbr.org/2019/01/what-will-software-look-like-once-anyone-can-create-it.

129 **NFL commissioner Roger Goodell:** The appearance of Google's CEO was never guaranteed; Page had a habit of never confirming his attendance on Google Calendar invites.

131 **internet generation's Siskel and Ebert:** Critics panned the movie. (It has a 0 percent rating on Rotten Tomatoes.) But it flourished on cable, debuting to more than seven million viewers as the top-ranked movie on TV.

132 **Kyncl told staff:** An interview with Jamie Byrne, YouTube director, September 2020. Kyncl did not recall saying this, according to a YouTube spokesperson.

132 **Smith cut in:** An interview with Patrick Walker, February 2021. In a statement a YouTube spokesperson said Kyncl did not recall this meeting and "does not gamble." A *Vice* representative did not return multiple requests for comment. Another YouTube employee recalled gambling in Las Vegas with Shane Smith on a separate occasion.

133 **"Cosby Show" room:** John Seabrook, "Streaming Dreams," *The New Yorker*, January 8, 2012, https://www.newyorker.com/magazine/2012/01/16/streaming-dreams.

Chapter 11: See It Now

136 **looking into the camera:** A. M. Sperber, *Murrow: His Life and Times* (New York: Fordham University Press, 1999), 355–6.

138 **a prominent Egyptian activist:** Another version of this event emerged in WikiLeaks cables, which claimed that the activist, Wael Abbas, after failing to contact Google, reached out to the U.S. embassy to resolve the issue.

141 **Unbeknownst to his company:** Wael Ghonim, *Revolution 2.0: The Power of the People Is Greater Than the People in Power* (Boston: Houghton Mifflin Harcourt, 2012).

142 **long rap sheet:** Ken Bensinger and Jeff Gottlieb, "Alleged Anti-Muslim Film Producer Has Drug, Fraud Convictions," *Los Angeles Times*, September 13, 2012, https://latimesblogs.latimes.com/lanow/2012/09/alleged-anti-muslim-film-producer-convictions-drugs-fraud.html.

142 **Muhammad as a pedophile and a brute:** Later, an actress in the movie would sue, claiming she was tricked into appearing in the movie and her lines were dubbed over. The full movie never came out.

142 **circulating it more widely:** Jillian C. York, *Silicon Values: The Future of Free Speech Under Surveillance Capitalism* (New York: Verso, 2022), 35.

143 **take the video trailer down:** The Department of State, under Hillary Clinton, was certainly not anti-Google. Jared Cohen, a staffer for Clinton, had called the Neda video "the most significant viral video of our lifetimes" and told YouTube's senior management that the site was "better than any intelligence we could get." Cohen then went to work for Google. Jessie Lichtenstein, "Digital Diplomacy," *The New York Times*, July 16, 2010, https://www.nytimes.com/2010/07/18/magazine/18web2-0-t.html.

Chapter 12: Will It Make the Boat Go Faster?

146 **"all over the map":** Walter Isaacson, "The Real Leadership of Steve Jobs," *Harvard Business Review*, April 2012, https://hbr.org/2012/04/the-real-leadership-lessons-of-steve-jobs.

147 **Everyone should be:** Doerr, *Measure What Matters*, 147.

151 **he told employees:** Doerr, *Measure What Matters*, 147.

152 **a business bestseller:** Ben Hunt-Davis, *Will It Make the Boat Go Faster* (Market Harborough: Troubador Publishing, 2011).

152 **YouTube's metrics dashboard:** This tool began as a too-long acronym, but someone liked the Rastafarian joke, so it stuck.

153 **a compelling subject line:** Doerr, *Measure What Matters*, 161–5.

153 **paid Google to run search ads:** Or, at least, Google once did. Arguments that Google later moved to favor keeping people on its own services in search results would become the subject of antitrust lawsuits.

Chapter 13: Let's Play

158 **"remarkably ugly":** Jason Kincaid, "YouTube Unveils Slick Experimental Redesign, Codenamed Cosmic Panda," *TechCrunch*, July 7, 2011, https://techcrunch.com/2011/07/07/youtube-unveils-slick-experimental-redesign-codenamed-cosmic-panda/.

159 **wrote an op-ed:** DeStorm Power, "Can I Count on YouTube?," *New Rockstars*, April 19, 2013, https://newmediarockstars.com/2013/04/destorm-power-can-i-count-on-youtube-op-ed/.

161 **he said in an interview:** Noreena Hertz, "Think Millenials Have It Tough? For 'Generation K,' Life is Even Harsher," *The Guardian*, March 19, 2016, https://www.theguardian.com/world/2016/mar/19/think-millennials-have-it-tough-for-generation-k-life-is-even-harsher.

Chapter 14: Disney Baby Pop-Up Pals Easter Eggs SURPRISE

168 **Kidvid rules:** Eventually, it would become three hours a week for "educational." Commercials could run for only twelve minutes every hour, less on weekends.

168 **threat of a license removal:** Kathryn C. Montgomery, *Generation Digital* (Cambridge: MIT Press, 2007). Portions of this chapter borrow from Montgomery's book, an excellent primer on the history of regulation in children's media.

168 **ignored other concerns:** "I want a country where children come first again and where virtue is honored," the head of the ultraconservative Family Research Council said in support of the Communications Decency Act, the 1996 law that created Section 230.

170 **upending higher education:** Khan, a college classmate of Mehrotra's, was a YouTube favorite.

170 **had christened YouTube:** Danielle Sacks, "How YouTube's Global Platform Is Redefining the Entertainment Business," *Fast Company*, January 31, 2011, https://www.fastcompany.com/1715183/how-youtubes-global-platform-redefining-entertainment-business.

171 **online *Billboard* for YouTube:** Joshua Cohen, "Top 100 Most Viewed YouTube Channels Worldwide: April 2014," *Tubefilter*, May 15, 2014, https://www.tubefilter.com/2014/05/15/top-100-most-viewed-youtube-channels-worldwide-april-2014/.

171 **one marketer told a reporter**: Hillary Reinsberg, "YouTube's Biggest Star Is an Unknown Toy-Reviewing Toddler Whisperer," *BuzzFeed*, July 18, 2014, https://www.buzzfeednews.com/article/hillaryreinsberg/youtubes-biggest-star-is-an-unknown-toy-reviewing-toddler-wh.

Chapter 15: The Five Families

180 **YouTube's rules on speech:** Around 2013, YouTube staffers met a Jewish nonprofit in Europe for the company's "trusted flaggers" program, which gave certain groups more tools to flag troubling content. The nonprofit begged YouTube to remove Holocaust denial, one attendee recalled. YouTube politely refused.

180 **Zunger proposed adding:** Mark Bergen, "YouTube Executives Ignored Warnings, Letting Toxic Videos Run Rampant," *Bloomberg*, April 2, 2019, https://www.bloomberg.com/news/features/2019-04-02/youtube-executives-ignored-warnings-letting-toxic-videos-run-rampant.

182 **a bumper sticker:** Robbie Brown, "Gun Enthusiast with Popular Online Videos Is Shot to Death in Georgia," *The New York Times*, January 10, 2013, https://www.nytimes.com/2013/01/11/us/keith-ratliff-gun-enthusiast-of-fpsrussia-is-shot-to-death.html.

184 **the industry's mafioso feel:** Strompolos had poached Maker's executive Ezra Cooperstein, who was also being recruited by AwesomenessTV, another MCN. Machinima's representative at Weinstein's meetings was the nephew of the firm's chair. Maker's representative, Chris Williams, was Weinstein's cousin.

185 **documenting the fight:** Ray William Johnson, "Why I Left Maker Studios," *New Rockstars*, December 11, 2012, https://newmediarockstars.com/2012/12/why-i-left-maker-studios/.

Chapter 16: Lean Back

188 **later told viewers:** Ingrid Nilsen, "This One Is For You (My Last Video)," YouTube video, June 30, 2020, 48:25, https://www.youtube.com/watch?v=zbuky-D7wy8.

189 **fashion observer explained:** Ruth LaFerla, "An Everywoman as Beauty Queen," *The New York Times*, August 5, 2009, https://www.nytimes.com/2009/08/06/fashion/06youtube.html.

191 **Engineers ran tests for Dallas:** YouTube, like most internet companies, ran these tests on subsets of its users without their knowledge. For years, YouTube showed a small percentage of viewers no ads at all, an unaware control group.

193 **search engine boilerplate:** Ben Collins, "Meet the 'Cult' Leader Stumping for Donald Trump," *The Daily Beast*, February 5, 2016, https://www.thedailybeast.com/meet-the-cult-leader-stumping-for-donald-trump.

193 **he told viewers:** Kevin Roose, "One: Wonderland," April 16, 2020, in *Rabbit Hole*, produced by *The New York Times*, podcast, 26:48, https://www.nytimes.com/2020/04/16/podcasts/rabbit-hole-internet-youtube-virus.html.

193 **If the subject came up:** In a statement, Molyneux wrote, "I am an advocate for a stateless society, since I accept that the foundation of moral philosophy is the *non-aggression principle*, which condemns the initiation of the use of force."

193 **an "anarcho-capitalist":** Mike Masnick, "'Anarcho-Capitalist' Stefan Molyneux, Who Doesn't Support Copyright, Abuses DMCA to Silence Critic," *Techdirt*, August 22, 2014, https://www.techdirt.com/2014/08/22/anarcho-capitalist-stefan-molyneux-who-doesnt-support-copyright-abuses-dmca-to-silence-critic/.

194 **suddenly left home:** Kate Hilpern, "You'll Never See Me Again," *The Guardian*, November 14, 2008, https://www.theguardian.com/lifeandstyle/2008/nov/15/family-relationships-fdr-defoo-cult.

194 **preached this online:** In a statement, Molyneux wrote that he has "never told adults to leave their families." He continued: "I remind them that it is an option under situations of continued and severe abuse, and my advice is to try to talk things out with their parents first, and engage with a professional counselor, either individually or with their family. This is neither radical nor unknown." He then cited similar advice from Dr. Phil.
Regarding charges against Papadopoulos, Molyneux wrote, "Eventually, my wife was reprimanded, the main issue being that, while family separation may be appropriate in cases of abuse, she did not get sufficient history from those sending in questions to discuss that."

194 **board later reprimanded:** Tu Thanh Ha, "Therapist Who Told Podcast Listeners to Shun Their Families Reprimanded," *The Globe and Mail*, November 1, 2012, https://www.theglobeandmail.com/news/toronto/therapist-who-told-podcast-listeners-to-shun-their-families-reprimanded/article4846791/.

194 **"Philosopher King":** This, Molyneux said in a statement, was a joke about Plato, "a totalitarian philosopher I strongly oppose."

194 **told a Canadian reporter:** Tu Thanh Ha, "How a Cyberphilosopher Convinced Followers to Cut Off Family," *The Globe and Mail*, December 12, 2008, https://www.theglobeandmail.com/technology/how-a-cyberphilosopher-convinced-followers-to-cut-off-family/article7511365/.

Chapter 17: The Mother of Google

196 **media visibility and stage presence:** For a counterexample, consider the Microsoft stagecraft with Bill Gates from the 1990s. You can find the cringey clips on YouTube.

196 **sanctity of its search rankings:** That is, except in China, where Google manipulated search to appease government censors before leaving the country.

197 **the child said:** Esther Wojcicki, *How to Raise Successful People: Simple Lessons for Radical Results* (New York: HarperCollins, 2018), 138.

197 **her book on parenting:** Wojcicki, *How to Raise Successful People*, 138.

197 **another magazine named her:** Elizabeth Murphy, "Inside 23andMe founder Anne Wojcicki's $99 DNA Revolution," *Fast Company*, October 14, 2013, https://www.fastcompany.com/3018598/for-99-this-ceo-can-tell-you-what-might-kill-you-inside-23andme-founder-anne-wojcickis-dna-r.

198 **a San Jose newspaper:** Mike Swift, "Susan Wojcicki: The Most Important Googler You've Never Heard Of," *The Mercury News*, February 3, 2011, https://www.mercurynews.com/2011/02/03/susan-wojcicki-the-most-important-googler-youve-never-heard-of/.

198 **dubbed her:** Patricia Sellers, "The New Valley Girls," *Fortune*, October 13, 2008, https://fortune.com/2008/10/13/the-new-valley-girls-2/.

198 **article remarked upon:** Robert Hof, "Look Out, Television: Google Goes for the Biggest Advertising Prize of All," *Forbes*, February 9, 2014, https://www.forbes.com/sites/roberthof/2014/01/22/look-out-television-already-the-king-of-online-ads-google-goes-for-the-big-prize/.

200 **Ramaswamy grew animated:** Amir Efrati, "The Ascension of Google's Sridhar Ramaswamy," *The Information*, April 6, 2015, https://www.theinformation.com/articles/the-ascension-of-google-s-sridhar-ramaswamy. Ramaswamy denied saying this.

200 **went public with their split:** Liz Gannes, "Google Co-Founder Sergey Brin and 23andMe Co-Founder Anne Wojcicki Have Split," *All Things D*, August 28, 2013, https://allthingsd.com/20130828/google-co-founder-sergey-brin-and-23andme-co-founder-anne-wojcicki-have-split/.

201 **Brin went to Burning Man:** Vanessa Grigoriadis, "O.K., Glass: Make Google Eyes," *Vanity Fair*, March 12, 2014, https://www.vanityfair.com/style/2014/04/sergey-brin-amanda-rosenberg-affair. After separating in 2013, Sergey Brin and Anne Wojcicki divorced in 2015. A YouTube representative said the company could not confirm Brin's attendance at Burning Man.

203 **said in a statement:** Claire Cain Miller, "Google Appoints Its Most Senior Woman to Run YouTube," *The New York Times*, February 5, 2014, https://bits.blogs.nytimes.com/2014/02/05/google-appoints-its-most-senior-woman-to-run-youtube/.

203 **Slide five showed:** Doerr, *Measure What Matters*, 166.

Chapter 18: Down the Tubes

207 **an anthropologist:** David Graeber, "On the Phenomenon of Bullshit Jobs," *Atlas of Places*, 2013, https://www.atlasofplaces.com/essays/on-the-phenomenon-of-bullshit-jobs/.

208 **her co-worker said:** Alex Morris, "When Google Walked," *New York*, February 5, 2019, https://nymag.com/intelligencer/2019/02/can-the-google-walkout-bring-about-change-at-tech-companies.html.

209 **she signed off:** Claire Stapleton, "Down the 'Tube: 10.3.2014," Tiny Letter, October 3, 2014, http://tinyletter.com/clairest/letters/down-the-tube-10-3-2014.

210 **potential antitrust scrutiny:** Ryan Mac, "Amazon Pounces On Twitch After Google Balks Due to Antitrust Concerns," *Forbes*, August 25, 2014, https://www.forbes.com/sites/ryanmac/2014/08/25/amazon-pounces-on-twitch-after-google-balks-due-to-antitrust-concerns/?sh=60d7c4865ab6.

210 **press would chide:** Rolfe Winkler, "YouTube: 1 Billion Viewers, No Profit," *The Wall Street Journal*, February 25, 2015, https://www.wsj.com/articles/viewers-dont-add-up-to-profit-for-youtube-1424897967.

211 **carried a long profile:** Jonathan Mahler, "YouTube's Chief, Hitting a New 'Play' Button," *The New York Times*, December 20, 2014, https://www.nytimes.com/2014/12/21/business/youtubes-chief-hitting-a-new-play-button.html.

212 **for "Googley" traits:** Laszlo Bock, "Here's Google's Secret to Hiring the Best People," *Wired*, April 7, 2015, https://www.wired.com/2015/04/hire-like-google/.

212 **her office bookshelf:** Mahler, "YouTube's Chief, Hitting a New 'Play' Button."

215 **"not a legal one":** Jeff Stone, "James Foley Execution Video Creates Editorial Questions for Twitter, YouTube," *International Business Times*, August 20, 2014, https://www.ibtimes.com/james-foley-execution-video-creates-editorial-questions-twitter-youtube-1664478.

215 **"dangerous precedent":** "Silicon Valley Firms Reacted Quickly to Halt Spread of Steven Sotloff Beheading Video," NBC, September 4, 2014, https://www.nbcbayarea.com/news/local/silicon-valley-firms-halted-spread-of-steven-sotloff-beheading-video/2085940/.

215 **told the press:** Barton Gellman and Ashkan Soltani, "NSA Infiltrates Links to Yahoo, Google Data Centers Worldwide, Snowden Documents Say," *The Washington Post*, October 30, 2013, https://www.washingtonpost.com/world/national-security/nsa-infiltrates-links-to-yahoo-google-data-centers-worldwide-snowden-documents-say/2013/10/30/e51d661e-4166-11e3-8b74-d89d714ca4dd_story.html.

216 **"before it's made":** Ben Quinn, "YouTube Staff Too Swamped to Filter Out All Terror-Related Content," *The Guardian*, January 28, 2015, https://www.theguardian.com/technology/2015/jan/28/youtube-too-swamped-to-filter-terror-content.

216 **"Sunlight is the best disinfectant":** A variation on the judicial aphorism from Justice Louis Brandeis.

216 **speech on censorship:** Mark Sweney, "Google Calls for Anti-Isis Push and Makes YouTube Propaganda Pledge," *The Guardian*, June 24, 2015, https://www.theguardian.com/media/2015/jun/24/google-youtube-anti-isis-push-inhuman-beheading-videos-censorship.

Chapter 19: True News

219 **grasp of digital stardom:** Brooks Barnes, "Disney Buys Maker Studios, Video Supplier for YouTube," *The New York Times*, March 24, 2014, https://www.nytimes.com/2014/03/25/business/media/disney-buys-maker-studios-video-supplier-for-youtube.html.

219 **who was displeased:** Kevin Roose, "What Does PewDiePie Really Believe?," *The New York Times*, October 9, 2019, https://www.nytimes.com/interactive/2019/10/09/magazine/PewDiePie-interview.html. A Disney spokesperson did not offer any additional comment.

220 **a *Variety* story:** Andrew Wallenstein, "If PewDiePie Is YouTube's Top Talent, We're All Doomed," *Variety*, September 11, 2013, https://variety.com/2013/biz/news/if-pewdiepie-is-youtubes-top-talent-were-all-doomed-1200607196/.

220 **"kind of scary":** Sven Grundberg and Jens Hansegard, "YouTube's Biggest Draw Plays Games, Earns $4 Million a Year," *The Wall Street Journal*, June 16, 2014, https://www.wsj.com/articles/youtube-star-plays-videogames-earns-4-million-a-year-1402939896.

222 **let him don headphones:** Kevin Roose, "The Making of a YouTube Radical," *The New York Times*, June 8, 2019, https://www.nytimes.com/interactive/2019/06/08/technology/youtube-radical.html.

222 **eloquent descriptor:** Roose, "What Does PewDiePie Really Believe?"

223 **woman's video diary:** Caitlin Dickson, "Richard Dawkins Gets into a Comments War with Feminists," *The Atlantic*, July 6, 2011, https://www.theatlantic.com/national/archive/2011/07/richard-dawkins-draws-feminist-wrath-over-sexual-harassment-comments/352530/.

224 **"talk-radio with video":** Brian Rosenwald, *Talk Radio's America: How an Industry Took Over a Political Party That Took Over the United States* (Cambridge: Harvard University Press, 2019). I relied on Rosenwald's excellent book throughout this chapter.

224 **clickable tabloid fare:** Some early hits included "Amy Winehouse's Disastrous Face" and "Vicious Girl Fight in the Bathroom."

225 **was not true:** Aja Romano, "What We Still Haven't Learned From Gamergate," *Vox*, January 7, 2021, https://www.vox.com/culture/2020/1/20/20808875/gamergate-lessons-cultural-impact-changes-harassment-laws.

225 **recalled later:** Kevin Roose, "Two: Looking Down," April 23, 2020, in *Rabbit Hole*, produced by *The New York Times*, podcast, 36:57, https://www.nytimes.com/2020/04/23/podcasts/rabbit-hole-internet-youtube-virus.html.

226 **neo-Nazi footage:** Brentin Mock, "Neo-Nazi Groups Share Hate via YouTube," *Southern Poverty Law Center Intelligence Report,* April 20, 2007, https://www.splcenter.org/fighting-hate/intelligence-report/2007/neo-nazi-groups-share-hate-youtube.

Chapter 20: Disbelief

230 **reports of sexual harassment:** Amy Brittain and Irin Carmon, "Charlie Rose's Misconduct Was Widespread at CBS and Three Managers Were Warned, Investigation Finds," *The Washington Post,* May 3, 2018, https://www.washingtonpost.com/charlie-roses-misconduct-was-widespread-at-cbs-and-three-managers-were-warned-investigation-finds/2018/05/02/80613d24-3228-11e8-94fa-32d48460b955_story.html.

231 **called itself Brain:** Gideon Lewis-Kraus, "The Great A.I. Awakening," *The New York Times,* December 14, 2016, https://www.nytimes.com/2016/12/14/magazine/the-great-ai-awakening.html.

232 **interview back in 2002:** Steve Levy, "Google Search Will Be Your Next Brain," *Wired,* January 16, 2015, https://www.wired.com/2015/01/google-search-will-be-your-next-brain/. (Levy cites his earlier interview with Page here.)

234 **network would recommend:** Casey Newton, "How YouTube Perfected the Feed," *The Verge,* August 30, 2017, https://www.theverge.com/2017/8/30/16222850/youtube-google-brain-algorithm-video-recommendation-personalized-feed.

Chapter 21: A Boy and His Toy

238 **admitted to a reporter:** Geoff Weiss, "This 4-Year-Old Has the Most-Viewed YouTube Channel in the World," *Tubefilter,* September 8, 2016, https://www.tubefilter.com/2016/09/08/ryan-toys-review-most-watched-youtube-channel-in-the-world/.

239 **find a list:** Joshua Cohen, "Meet the Top 1% of YouTube's 'Google Preferred' Channels for Advertisers," *Tubefilter,* April 18, 2014, https://www.tubefilter.com/2014/04/18/youtube-google-preferred-channels-top-1-percent-advertisers/.

239 **obsessed offspring:** Laura T. Coffey, "'Crack for Toddlers': Mysterious Toy Review Videos Enchant Kids, Bring in Millions of Dollars," *Today,* July 31, 2014, https://www.today.com/parents/disneycollector-toy-reviewer-enchants-toddlers-youtube-1D80001314.

240 **"an acid trip":** Chavie Lieber, "How L.O.L. Dolls Became the Dopamine Hit of a Generation," *The New York Times,* April 16, 2020, https://www.nytimes.com/2020/04/16/parenting/lol-surprise-doll-isaac-larian.html.

241 **"To her, it's music":** Ben Popper, "Red Dawn: An Inside Look at YouTube's New Ad-Free Subscription Service," *The Verge,* October 15, 2015, https://www.theverge.com/2015/10/21/9566973/youtube-red-ad-free-offline-paid-subscription-service.

242 **the *Daily Mail* outed:** Paul Thompson, "Mystery Woman Behind the 'Richest Hands on the Internet' Revealed: Former Pornstar Makes $5m a Year Unwrapping Disney Toys on YouTube," *Daily Mail,* February 24, 2015, https://www.dailymail.co.uk/news/article-2958242/Brazilian-former-porn-star-Diane-DeJesus-mystery-figure-5million-year-YouTube-sensation-DC-Toys-Collector.html.

244 **The party featured**: Belinda Luscombe, "Meet YouTube View Master," *Time,* August 17, 2015, https://time.com/4012832/meet-youtubes-view-master/.

245 **service inside schools didn't take:** In 2012, YouTube convened education officials and broadcasters for an education event where a teacher complained about seeing tequila ads on YouTube with her class. "If my kid is forced to watch a Doritos commercial while at school," one attendee griped, "I swear I will sue that district."

Chapter 22: Spotlight

248 **published a survey:** Susanne Ault, "Survey: YouTube Stars More Popular Than Mainstream Celebs Among U.S. Teens," *Variety,* August 5, 2014, https://variety.com/2014

/digital/news/survey-youtube-stars-more-popular-than-mainstream-celebs-among-u-s-teens-1201275245/.

252 **gaming lore, science, and YouTube:** Patrick also worked for Big Frame, the early management company that represented Nilsen, DeStorm Power, and MysteryGuitarMan.

254 **lackluster reviews and interest:** Brian Lowry, "Review: 'Scare PewDiePie' on YouTube Red," *Variety*, February 10, 2016, https://variety.com/2016/digital/reviews/scare-pewdiepie-review-youtube-red-1201701504/.

254 **YouTube Space:** Those with more than ten thousand subscribers could apply to film at the production space.

Chapter 23: Joke, Threat, Obvious

259 **breaking into joyful sobs:** Ingrid Nelson, "Something I Want You to Know (Coming Out)," YouTube video, June 9, 2015, 19:12, https://www.youtube.com/watch?v=Eh7WRYXVh9M.

260 **befitting a politician:** Three years later, Grove would leave Google to become an elected official in his home state of Minnesota.

260 **Jimmy Kimmel's talk show:** Later, he would go on the talk show with Jimmy Fallon, who tousled his hair.

261 **Southern Poverty Law Center described:** "Extremist Info: Stefan Molyneux," Southern Poverty Law Center, undated, https://www.splcenter.org/fighting-hate/extremist-files/individual/stefan-molyneux.

261 **Molyneux would refute:** That year, Molyneux hosted the Danish academic Helmuth Nyborg and the author Jared Taylor. In a statement, Molyneux wrote, "I have had many people on my show that I disagree with—in fact, no human belief systems overlap perfectly, nor should they. . . . The phrase 'white supremacist' refers to the belief that whites should violently rule over all other races, which would be a gruesome (and most likely genocidal) violation of the non-aggression principle I hold sacred. To my knowledge, I have never hosted anyone who holds such vile beliefs. I certainly have never heard either Jared Taylor or Dr. Nyborg express such a thirst for violence." The Southern Poverty Law Center defines white nationalism as a belief that "white identity should be the organizing principle of the countries that make up Western civilization."

262 **Black YouTube animator:** Viewers adored Thorne too, although he later confessed to YouTube brass that his videos with animated thumbnails did better than those showing his face. It seemed people clicked less on Black faces and the algorithm followed suit.

263 **told the journalist:** Joshua Green, *Devil's Bargain: Steve Bannon, Donald Trump, and the Storming of the Presidency* (New York: Penguin Press, 2017).

264 **A later study:** Bernhard Rieder, Ariadna Matamoros-Fernández, and Òscar Coromina, "From Ranking Algorithms to 'Ranking Cultures': Investigating the Modulation of Visibility in YouTube Search Results," *Convergence: The International Journal of Research into New Media Technologies* 24, no. 1 (January 10, 2018): 50–68.

264 **noting the success media companies:** Mike Shields, "Some Media Companies Cool on YouTube Distribution," *The Wall Street Journal*, July 21, 2016, https://www.wsj.com/articles/some-media-companies-cool-on-youtube-distribution-1469095200.

265 **amount of YouTube's time and attention:** A YouTube spokesperson said the company has "always focused on YouTube" and was not concerned with Facebook at the time.

268 **DeFranco countered on Twitter:** Aja Romano, "YouTube's 'Ad-Friendly' Content Policy May Push One of Its Biggest Stars Off the Website," *Vox*, September 2, 2016, https://www.vox.com/2016/9/2/12746450/youtube-monetization-phil-defranco-leaving-site.

270 **devoted to Trump:** Ben Collins, "'Is Hillary Dying' Hoax Started by Pal of Alex Jones," *The Daily Beast,* August 9, 2016, https://www.thedailybeast.com/is-hillary-dying-hoax-started-by-pal-of-alex-jones.

270 **showed a slip in trajectory:** Doerr, *Measure What Matters,* 168.

Chapter 24: The Party Is Over

273 **leaked to *Breitbart News*:** Allum Bokhari, "LEAKED VIDEO: Google Leadership's Dismayed Reaction to Trump Election," *Breitbart,* September 12, 2018, https://www.breitbart.com/tech/2018/09/12/leaked-video-google-leaderships-dismayed-reaction-to-trump-election/.

274 **a photo of him:** Trey Parker, "The 100 Most Influential People: Felix Kjellberg (a.k.a. PewDiePie)," *Time,* April 21, 2016, https://time.com/collection-post/4302406/felix-kjellberg-pewdiepie-2016-time-100/.

275 **Nussbaum would write:** Emily Nussbaum, *I Like to Watch: Arguing My Way Through the TV Revolution* (New York: Random House, 2019).

276 **Kjellberg later recalled:** Kevin Roose, "Six: Impasse," April 21, 2020, in *Rabbit Hole,* produced by *The New York Times,* podcast, 24:25, https://www.nytimes.com/2020/05/21/podcasts/rabbit-hole-PewDiePie-youtube-virus.html.

277 **no mention of Jews:** Abby Phillip, "Trump's Statement Marking Holocaust Remembrance Leaves Out Mention of Jews," *The Washington Post,* January 27, 2017, https://www.washingtonpost.com/politics/trumps-statement-marking-holocaust-remembrance-leaves-out-mention-of-jews/2017/01/27/0886d3c2-e4bd-11e6-a547-5fb9411d332c_story.html.

278 **The *Journal* article:** Rolfe Winkler, Jack Nicas, and Ben Fritz, "Disney Severs Ties with YouTube Star PewDiePie After Anti-Semitic Posts," *The Wall Street Journal,* February 14, 2017, https://www.wsj.com/articles/disney-severs-ties-with-youtube-star-pewdiepie-after-anti-semitic-posts-1487034533.

279 **in a later interview:** Roose, "Six: Impasse."

279 **"My Response":** This video has since been deleted.

280 **released a video:** The Film Theorists, "Film Theory: Why Pewdiepie's Fiverr Joke Backfired," YouTube video, February 25, 2017, 20:30, https://www.youtube.com/watch?v=DxphJ-dnX2Y.

280 **"It's often hard to see":** For good measure, Patrick also noted that old media has "generally been a huge steaming pile of turd to online personalities."

281 **Kyncl wrote:** Robert Kyncl with Maany Peyvan, *Streampunks: YouTube and the Rebels Remaking Media* (New York: Harper Business, 2017).

Chapter 25: Adpocalypse

285 **this front-page headline:** Alexi Mostrous, "Big Brands Fund Terror Through Online Adverts," *The Times,* February 9, 2017, https://www.thetimes.co.uk/article/big-brands-fund-terror-knnxfgb98.

285 **told *The Guardian*:** Jane Martinson, "Guardian Pulls Ads from Google After They Were Placed Next to Extremist Material," *The Guardian,* March 16, 2017, https://www.theguardian.com/media/2017/mar/16/guardian-pulls-ads-google-placed-extremist-material.

286 **identified several brand names:** Jack Nicas, "Google's YouTube Has Continued Showing Brands' Ads with Racist and Other Objectionable Videos," *The Wall Street Journal,* March 24, 2017, https://www.wsj.com/articles/googles-youtube-has-continued-showing-brands-ads-with-racist-and-other-objectionable-videos-1490380551.

286 **reminded marketers in a note:** Jennifer Faull, "The Screw Tightens on Google as Holding Groups Advise Advertisers to Reassess the Risks," *The Drum*, March 17, 2017, https://www.thedrum.com/news/2017/03/17/the-screw-tightens-google-holding-groups-advise-advertisers-reassess-the-risks.

287 **explained in an interview:** Peter Kafka, "Google Says Its YouTube Ad Problem Is 'Very Very Very Small' But It's Getting Better at Fixing It Anyway," *Recode*, April 3, 2017, https://www.vox.com/2017/4/3/15157654/google-youtube-advertising-controversy-interview-philipp-schindler.

287 **Wojcicki insisted:** Lucas Shaw and Mark Bergen, "YouTube's Plan to Clean Up the Mess That Made It Rich," *Bloomberg Businessweek*, April 26, 2018, https://www.bloomberg.com/news/features/2018-04-26/youtube-may-be-a-horror-show-but-no-one-can-stop-watching.

288 **Internet Creators Guild:** Green's colleague Laura Chernikoff ran the guild.

Chapter 26: Reinforce

292 **an American cleric:** Martin Evans, Nicola Harley, and Harry Yorke, "London Terrorist Had Twice Been Referred to Police Over His Extremist Views," *The Telegraph*, June 4, 2017, https://www.telegraph.co.uk/news/2017/06/04/london-terrorist-had-twice-referred-police-extremist-views/.

298 **abundance of *Minecraft*:** Another data point in this theory: Microsoft, Google's rival, bought the *Minecraft* studio in 2014 for $2.5 billion.

298 **researcher described Reinforce:** Kevin Roose, "The Making of a YouTube Radical," *The New York Times*, June 8, 2019, https://www.nytimes.com/interactive/2019/06/08/technology/youtube-radical.html.

298 **allowed several staffers:** Casey Newton, "How YouTube Perfected the Feed," *The Verge*, August 30, 2017, https://www.theverge.com/2017/8/30/16222850/youtube-google-brain-algorithm-video-recommendation-personalized-feed.

299 **a curious BBC reporter:** "Why Do People Still Think the Earth Is Flat?," BBC News, November 14, 2017, https://www.bbc.com/news/av/41973119.

299 **conference speaker recounted:** Alan Burdick, "Looking for Life on a Flat Earth," *The New Yorker*, May 30, 2018, https://www.newyorker.com/science/elements/looking-for-life-on-a-flat-earth.

300 **described the scene thusly:** Harry McCracken, "Susan Wojcicki Has Transformed YouTube—But She Isn't Done Yet," *Fast Company*, June 18, 2017, https://www.fastcompany.com/40427026/susan-wojcickis-youtube-isnt-tv-but-its-tvs-biggest-rival.

300 **her service's blemishes:** Although Wojcicki told the magazine she had no plans to change the ad commission terms with creators.

301 **ten-page memorandum:** Kate Conger, "Exclusive: Here's the Full 10-Page Anti-Diversity Screed Circulating Internally at Google," *Gizmodo*, August 5, 2017, https://gizmodo.com/exclusive-heres-the-full-10-page-anti-diversity-screed-1797564320.

301 **was on vacation:** Mark Bergen and Brad Stone, "Everyone's Mad at Google and Sundar Pichai Has to Fix It," *Bloomberg Businessweek*, October 19, 2017, https://www.bloomberg.com/news/features/2017-10-19/everyone-s-mad-at-google-and-sundar-pichai-has-to-fix-it.

302 **"Google Memo Fired Employee Speaks Out!":** Exact date unknown because the original video has since been removed.

302 ***Wired* wrote:** Megan Molteni and Adam Rogers, "The Actual Science of James Damore's Google Memo," *Wired*, August 15, 2017, https://www.wired.com/story/the-pernicious-science-of-james-damores-google-memo/.

302 **the company shared publicly:** Susan Wojcicki, "Read YouTube CEO Susan Wojcicki's Response to the Controversial Google Anti-Diversity Memo," *Fortune*, August 9, 2017, https://fortune.com/2017/08/09/google-diversity-memo-wojcicki/.

303 **in a subsequent interview:** Eric Johnson, "YouTube's Susan Wojcicki Explains Why the 'Google Memo' Author Had to Be Fired," *Recode*, October 16, 2017, https://www.vox.com/2017/10/16/16479486/youtube-susan-wojcicki-james-damore-google-memo-diversity-gender-kara-swisher-podcast.

Chapter 27: Elsagate

304 **told a fellow YouTuber:** Stanley "Dirt Monkey" Genadek, "Interview with Geek to Freak Greg Chism," YouTube video, May 23, 2015, 37:06, https://www.youtube.com/watch?v=_vFDsw9a3Ho.

305 **Klein told viewers:** Ethan and Hila, "How to Traumatize Your Children PRANK," YouTube video, March 3, 2016, 13:00, https://www.youtube.com/watch?v=7bCzbUiB87M.

306 **told a reporter:** Ben Popper, "Adults Dressed as Superheroes Is YouTube's New, Strange, and Massively Popular Genre," *The Verge*, February 20, 2017, https://www.theverge.com/2017/2/20/14489052/youtube-kids-videos-superheroes-disney-characters-fart-jokes.

307 **chosen trade was exceptionally weird:** Many who jumped on the "Finger Family" trend stayed anonymous because they suspected the trend's flash in the pan would end. "They were grabbing cash while they could," Ranta recalled.

307 **wrote about the trend:** Rachel Deal, "The Ballad of Elsa and Spiderman," *The Awl*, February 23, 2017, https://www.theawl.com/2017/02/the-ballad-of-elsa-and-spiderman/.

307 **a damning story:** "The Disturbing YouTube Videos That Are Tricking Children," BBC, March 27, 2017, https://www.bbc.com/news/blogs-trending-39381889.

311 **on "startling" videos:** Sapna Maheshwari, "On YouTube Kids, Startling Videos Slip Past Filters," *The New York Times*, November 4, 2017, https://www.nytimes.com/2017/11/04/business/media/youtube-kids-paw-patrol.html.

312 **on the blogging site Medium:** James Bridle, "Something Is Wrong on the Internet," Medium, November 6, 2017, https://medium.com/@jamesbridle/something-is-wrong-on-the-internet-c39c471271d2.

313 **Article title:** Mark Bridge and Alexi Mostrous, "Child Abuse on YouTube," *The Times*, November 18, 2017, https://www.thetimes.co.uk/article/child-abuse-on-youtube-q3x9zfkch.

313 **he later confessed:** Daisuke Wakabayashi, "A Former Google Executive Takes Aim at His Old Company with a Start-Up," *The New York Times*, June 19, 2020, https://www.nytimes.com/2020/06/19/technology/google-neeva-executive.html.

314 **April told viewers:** April and Davey, "WHY WE STOPPED MAKING SUPERHERO VIDEOS . . . ," YouTube video, August 16, 2017, 14:50, https://www.youtube.com/watch?v=N0gyiSYwYgs.

314 **"created a monster":** Charlie Warzel and Remy Smidt, "YouTubers Made Hundreds of Thousands Off of Bizarre and Disturbing Child Content," *BuzzFeed*, December 11, 2017, https://www.buzzfeednews.com/article/charliewarzel/youtubers-made-hundreds-of-thousands-off-of-bizarre-and.

315 **told *BuzzFeed News*:** Remy Smidt, "Authorities Say YouTube's 'Toy Freaks' Dad Is Under Investigation—But They Won't Say Who's in Charge," *BuzzFeed*, November 30, 2017, https://www.buzzfeednews.com/article/remysmidt/toy-freaks-youtube.

Chapter 28: Bad Actors

317 **was given little:** Representatives for Accenture declined to comment.

319 **in damning investigations:** Casey Newton, "The Terror Queue," *The Verge*, December 16, 2019, https://www.theverge.com/2019/12/16/21021005/google-youtube-moderators-ptsd-accenture-violent-disturbing-content-interviews-video.

320 **remove cartoons of Peppa Pig:** A YouTube spokesperson said the company's systems would not be able to do this.

320 **told a reporter:** Newton, "The Terror Queue."

321 **child abuse in videos:** "I remember that I was quite confused," Sjøberg said, recalling "a series of videos with children and fireworks or painful pranks, which I did not know how to tackle."

322 **told an advertising magazine:** T. L. Stanley, "How Vine's Hunky Goofball Logan Paul Plans to Become a Mainstream Superstar," *Adweek*, January 24, 2016, https://www.adweek.com/brand-marketing/how-vines-hunky-goofball-logan-paul-plans-become-mainstream-superstar-169152/.

324 **analyzed fifty videos:** Chris Stokel-Walker, *YouTubers: How YouTube Shook Up TV and Created a New Generation of Stars* (Kingston, Ontario: Canbury Press, 2019), 22.

325 ***Guardian* headline:** Paul Lewis, "'Fiction Is Outperforming Reality': How YouTube's Algorithm Distorts Truth," *The Guardian*, February 2, 2018, https://www.theguardian.com/technology/2018/feb/02/how-youtubes-algorithm-distorts-truth.

325 **Chaslot's research appeared:** Jack Nicas, "How YouTube Drives People to the Internet's Darkest Corners," *The Wall Street Journal*, February 7, 2018, https://www.wsj.com/articles/how-youtube-drives-viewers-to-the-internets-darkest-corners-1518020478.

326 **filled the data void:** John Hermann, "The Making of a No. 1 YouTube Conspiracy Video After the Parkland Tragedy," *The New York Times*, February 21, 2018, https://www.nytimes.com/2018/02/21/business/media/youtube-conspiracy-video-parkland.html.

328 **said it had not been informed:** Megan Farokhmanesh, "YouTube Didn't Tell Wikipedia About Its Plans for Wikipedia," *The Verge*, March 14, 2018, https://www.theverge.com/2018/3/14/17120918/youtube-wikipedia-conspiracy-theory-partnerships-sxsw.

329 **global pandemic struck:** A YouTube spokesperson said the company "took a strong supportive stance on vaccines from the very beginning" of the COVID-19 pandemic "and was one of the first companies to implement a COVID-vaccine misinfo policy."

Chapter 29: 901 Cherry Avenue

331 **or Green Nasim:** Daisuke Wakabayashi, Thomas Erdbrink, and Matthew Haag, "'Vegan Bodybuilder': How YouTube Attacker, Nasim Aghdamn, Went Viral in Iran," *The New York Times*, April 4, 2018, https://www.nytimes.com/2018/04/04/technology/nasim-aghdam-youtube-shooter.html.

331 **moved to California:** Kristina Davis, "PETA Protests Military's Use of Pigs in Training," *The San Diego Union-Tribune*, August 13, 2009, https://www.sandiegouniontribune.com/military/sdut-peta-protests-militarys-use-pigs-training-2009aug13-htmlstory.html.

333 **her grandmother near San Diego:** Ethan Baron, "YouTube Shooter's Father Says She Was Angry at Company," *East Bay Times*, April 3, 2018, https://www.eastbaytimes.com/2018/04/03/youtube-shooters-father-says-she-was-angry-at-company/.

333 **she purchased a 9 mm:** "Store Worker Says YouTube Shooter Gun Buy Did Not Stand Out," Associated Press, April 6, 2018, https://apnews.com/article/a40fc8ef512549bda1ffd71075a2eed1.

333 **he called police back:** Anthony Pura, "Shooter's Family Warned Police About YouTube Grudge," KGTV San Diego, April 3, 2018, https://www.10news.com/news/shooters-family-warned-police-about-youtube-grudge.

Chapter 30: Boil the Ocean

336 **would directly threaten:** Joseph Bernstein, "YouTube's Newest Far-Right, Foul-Mouthed, Red-Pilling Star Is a 14-Year-Old Girl," *BuzzFeed*, May 13, 2019, https://www.buzzfeednews.com/article/josephbernstein/youtubes-newest-far-right-foul-mouthed-red-pilling-star-is.

338 **March *New York Times* op-ed:** Zeynep Tufekci, "YouTube, the Great Radicalizer," *The New York Times*, March 10, 2018, https://www.nytimes.com/2018/03/10/opinion/sunday/youtube-politics-radical.html.

339 **more complaints came:** A YouTube spokesperson said the commercial YouTube contracted was not dismissed for being too polarizing but because the advertising "agency didn't deliver what was asked."

341 **praising RT:** Daisuke Wakabayashi and Nicholas Confessore, "Russia's Favored Outlet Is an Online News Giant. YouTube Helped," *The New York Times*, October 23, 2017, https://www.nytimes.com/2017/10/23/technology/youtube-russia-rt.html.

342 **the law did not dictate this:** Catherine Padhi, "Ted Cruz vs. Section 230: Misrepresenting the Communications Decency Act," *Lawfare*, April 20, 2018, https://www.lawfareblog.com/ted-cruz-vs-section-230-misrepresenting-communications-decency-act.

342 **blustering threat still worked:** The law actually dictates that websites are clear to make "good faith" efforts to restrict material considered "obscene, lewd, lascivious, filthy, excessively violent, harassing, or otherwise objectionable."

343 **Lewis found that:** Becca Lewis, "Alternative Influence: Broadcasting the Reactionary Right on YouTube," *Data & Society*, September 18, 2018, https://datasociety.net/library/alternative-influence/.

343 **gave airtime to fringier figures:** Academia did relatively little work on YouTube, in part due to its copious volume. Lewis's report, for the NGO Data & Society, shed light for her insight and because she did something few researchers and YouTube staff did: she watched a ton of YouTube.

343 **told *The Guardian*:** Emine Saner, "YouTube's Susan Wojcicki: 'Where's the Line of Free Speech—Are You Removing Voices That Should Be Heard?,'" *The Guardian*, August 10, 2019, https://www.theguardian.com/technology/2019/aug/10/youtube-susan-wojcicki-ceo-where-line-removing-voices-heard.

344 **visited her Toronto home:** Daniel Lombroso, "Why the Alt-Right's Most Famous Woman Disappeared," *The Atlantic*, October 16, 2020, https://www.theatlantic.com/politics/archive/2020/10/alt-right-star-racist-propagandist-has-no-regrets/616725/.

344 **"the lowest rungs of civilization":** In a statement, Molyneux wrote, "I don't recall using this specific phrase, but it would not surprise me. In my speeches in Australia, using first-hand accounts as well as published academic sources, I detailed the horrific abuses against children—and women—performed in the past by Aboriginal tribes.

The arguments Molyneux cites have been widely debunked. The Australian government released a report that said family and sexual violence was more common among Aboriginal woman but wrote that the issue "must be understood in the historical context of white settlement and [colonization]." See: "Family Violence and Aboriginal and Torres Strait Islander Victim-Survivors," https://www.vic.gov.au/victorian-family-violence-research-agenda-2021-2024.

346 **prior day's *Washington Post*:** Craig Timberg, Elizabeth Dwoskin, Tony Romm, and Andrew Ba Tran, "Two Years After #Pizzagate Showed the Dangers of Hateful Conspiracies, They're Still Rampant on YouTube," *The Washington Post*, December 10, 2018, https://www.washingtonpost.com/business/technology/hateful-conspiracies-thrive-on-youtube-despite-pledge-to-clean-up-problematic-videos/2018/12/10/625730a8-f3f8-11e8-9240-e8028a62c722_story.html. Guillaume Chaslot, the former YouTube engineer, provided research for this article with his new group, AlgoTransparency.

348 **opened her laptop:** Alex Morris, "When Google Walked," *New York*, February 5, 2019, https://nymag.com/intelligencer/2019/02/can-the-google-walkout-bring-about-change-at-tech-companies.html. I used details from Morris's great account in the remainder of the chapter.

348 **found a link to a morning headline:** Daisuke Wakabayashi and Katie Benner, "How Google Protected Andy Rubin, the 'Father of Android,'" *The New York Times*, October 25, 2018, https://www.nytimes.com/2018/10/25/technology/google-sexual-harassment-andy-rubin.html.

349 **"like Joan of Arc or something":** Morris, "When Google Walked."
349 **would later write:** Claire Stapleton, "Google Loved Me, Until I Pointed Out Everything That Sucked About It," *Elle*, December 19, 2019, https://www.elle.com/culture/tech/a30259355/google-walkout-organizer- claire-stapleton/.

Chapter 31: The Master's Tools

352 **dropped the n-word:** Aja Romano, "YouTube Star PewDiePie Used the N-word in a Live Stream, After Months of Denying He's Racist," *Vox*, September 11, 2017, https://www.vox.com/culture/2017/9/11/16288826/pewdiepie-n-word-playerunknown-battlegrounds.
355 **"just lip service":** Wojcicki, through a YouTube spokesperson, declined to comment on this interaction.
357 **north of Sydney:** Details on the Christchurch terrorist and event came from contemporaneous accounts, news reports, and the Royal Commission of Inquiry into the Terrorist Attack of Christchurch Mosques, https://christchurchattack.royalcommission.nz/.
358 **"immediately condemned the New Zealand terrorist":** Molyneux also wrote, "The shooter in New Zealand apparently did donate to my show, long before he committed his evil actions. I have always advocated for free speech, peace and reason for the resolution of disputes—and vehemently condemned terrorism, since it is an utterly immoral violation of the non-aggression principle, which states that violence—outside of immediate self-defense—is evil. . . . Clearly he rejected my staunch advocacy of the non-aggression principle, and did the exact opposite of everything I argue for—thus rejecting everything I stand for."
358 **before his attack:** These Facebook posts came courtesy of Matt Nippert, an investigative reporter with the New Zealand Herald, who conducted his research into the Christ church shooting as a fellow at the University of Cambridge.
359 **a friend remembered:** Shamim Homayun, "Remembering My Friend, and Why There Is No Right Way to Mourn the Christchurch Attacks," *The Conversation*, March 12, 2020, https://theconversation.com/remembering-my-friend-and-why-there-is-no-right-way-to-mourn-the-christchurch-attacks-133239.
359 **stood at the door of his mosque:** Charlie Mitchel, "'Welcome, Brother': A Community That Stressed Peace Is Undone by Violence," *The Sydney Morning Herald*, March 18, 2019, https://www.smh.com.au/world/oceania/welcome-brother-a-community-that-stressed-peace-is-undone-by-violence-20190318-p5152x.html.
360 **would describe it:** Elizabeth Dwoskin and Craig Timberg, "Inside YouTube's Struggles to Shut Down Video of the New Zealand Shooting—and the Humans Who Outsmarted Its Systems," *The Washington Post*, March 18, 2019, https://www.washingtonpost.com/technology/2019/03/18/inside-youtubes-struggles-shut-down-video-new-zealand-shooting-humans-who-outsmarted-its-systems/.

Chapter 32: Roomba

365 **the Davos set:** Clare Duffy, "Marc Benioff Says It's Time to Break Up Facebook," CNN, October 17, 2019, https://www.cnn.com/2019/10/16/tech/salesforce-marc-benioff-break-up-facebook-boss-files/index.html.
366 **investigation in 2019:** Max Fisher and Amanda Taub, "How YouTube Radicalized Brazil," *The New York Times*, August 11, 2019, https://www.nytimes.com/2019/08/11/world/americas/youtube-brazil.html.
366 **he told a reporter:** (Me.) Bergen, "YouTube Executives Ignored Warnings, Letting Toxic Videos Run Rampant."
366 **"alleging that a group":** Specifically, YouTube prohibited exclusion "based on qualities like age, gender, race, caste, religion, sexual orientation or veteran status."

369 **Wojcicki replied:** MostlySane, "In Conversation with CEO, YouTube—Susan Wojcicki," YouTube video, April 16, 2019, 26:30, https://www.youtube.com/watch?v=6P-9uEvKD0o.

372 **"jacksepticeye and Elon Musk":** jacksepticeye was a top creator; Elon Musk appeared on YouTube often but didn't have his own channel.

373 **one email newsletter:** Claire Stapleton, "Down the 'Tube: (no subject)," Tiny Letter, February 15, 2019, https://tinyletter.com/clairest/letters/down-the-tube-no-subject.

373 **one tormented conversation:** Stapleton, "Google Loved Me, Until I Pointed Out Everything That Sucked About It."

374 **a loaded word:** Nitasha Tiku, "Google Walkout Organizers Say They're Facing Retaliation," *Wired*, April 22, 2019, https://www.wired.com/story/google-walkout-organizers-say-theyre-facing-retaliation/.

375 **Green began:** hankschannel, "YouTube, Pandemics, Creators, and Power: An Interview with Susan Wojcicki and Hank Green," YouTube video, May 6, 2020, 54:38, https://www.youtube.com/watch?v=_XPXht-gyj4.

Chapter 33: Which YouTube?

379 **"My account was in perfectly good standing":** In a statement, Molyneux said he retired from political commentary in 2020 to work on a book about "peaceful parenting."

381 **"a dirty word inside YouTube":** A YouTube spokesperson said that the SPLC is "not widely accepted as an authoritative voice on hate groups."

383 **Press dubbed her:** Katherine Cross, "The Oscar Wilde of YouTube Fights the Alt-Right with Decadence and Seduction," *The Verge*, August 24, 2018, https://www.theverge.com/tech/2018/8/24/17689090/contrapoints-youtube-natalie-wynn.

384 **promoting coronavirus hoaxes:** Marlow Stern, "Jake Paul Believes COVID Is 'a Hoax' and '98% of News Is Fake,'" *The Daily Beast*, November 25, 2020, https://www.thedailybeast.com/youtuber-jake-paul-believes-covid-is-a-hoaxand-i-am-fake-news.

384 **looting during summer:** Jake Paul was charged with misdemeanor trespassing at a shopping mall but he denied participating in looting or vandalism.

387 **YouTube managers argued:** Mike Isaac, Kate Conger, and Daisuke Wakabayashi, "What to Expect from Facebook, Twitter and YouTube on Election Day," *The New York Times*, November 2, 2020, https://www.nytimes.com/2020/11/02/technology/facebook-twitter-youtube-election-day.html.

388 **sent out a statement:** Mark Bergen, "YouTube Election Loophole Lets Some False Trump-Win Videos Spread," *Bloomberg*, November 10, 2020, https://www.bloomberg.com/news/articles/2020-11-10/youtube-election-loophole-lets-some-false-trump-win-videos-spread.

Epilogue

389 **he admitted:** Neima Jahromi, "The Fight for the Future of YouTube," *The New Yorker*, July 18, 2019, https://www.newyorker.com/tech/annals-of-technology/the-fight-for-the-future-of-youtube.

393 **admitted about TikTok:** hankschannel, "YouTube, Pandemics, Creators, and Power: An Interview with Susan Wojcicki and Hank Green," YouTube video, May 6, 2020, 54:38, https://www.youtube.com/watch?v=_XPXht-gyj4.

393 **would become TikTok:** ByteDance, a tech firm based in Beijing, acquired Musical.ly in 2017 and later refashioned that company's app as TikTok. A Google spokesperson declined to comment on its talks with Musical.ly.

393 **Americans watched more TikTok:** The study covered only Android phones. On iPhones, TikTok ranked as the most popular app in the world.

394 **were preschooler fare:** Joshua Cohen, "Top 50 Most Viewed US YouTube Channels," *Tubefilter*, December 31, 2020, https://www.tubefilter.com/2020/12/31/top-50-most-viewed-us-youtube-channels-2020-12-28/.

395 **that "drive outcomes":** Lucas Shaw, "YouTube Will Fund Kids Shows Based on These 12 Words," *Bloomberg*, February 4, 2020, https://www.bloomberg.com/news/articles/2020-02-04/youtube-will-fund-kids-shows-based-on-these-12-words.

395 **Moonbug Entertainment:** Lucas Shaw, "Studio Behind 'Cocomelon' Acquired in $3 Billion Deal," *Bloomberg*, November 4, 2021, https://www.bloomberg.com/news/articles/2021-11-04/-cocomelon-studio-fetches-3-billion-in-blackstone-backed-deal. Cocomelon, Little Baby Bum, and Blippi, a manic heir of *Blue's Clues* with an enormous toddler following.

395 **2020 *New York Times* headline:** Anahad O'Connor, "Are 'Kidfluencers' Making Our Kids Fat?," *The New York Times*, October 26, 2020, https://www.nytimes.com/2020/10/26/well/family/Youtube-children-junk-food-child-obesity.html.

396 **starting in 2018:** Madeline Berg, "How This 7-Year-Old Made $22 Million Playing with Toys," *Forbes*, December 3, 2018, https://www.forbes.com/sites/maddieberg/2018/12/03/how-this-seven-year-old-made-22-million-playing-with-toys-2/.

396 **banned videos promoting QAnon:** YouTube said it had categorized pro-QAnon videos as "borderline," in its penalty box, since early 2019.

397 **Facebook's popularity charts:** Facebook often disputed these measurements.

398 **low and falling:** In 2021 YouTube shared a metric called Violative View Rate: for every ten thousand views, roughly eighteen came from videos eventually removed for breaking rules. This rate had declined 70 percent since 2017, according to the company.

398 **did not share comparable data:** Mark Bergen, "YouTube Avoids Facebook-Level Criticism from Biden," *Bloomberg*, July 22, 2021, https://www.bloomberg.com/news/newsletters/2021-07-22/google-s-youtube-avoids-facebook-level-criticism-from-biden-on-vaccine-misinfo.

399 **In the fall of 2021:** The vaccine ban included two caveats: scientific discussions and "personal testimonials" about vaccines were still permitted.

399 **authored an op-ed:** Susan Wojcicki, "Free Speech and Corporate Responsibility Can Coexist Online," *The Wall Street Journal*, August 1, 2021, https://www.wsj.com/articles/free-speech-youtube-section-230-censorship-content-moderation-susan-wojcicki-social-media-11627845973.

399 **he blogged:** Neal Mohan, "Perspective: Tackling Misinformation on YouTube," YouTube Official Blog, August 25, 2021, https://blog.youtube/inside-youtube/tackling-misinfo/.

402 **sat down with Wojcicki:** hankschannel, "YouTube, Pandemics, Creators, and Power."

Index